LEON COOPER
Professor of Operations Research
Southern Methodist University

DAVID STEINBERG
Assistant Professor of Applied Mathematics
Washington University

Introduction to
METHODS OF OPTIMIZATION

W. B. SAUNDERS COMPANY
Philadelphia · London · Toronto

W. B. Saunders Company: West Washington Square
Philadelphia, Pa. 19105

12 Dyott Street
London, WC1A 1DB

833 Oxford Street
Toronto 18, Ontario

Introduction to Methods of Optimization ISBN 0-7216-2693-9

Press of W. B. Saunders Company. Library of Congress catalog card number 72-118583.

Print No.: 9 8 7 6 5 4 3

PREFACE

The past few decades have seen an ever burgeoning growth in the application of the scientific method and, more particularly, of quantitative methods to aspects and areas of practical affairs from which they were virtually absent previously. One of the most important contributing factors to this development has been the existence and availability of digital computers. However, even if computers were not extensively used, it is evident that a new spirit of quantitative inquiry is emerging in disciplines as varied as political science, linguistic analysis, mathematical biology, management science, sociology, and other fields.

One of the important goals or objectives common to many diverse disciplines and problem areas is optimization. Loosely speaking, optimization can be defined as finding the best way to carry out an action or produce a result. More accurately, we may say that we wish to define a system of some sort, identify its variables and the conditions they must satisfy, define a measure of effectiveness of the system, and then seek the state of the system (values of the variables) that gives the most desirable (largest or smallest) measure of effectiveness.

The purpose of this book is to provide an introductory, yet reasonably rigorous, treatment of the basic methods and techniques available for the solution of optimization problems. The book does not pretend to provide encyclopedic completeness, but the authors do believe that many of the most important methods have been covered adequately. Because of limitations of space, we have not been able to consider such topics as the calculus of variations, the maximum principle, and various topics in stochastic optimization.

The basic orientation of the book is to provide a background for problem-solvers. We have not striven for mathematical elegance or extreme rigor. However, neither have we ignored the necessity to provide reasonably complete and accurate mathematical proofs.

The only specific mathematical background assumed is a first-year college course in calculus. Chapters 2 and 3 provide the background required in linear algebra and n-dimensional geometry, respectively. Hence this text can be used by engineers, business students, economists, and social scientists. It was designed for use primarily in a two-semester undergraduate course, or perhaps a first-year graduate course in some subject areas. However, by the selection of specific chapters a useful one-semester course can also be taught. For example, Chapters 1 to 4, Chapter 6, and topics in Chapter 7 could be taught in a one-semester course.

The two chapters following the Introduction deal with aspects of linear algebra and with n-dimensional geometry. If the text is used by students who possess a background in one or both of these subjects, these chapters may be omitted. In Chapter 4 the underlying concepts of classical optimization are presented and their use in calculation is discussed. Following this, a fairly thorough discussion of search methods and their use is presented in Chapter 5.

Chapters 6 and 7 are devoted to the development, understanding, and use of linear programming as a computational method. Special topics, such as the transportation model and the Dantzig-Wolfe decomposition principle, are also discussed. In Chapters 8 and 9 various algorithms, both exact and approximate, are discussed for solving nonlinear programming problems, including integer programming problems. There is also an introduction to the use of dynamic programming methodology.

The authors wish to express their debt to many of their students who have been taught some of this material. Their comments and reactions have been very helpful.

LEON COOPER
DAVID I. STEINBERG

CONTENTS

Chapter 1

INTRODUCTION

1.1 INTRODUCTION

In a vague, often ill-defined way, mankind has long sought or professed to seek, better ways to carry out the tasks of daily life. Most of the time we seek merely to obtain some improvement in the level of performance. Throughout human history the long quest for, first, more effective sources of food and, later, materials, power, and mastery of the physical environment attests to this general desire. It is not our purpose here to describe this process or the complex motivations behind it; it is also certainly not clear that we could do so, even if we wanted to.

Relatively late in human history, certain kinds of general questions began to be formulated quantitatively, first in words, and later in symbolic notations. One pervasive aspect of these general questions was a seeking after the "best" or "optimum." It should be emphasized that these words do not usually have precise meaning, even today, after massive efforts have been made to describe complex human and social situations, unless one can write down, in effect, a mathematical expression containing one or more variables, the value of which is to be determined. The question that is then asked, in general terms, is what values should these variables have so that the mathematical expression has the greatest possible numerical value (maximization) or the least possible numerical value (minimization). This general process of maximizing or minimizing is referred to as optimization.

In this book we shall not be concerned with the problem of how well any given mathematical expression describes the portion of reality with which it is supposed to deal. That is the central problem of the various disciplines which either are or are struggling to become "sciences." Our concern here is, given a meaningful mathematical function of one or more variables, what value(s) of the variables, within certain allowable limits, will make the function take on its maximum or minimum value.

This process may be described mathematically as

$$
\begin{aligned}
\text{Maximize} \quad & f(x_1, x_2, \ldots, x_n) \\
\text{Subject to} \quad & g_1(x_1, x_2, \ldots, x_n) \leq b_1 \\
& g_2(x_1, x_2, \ldots, x_n) \leq b_2 \\
& \quad \vdots \\
& g_m(x_1, x_2, \ldots, x_n) \leq b_m
\end{aligned} \tag{1-1}
$$

By $f(x_1, x_2, \ldots, x_n)$ we mean some function of the variables $x_1, x_2, \ldots, x_n$. It may or may not be continuous. It may or may not be differentiable, and so forth. It may only be a table of numbers giving the correspondence between the various values of the x_j and the value of f. This is precisely what one would obtain if the numbers originated in a laboratory or in other observations of natural phenomena. For purposes of mathematical and computational convenience one often connects a set of such points in a $(n + 1)$-dimensional space by a "smooth surface" and pretends that a mathematical function exists which describes the phenomenon in question. This is often the first introduction of disparity between a mathematical model and a physical situation.

The $g_i(x_1, x_2, \ldots, x_n)$ may also be functions similar to f and the preceding remarks may often apply to these constraints or restrictions on the variables. Often, the $g_i(x_1, x_2, \ldots, x_n)$ are very simple functions. For example, nonnegativity restrictions of the type

$$x_j \geq 0 \tag{1-2}$$

are often included in the set of constraints of (1-1). Sometimes in the statement of an optimization problem, they are expressed explicitly. However, the general statement (1-1) encompasses all optimization problems.

A very important special case of problem (1-1) is the problem in which each of the functions $f(x_1, x_2, \ldots, x_n)$ and $g_i(x_1, x_2, \ldots, x_n)$, $i = 1, 2, \ldots, m$, is a linear function. Such a problem may be described mathematically in the following form:

$$
\begin{aligned}
\text{Maximize} \quad & z = c_1x_1 + c_2x_2 + \dots + c_nx_n \\
\text{Subject to} \quad & a_{11}x_1 + a_{12}x_2 + \dots + a_{1n}x_n \le b_1 \\
& a_{21}x_1 + a_{22}x_2 + \dots + a_{2n}x_n \le b_2 \\
& \quad \vdots \\
& a_{m1}x_1 + a_{m2}x_2 + \dots + a_{mn}x_n \le b_m \\
& x_j \ge 0, \quad j = 1, 2, \dots, n
\end{aligned} \tag{1-3}
$$

This problem is known as a *linear programming problem.* The function to be maximized, $z = c_1x_1 + c_2x_2 + \dots + c_nx_n$, is called the *objective function;* any or all of the inequalities in (1-3) may sometimes be replaced by equalities. A great variety of economic, social, and industrial problems can be described in terms of a linear programming problem. It is perhaps the single most important (or, at least, the most widely used) form of mathematical optimization model. All of Chapter 6 and much of Chapter 7 are devoted to a study of different methods for solving linear programming problems and of special types of linear programming problems.

It is alleged by Virgil that Queen Dido was an early practitioner of the art of optimization. Legend has it that she was allowed to have the largest area of land that could be "surrounded" by the hide of a bull, for what was to become the city of Carthage. The land was to be adjacent to the sea. Her solution was to make a rope from the hide. (How fine this rope was is not certain, but is most important.) She then formed a semicircle with the rope, with the sea shore as the diameter. Obviously, she had concluded that a semicircle gave the largest area for a fixed length of rope. This is a relatively simple optimization problem for a student today. Queen Dido's solution, which was correct, was not proven until many hundreds of years later. It involves the calculus of variations, a subject not treated in this book. It has an elegant elementary theory but is beset with exceedingly difficult computational problems, for any but the most simple problems.

The case of the calculus of variations illustrates one of the characteristic aspects of the entire subject of mathematical optimization. It is relatively easy to formulate problems and, in most cases, to demonstrate the *existence* of a solution to the problems. However, in all but a few simple generic types, there are no exact computational methods which are guaranteed to obtain a solution.

In this book we are presenting a selection, by no means complete, of some of the better known methods and methods which the authors feel are useful. First, we have provided some necessary mathematical background. Then we have considered classical optimization methods for functions of one or more variables without constraints on the variables. Following this we have considered constrained problems by classical methods. From a computational point of view, these methods are of limited utility. Next we have considered search methods which are very useful computationally. The latter chapters of the book deal with mathematical programming, one of the most important developments in contemporary applied mathematics. Some of the material in this book has been developed by the authors and the students of one of the authors.

The emphasis in this text is to present an *introductory* picture of the state of optimization as it is practiced computationally at present. Theory is either presented in detail or references are given. It has not been possible to include all that we might have wished. However, in all cases, the interested reader may deepen his understanding by consulting the references which are given.

1.2 SET NOTATION

At various places in the text we shall be discussing "sets" of points with certain properties. We shall close this chapter with a brief discussion of the notation which will be used in connection with such sets.

By a *point* we mean an ordered list of numbers. Thus, (x_1, x_2) is a point consisting of the two numbers x_1, x_2; $(x_1, x_2, \ldots, x_n)$ is a point consisting of the n numbers $x_1, x_2, \ldots, x_n$. Such a point is often called an n-tuple. Each of the numbers which comprise a given point are called the *components* of the point. Thus, x_1 is the first component of the point (x_1, x_2, x_3), x_2 is its second component, and so forth.

The set of all points with three components consists of all the points (x_1, x_2, x_3), where each of the three components can have any numeric value.

We shall often be concerned with sets in which we have imposed certain conditions on some or all of the components of the points in the set. For example, we might wish to consider the set of all points with three components such that the last component is zero. If we denote† this set by S, then we shall write

$$S = \{(x_1, x_2, x_3) \mid x_3 = 0\} \tag{I-4}$$

† Note: Sets are usually denoted by capital letters; in the next chapter we shall see that *matrices* are also denoted by capital letters; this fact should cause no confusion, however, since the meaning shall be quite clear from the context.

Equation (1-4) is read "S equals the set of all points (x_1, x_2, x_3) such that $x_3 = 0$." The vertical bar divides the information about the set S which is contained in the braces { } into two parts: to the left of the bar, we describe the general form of the points of S; to the right of the bar, we describe any special characteristics of these points which distinguishes them from points not necessarily in S (e.g., characteristics which a point must possess in order to be a point in the set S).

Members of sets are sometimes called *elements* of the set; if a point† $\bar{x}$ is an element of the set S, then we write

$$\bar{x} \in S \tag{1-5}$$

The expression "$\bar{x} \in S$" is read "$\bar{x}$ is an element of S." Conversely, the expression "$\bar{y} \notin S$" is read "$\bar{y}$ is not an element of S."

If all the elements of a set S are also elements of another set T, then S is said to be a *subset* of T. Symbolically, we write

$$S \subset T \tag{1-6}$$

The expression (1-6) is read "S is a subset of T" or "S is contained in T." Alternatively, we often write

$$T \supset S \tag{1-7}$$

which is equivalent to (1-6).

Finally, we conclude this section by defining two relationships concerning groups of sets:

Intersection: The intersection of two sets S_1 and S_2 is the set of all points $\bar{x}$ such that $\bar{x}$ is an element of *both* S_1 and S_2.
If we denote the intersection of the two sets S_1 and S_2 by T, then the definition states that

$$T = \{\bar{x} \mid \bar{x} \in S_1 \quad \textit{and} \quad \bar{x} \in S_2\} \tag{1-8}$$

Symbolically, we may also describe T by the following expression:

$$T = S_1 \cap S_2 \tag{1-9}$$

Equation (1-9) is read, "T equals S_1 intersected with S_2," or "T equals the intersection of S_1 and S_2."

The intersection of n sets $S_1, S_2, \ldots, S_n$ is defined by

$$T \equiv S_1 \cap S_2 \cap \ldots \cap S_n \equiv \bigcap_{i=1}^{n} S_i \tag{1-10}$$

or

$$T = \{\bar{x} \mid \bar{x} \in S_1 \quad \textit{and} \quad \bar{x}_2 \in S_2 \quad \textit{and} \quad \ldots \quad \textit{and} \quad \bar{x} \in S_n\} \tag{1-11}$$

† A point is often denoted by a lower case letter with a "bar" over it.

Union: The union of two sets S_1 and S_2 is the set of all points $\bar{x}$ such that $\bar{x}$ is an element of *either* S_1 or S_2.
Thus, if P denotes the union of the two sets S_1 and S_2, then

$$P = \{\bar{x} \mid \bar{x} \in S_1 \quad or \quad \bar{x} \in S_2\} \tag{1-12}$$

Symbolically, we write that

$$P = S_1 \cup S_2 \tag{1-13}$$

Equation (1-13) is read, "P equals the union of S_1 and S_2."

The union of n sets $S_1, S_2, \ldots, S_n$ is defined by

$$P \equiv S_1 \cup S_2 \cup \ldots \cup S_n \equiv \bigcup_{i=1}^{n} S_i \tag{1-14}$$

or equivalently,

$$T = \{\bar{x} \mid \bar{x} \in S_1 \quad or \quad \bar{x} \in S_2 \quad or \quad \ldots \quad or \quad \bar{x} \in S_n\} \tag{1-15}$$

EXAMPLES

Let
$$S_1 = \{(x_1, x_2, x_3) \mid x_3 = 0\}$$
$$S_2 = \{(x_1, x_2, x_3) \mid x_1 = x_2\}$$
$$S_3 = \{(x_1, x_2, x_3) \mid x_2 = x_3 = 0\}$$
$$S_4 = \{(x_1, x_2, x_3) \mid x_3 = 1\}$$

(a) Since every element of S_3 has the property that $x_3 = 0$, each such element is also in S_1; thus, $S_3 \subset S_1$.

(b) The intersection of S_1 with S_2 is the set

$$T \equiv S_1 \cap S_2 = \{(x_1, x_2, x_3) \mid x_1 = x_2 \quad \text{and} \quad x_3 = 0\}$$

The element $(1, 1, 0) \in T$.

(c) The intersection of S_2 with S_3 is the set

$$\begin{aligned} Q \equiv S_2 \cap S_3 &= \{(x_1, x_2, x_3) \mid x_1 = x_2 \quad and \quad x_2 = x_3 = 0\} \\ &= \{(x_1, x_2, x_3) \mid x_1 = x_2 = x_3 = 0\} \\ &= \{(0, 0, 0)\}. \end{aligned}$$

That is, Q contains only the point $(0, 0, 0)$.

(d) The intersection of S_1 with S_4 is the set

$$N \equiv S_1 \cap S_4 = \{(x_1, x_2, x_3) \mid x_3 = 0 \quad and \quad x_3 = 1\}$$

Since there are obviously no points (x_1, x_2, x_3) such that $x_3 = 0$ and $x_3 = 1$, N contains no elements. Such a set is called the *empty set*, or *null set*, and is denoted by $\varnothing$.

(e) The union of S_1 with S_4 is the set

$$R \equiv S_1 \cup S_4 = \{(x_1, x_2, x_3) \mid x_3 = 0 \quad \text{or} \quad x_3 = 1\}$$

Thus, the point $(1, 1, 0) \in R$ and also the point $(1, 1, 1) \in R$.

Chapter 2

MATRIX ALGEBRA

2.1 INTRODUCTION

Many mathematical models in both the physical sciences and the social sciences involve a system of simultaneous linear equations. Matrix algebra developed from studies of such systems of equations. Originally, matrix algebra was primarily a notational convenience, useful in the derivation of complicated relationships between different sets of equations. However, it soon became apparent that many of the properties of matrices and matrix algebra could be extended to more general concepts; this led to a branch of mathematics called linear algebra which today has become in itself a fundamental tool for the formulation and solution of a wide variety of mathematical models. In this chapter, only the most fundamental properties of matrix algebra will be investigated, primarily so that the notational advantages may be made use of in later chapters.

2.2 MATRICES

Let us look at a typical system of simultaneous equations:

$$\begin{Bmatrix} 2x_1 + 3x_2 + 4x_3 = 1 \\ -x_1 + 2x_2 - 2x_3 = 3 \end{Bmatrix} \qquad \textbf{(2-1)}$$

In such a system, we wish to determine a set of values for x_1, x_2, x_3 which satisfies each of the equations in (2-1). In some instances we might wish to determine *all possible* sets of values of x_1, x_2, x_3 which simultaneously satisfy the equations in (2-1). In any case, the symbols x_1, x_2, x_3 represent the *variables* of the system of equations. The coefficients of x_1, x_2, x_3 in the first equation are, respectively, 2, 3, and 4; the coefficients of x_1, x_2, x_3 in the second equation are, respectively, -1, 2, and -2. Note that it is these coefficients, along with the numbers on the right-hand side of each equation, that actually specify the system of equations. In other words, no new information is obtained from the presence of the symbols "x_1, x_2, x_3." We could just as easily have written

$$\left\{\begin{aligned} 2y_1 + 3y_2 + 4y_3 &= 1 \\ -y_1 + 2y_2 - 2y_3 &= 3 \end{aligned}\right\} \tag{2-2}$$

The systems (2-1) and (2-2) are completely equivalent: they have the same solutions. Suppose we write the coefficients of the variables in a rectangular table, or array:

$$\begin{matrix} 2 & 3 & 4 \\ -1 & 2 & -2 \end{matrix}$$

If we specify that the elements in the first row of this array represent the coefficients of the variables in the first equation and that the elements in the second row represent the coefficients of the variables in the second equation, then the entire system (2-1) can be represented as a rectangular array, by merely adding a column to the preceding array, consisting of the numbers on the right-hand side of each equation.

$$\begin{matrix} 2 & 3 & 4 & 1 \\ -1 & 2 & -2 & 3 \end{matrix}$$

Any rectangular array of numbers is called a *matrix;* a matrix is usually enclosed by square brackets:

$$\begin{bmatrix} 2 & 3 & 4 & 1 \\ -1 & 2 & -2 & 3 \end{bmatrix}$$

In some texts, double vertical bars, ‖ ‖, or parentheses, ( ), are used instead of the brackets. Throughout this text, capital letters will be used to represent matrices. Thus, for example, we could write

$$A = \begin{bmatrix} 2 & 3 & 4 & 1 \\ -1 & 2 & -2 & 3 \end{bmatrix}$$

A matrix with m rows and n columns is called *a matrix of order* (m, n) or *an* $m \times n$ (read "m by n") *matrix*. Thus, a general $m \times n$

matrix could be written symbolically as

$$A = \begin{bmatrix} a_{11} & a_{12} & \cdots & a_{1n} \\ a_{21} & a_{22} & \cdots & a_{2n} \\ \cdot & \cdot & & \\ \cdot & \cdot & & \\ \cdot & \cdot & & \\ a_{m1} & a_{m2} & \cdots & a_{mn} \end{bmatrix} \tag{2-3}$$

The symbols a_{ij} in (2-3) are called the *elements* of the matrix. Usually, these elements will be real numbers, but they may also be complex numbers or functions. Such numbers will be called *scalars*, to distinguish them from matrices. The subscripts i and j of the element a_{ij} indicate the row and column in which a_{ij} is located (e.g., a_{23} represents the element in the second row and third column).

It will sometimes be convenient to abbreviate (2-3) as

$$A = \|a_{ij}\|_{(m\times n)} = [a_{ij}]_{(m\times n)} \tag{2-4}$$

which indicates that A is an $m \times n$ matrix whose elements are the a_{ij}'s. When the order of the matrix is unimportant or is clear from the context, (2-4) may be abbreviated further to

$$A = \|a_{ij}\| = [a_{ij}] \tag{2-5}$$

Two matrices are said to be *equal* if and only if they have the same order and if their corresponding elements are equal; that is, if $A = [a_{ij}]_{(m\times n)}$ and $B = [b_{ij}]_{(m\times n)}$, then $A = B$ if and only if $a_{ij} = b_{ij}$ for all i and j.

2.3 OPERATIONS WITH MATRICES: ADDITION, SCALAR MULTIPLICATION, AND SUBTRACTION

In order to enhance the notational conveniences afforded by the use of matrices, it is desirable to define a set of operational rules for combining matrices. Before doing so, let's consider the following example:

Suppose we are conducting a poll of voters within a state; a summary of the information collected might have the following form:

Numbers of Voters (in units of 10,000)
for Candidate 1, 2, 3

	Urban Areas			Rural Areas		
Categories	1	2	3	1	2	3
I	2	1	3	3	2	1
II	3	1	2	4	3	1
III	3	1	1	4	2	2
IV	1	2	2	3	2	1

The categories I, II, III, and IV might represent, for example, some sort of socioeconomic breakdown. There are several ways that this data could be represented in matrix form:

$$M = \begin{bmatrix} 2 & 1 & 3 & 3 & 2 & 1 \\ 3 & 1 & 2 & 4 & 3 & 1 \\ 3 & 1 & 1 & 4 & 2 & 2 \\ 1 & 2 & 2 & 3 & 2 & 1 \end{bmatrix} \tag{2-6}$$

Equation (2-6) represents the data of the table in one 4×6 matrix. We might also want to separate the data into two matrices:

$$A = \begin{bmatrix} 2 & 1 & 3 \\ 3 & 1 & 2 \\ 3 & 1 & 1 \\ 1 & 2 & 2 \end{bmatrix} \qquad B = \begin{bmatrix} 3 & 2 & 1 \\ 4 & 3 & 1 \\ 4 & 2 & 2 \\ 3 & 2 & 1 \end{bmatrix} \tag{2-7}$$

Matrix A contains the data for voters in urban areas, matrix B for voters in rural areas. If we did not wish to distinguish between the voting preferences of urban voters and rural voters, we could combine these data to obtain a new table:

	Numbers of Voters (in units of 10,000)		
Categories	1	2	3
I	5	3	4
II	7	4	3
III	7	3	3
IV	4	4	3

This table can be represented in matrix form as

$$C = \begin{bmatrix} 5 & 3 & 4 \\ 7 & 4 & 3 \\ 7 & 3 & 3 \\ 4 & 4 & 3 \end{bmatrix}$$

Observe that if corresponding elements of matrix A and matrix B are added together to form a new matrix, this new matrix will be matrix C. This fact leads us to the definition of matrix addition:

If $A = [a_{ij}]_{(m \times n)}$ and $B = [b_{ij}]_{(m \times n)}$, then the *sum* $A + B$ is the matrix $[(a_{ij} + b_{ij})]_{(m \times n)}$. Only matrices of the same order may be added together. Thus, we have

$$\begin{bmatrix} 2 & 1 & 3 \\ 3 & 1 & 2 \\ 3 & 1 & 1 \\ 1 & 2 & 2 \end{bmatrix} + \begin{bmatrix} 3 & 2 & 1 \\ 4 & 3 & 1 \\ 4 & 2 & 2 \\ 3 & 2 & 1 \end{bmatrix} = \begin{bmatrix} 5 & 3 & 4 \\ 7 & 4 & 3 \\ 7 & 3 & 3 \\ 4 & 4 & 3 \end{bmatrix} \quad \textbf{(2-9)}$$

or, $A + B = C$.

Sometimes, it might be preferable to treat the data in terms of a per cent of the total population. If the categories I, II, III, and IV are mutually exclusive, then we can obtain the total population by merely summing all the elements of C, yielding a total population of 50,000. Thus, in order to convert the data into per cent form, we must multiply each entry in the table by $\frac{1}{50}$; this corresponds to multiplying each element in matrix C by $\frac{1}{50}$, yielding a new matrix D.

$$D = \begin{bmatrix} 0.10 & 0.06 & 0.08 \\ 0.14 & 0.08 & 0.06 \\ 0.14 & 0.06 & 0.06 \\ 0.08 & 0.08 & 0.06 \end{bmatrix} \quad \textbf{(2-10)}$$

The operation of multiplying each element of a matrix by a scalar is called *scalar multiplication.* If $A = [a_{ij}]$ and α is a scalar, then the multiplication of A by α is written αA (or $A\alpha$), and

$$\alpha A = [\alpha a_{ij}] \quad \textbf{(2-11)}$$

The difference of two matrices (matrix subtraction) A, B, both of the same order, is denoted by A-B, and is defined by

$$A + (-1)B \quad \textbf{(2-12)}$$

where in (2-12), B is first multiplied by the scalar -1, and the resulting matrix is added to A.

2.4 MULTIPLICATION OF MATRICES

The definition of matrix multiplication is not quite as simple as that of matrix addition or subtraction. In order to determine a definition that will be meaningful, let's return to the system of simultaneous equations (2-1) of Section 2.2:

$$2x_1 + 3x_2 + 4x_3 = 1$$

$$-x_1 + 2x_2 - 2x_3 = 3$$

The matrix of coefficients of this system is

$$A = \begin{bmatrix} 2 & 3 & 4 \\ -1 & 2 & -2 \end{bmatrix} \tag{2-13}$$

Let's consider a (3×1) matrix $\begin{bmatrix} x_1 \\ x_2 \\ x_3 \end{bmatrix}$ and a (2×1) matrix $\begin{bmatrix} 1 \\ 3 \end{bmatrix}$; a matrix with only one column is called a *column vector*, or simply a vector. (Similarly, a matrix with only one row is called a *row* vector.) A vector with n elements is sometimes called an *n-component* vector or an *n-vector*. In this text a vector will be denoted by a lower case letter with a "bar" over it, and all other matrices by capital letters.

We would like to define matrix multiplication in such a way that we may write equation (2-1) as

$$\begin{bmatrix} 2 & 3 & 4 \\ -1 & 2 & -2 \end{bmatrix} \begin{bmatrix} x_1 \\ x_2 \\ x_3 \end{bmatrix} = \begin{bmatrix} 1 \\ 3 \end{bmatrix} \tag{2-14}$$

or

$$A\bar{x} = \bar{b} \tag{2-15}$$

where

$$\bar{x} = \begin{bmatrix} x_1 \\ x_2 \\ x_3 \end{bmatrix} \qquad \bar{b} = \begin{bmatrix} 1 \\ 3 \end{bmatrix}$$

In equation (2-15) we are in some sense "multiplying" the matrix A by the column vector $\bar{x}$. We want the result of this multiplication to yield the original system of equations:

$$\begin{bmatrix} 2 & 3 & 4 \\ -1 & 2 & -2 \end{bmatrix} \begin{bmatrix} x_1 \\ x_2 \\ x_3 \end{bmatrix} = \begin{bmatrix} (2x_1 + 3x_2 + 4x_3) \\ (-x_1 + 2x_2 - 2x_3) \end{bmatrix} = \begin{bmatrix} 1 \\ 3 \end{bmatrix} \tag{2-16}$$

Thus, the product of the (2×3) matrix A and the (3×1) vector $\bar{x}$ yields a (2×1) vector $\bar{b}$. Now, the question is, "What *rule* for multiplying A by $\bar{x}$ will yield $\bar{b}$?" The first element of $\bar{b}$ is equal to $(2x_1 + 3x_2 + 4x_3)$, and this element is obtained by taking the *first* element of the *first row* of A and multiplying it by the *first element* of $\bar{x}$ and then adding the result to the product of the *second* element in the *first row* of A and the *second* element of $\bar{x}$, and, finally, adding this result to the product of the *third* element of the *first row* of A and the *third* element of $\bar{x}$. Similarly, the second element of $\bar{b}$ is obtained by repeating the foregoing procedure with the elements in the *second row* of A replacing those in the first row.

In general, multiplying an $(m \times n)$ matrix A by an n-component vector $\bar{x}$ will result in an m-component vector, say

$$\bar{b} = \begin{bmatrix} b_1 \\ b_2 \\ \cdot \\ \cdot \\ \cdot \\ b_m \end{bmatrix}$$

To obtain b_i (the ith *component*, or element, of $\bar{b}$), multiply the corresponding elements of the ith row of A and of $\bar{x}$ and add the resulting products; symbolically,

$$b_i = a_{i1}x_1 + a_{i2}x_2 + \dots + a_{in}x_n \tag{2-17}$$

Multiplication of a matrix by a vector is only defined when the number of components of the vector equals the number of *columns* in the matrix.

EXAMPLE

$$A = \begin{bmatrix} 3 & 2 & 1 & 0 \\ 4 & 1 & 2 & 1 \\ 1 & 2 & 2 & 3 \end{bmatrix} \qquad \bar{x} = \begin{bmatrix} 2 \\ 5 \\ 4 \\ 2 \end{bmatrix}$$

$$A\bar{x} = \begin{bmatrix} 3 & 2 & 1 & 0 \\ 4 & 1 & 2 & 1 \\ 1 & 2 & 2 & 3 \end{bmatrix} \begin{bmatrix} 2 \\ 5 \\ 4 \\ 2 \end{bmatrix}$$

$$= \begin{bmatrix} (3 \times 2 + 2 \times 5 + 1 \times 4 + 0 \times 2) \\ (4 \times 2 + 1 \times 5 + 2 \times 4 + 1 \times 2) \\ (1 \times 2 + 2 \times 5 + 2 \times 4 + 3 \times 2) \end{bmatrix} = \begin{bmatrix} 20 \\ 23 \\ 26 \end{bmatrix}$$

The generalization of matrix multiplication to the product of two matrices is made by considering each column of the second matrix in the product as a column vector, and computing the column vector resulting from multiplying the first matrix by this column. The result of multiplying the first matrix by the kth column of the second matrix is the kth column of the product of the two matrices. Symbolically, if A is an $(m \times n)$ matrix and B is an $(n \times p)$ matrix, then the *product* AB is an $(m \times n)$ matrix. In the product AB the matrix A is said to be *postmultiplied* by B, or B is said to be *premultiplied* by A. Matrix multiplication is defined only when the number of *columns* of the first matrix (A) is equal to the number of *rows* of the second matrix (B). Thus,

$$\begin{bmatrix} a_{11} & a_{12} & \cdots & a_{1n} \\ a_{21} & a_{22} & \cdots & a_{2n} \\ \cdot & & & \cdot \\ \cdot & & & \cdot \\ \cdot & & & \cdot \\ a_{m1} & a_{m2} & \cdots & a_{mn} \end{bmatrix} \begin{bmatrix} b_{11} & b_{12} & \cdots & b_{1p} \\ b_{21} & b_{22} & \cdots & b_{2p} \\ \cdot & \cdot & & \cdot \\ \cdot & \cdot & & \cdot \\ \cdot & \cdot & & \cdot \\ b_{n1} & b_{n2} & \cdots & b_{np} \end{bmatrix}$$

$$= \begin{bmatrix} c_{11} & c_{12} & \cdots & c_{1p} \\ c_{21} & c_{22} & \cdots & c_{2p} \\ \cdot & \cdot & & \\ \cdot & \cdot & & \\ \cdot & \cdot & & \\ c_{m1} & c_{m2} & \cdots & c_{mp} \end{bmatrix} \qquad \textbf{(2-18)}$$

where†

$$c_{ij} = a_{i1}b_{1j} + a_{i2}b_{2j} + \ldots + a_{in}b_{nj} = \sum_{k=1}^{n} a_{1k}b_{kj} \qquad \textbf{(2-19)}$$

When the number of columns of a matrix A is the same as the number of rows of a matrix B, *A is said to be conformable to B* (for multiplication).

---

† The $\sum$ is the symbol for summation, and the subscript k in (2-19) is called the "dummy variable" of summation; the symbol $\sum_{k=1}^{n} t_k$ is read, "Sum the numbers t_k, from $k = 1$ to $k = n$;" thus, $\sum_{k=1}^{n} t_k = t_1 + t_2 + \ldots + t_n$.

EXAMPLE

$$\begin{bmatrix} 1 & 3 & 2 \\ 0 & 5 & 4 \\ 2 & 1 & 0 \\ 6 & 3 & 1 \end{bmatrix}_{(4\times3)} \begin{bmatrix} 1 & 2 \\ 3 & 4 \\ 2 & 1 \end{bmatrix}_{(3\times2)}$$

$$= \begin{bmatrix} (1 \times 1 + 3 \times 3 + 2 \times 2) & (1 \times 2 + 3 \times 4 + 2 \times 1) \\ (0 \times 1 + 5 \times 3 + 4 \times 2) & (0 \times 2 + 5 \times 4 + 4 \times 1) \\ (2 \times 1 + 1 \times 3 + 0 \times 2) & (2 \times 2 + 1 \times 4 + 0 \times 1) \\ (6 \quad 1 + 3 \quad 3 + 1 \quad 2) & (6 \quad 2 + 3 \quad 4 + 1 \quad 1) \end{bmatrix}$$

$$= \begin{bmatrix} 14 & 16 \\ 23 & 24 \\ 5 & 8 \\ 17 & 25 \end{bmatrix}_{(4\times2)}$$

It should be clear that matrix addition has the property that for any two $m \times n$ matrices A and B

$$A + B = B + A \tag{2-20}$$

Equation (2-20) is sometimes called the *commutative law of matrix addition*. (That matrix addition has the property (2-20) is referred to as "matrix addition is commutative.")

Matrix multiplication, however, is not in general commutative. That is, it is not true in general that $AB = BA$. First of all, even if A is conformable to B for multiplication, it may not be true that B is conformable to A for multiplication, so that the product BA may not be defined.

EXAMPLE

$$A = \begin{bmatrix} 1 & 3 & 2 \\ 0 & 5 & 4 \\ 2 & 1 & 0 \\ 6 & 3 & 1 \end{bmatrix} \qquad B = \begin{bmatrix} 1 & 2 \\ 3 & 4 \\ 2 & 1 \end{bmatrix} \qquad AB = \begin{bmatrix} 14 & 16 \\ 23 & 24 \\ 5 & 8 \\ 17 & 25 \end{bmatrix}$$

The product

$$BA = \begin{bmatrix} 1 & 2 \\ 3 & 4 \\ 2 & 1 \end{bmatrix} \begin{bmatrix} 1 & 3 & 2 \\ 0 & 5 & 4 \\ 2 & 1 & 0 \\ 6 & 3 & 1 \end{bmatrix}$$

is not defined, since the number of columns in B is not equal to the number of rows in A.

Moreover, even if AB and BA are both defined, it is still possible (in fact, likely) that $AB \neq BA$.

EXAMPLE

$$A = \begin{bmatrix} 1 & 2 \\ -1 & 0 \\ 3 & 1 \end{bmatrix} \qquad B = \begin{bmatrix} -1 & 0 & 6 \\ 2 & 4 & -1 \end{bmatrix}$$

$$AB = \begin{bmatrix} 1 & 2 \\ -1 & 0 \\ 3 & 1 \end{bmatrix} \begin{bmatrix} -1 & 0 & 6 \\ 2 & 4 & -1 \end{bmatrix} = \begin{bmatrix} 3 & 8 & 4 \\ 1 & 0 & -6 \\ -1 & 4 & 17 \end{bmatrix}$$

$$BA = \begin{bmatrix} -1 & 0 & 6 \\ 2 & 4 & -1 \end{bmatrix} \begin{bmatrix} 1 & 2 \\ -1 & 0 \\ 3 & 1 \end{bmatrix} = \begin{bmatrix} 17 & 4 \\ -5 & 3 \end{bmatrix}$$

From this example, and from the definition of matrix multiplication, it should now be clear that in order for the products AB and BA to be matrices of the same order, A and B must be square matrices; however, even if this is the case, it is still in general not true that $AB = BA$, as the following example illustrates.

EXAMPLE

$$A = \begin{bmatrix} 1 & 2 \\ 0 & -1 \end{bmatrix} \qquad B = \begin{bmatrix} 3 & -2 \\ 5 & 4 \end{bmatrix}$$

$$AB = \begin{bmatrix} 1 & 2 \\ 0 & -1 \end{bmatrix} \begin{bmatrix} 3 & -2 \\ 5 & 4 \end{bmatrix} = \begin{bmatrix} 13 & 6 \\ -5 & -4 \end{bmatrix}$$

$$BA = \begin{bmatrix} 3 & -2 \\ 5 & 4 \end{bmatrix} \begin{bmatrix} 1 & 2 \\ 0 & -1 \end{bmatrix} = \begin{bmatrix} 3 & 8 \\ 5 & 6 \end{bmatrix}$$

Another rather unfortunate property of matrix multiplication is the fact that the "cancellation law"† does *not* hold, in general; that is, given three matrices A, B, C, if $AB = AC$ and $A \neq O$,‡ it is not necessarily true that $B = C$.

EXAMPLE

$$A = \begin{bmatrix} 3 & 1 & 0 \\ -1 & 2 & 0 \\ 4 & -5 & 0 \end{bmatrix} \qquad B = \begin{bmatrix} 1 & 2 & 3 \\ -1 & -2 & 1 \\ 0 & 4 & 6 \end{bmatrix}$$

$$C = \begin{bmatrix} 1 & 2 & 3 \\ -1 & -2 & 1 \\ 2 & 0 & -8 \end{bmatrix}$$

$$AB = \begin{bmatrix} 3 & 1 & 0 \\ -1 & 2 & 0 \\ 4 & -5 & 0 \end{bmatrix} \begin{bmatrix} 1 & 2 & 3 \\ -1 & -2 & 1 \\ 0 & 4 & 6 \end{bmatrix} = \begin{bmatrix} 2 & 4 & 10 \\ -3 & -6 & -1 \\ 9 & 18 & 7 \end{bmatrix}$$

$$= \begin{bmatrix} 3 & 1 & 0 \\ -1 & 2 & 0 \\ 4 & -5 & 0 \end{bmatrix} \begin{bmatrix} 1 & 2 & 3 \\ -1 & -2 & 1 \\ 2 & 0 & -8 \end{bmatrix} = AC$$

Another law of scalar algebra which does not hold for matrices is the following: if a, b are two scalars and if $ab = 0$, then either $a = 0$ or $b = 0$ or both a, b equal zero. In the case of matrices, it is possible to multiply two nonzero matrices together and obtain a zero matrix. Consider the following:

EXAMPLE

$$A = \begin{bmatrix} 3 & 1 & 0 \\ 6 & 2 & 0 \\ -3 & -1 & 0 \end{bmatrix} \qquad B = \begin{bmatrix} 2 & -1 & 3 \\ -6 & 3 & -9 \\ 5 & 4 & 5 \end{bmatrix}$$

$$AB = \begin{bmatrix} 3 & 1 & 0 \\ 6 & 2 & 0 \\ -3 & -1 & 0 \end{bmatrix} \begin{bmatrix} 2 & -1 & 3 \\ -6 & 3 & -9 \\ 5 & 4 & 5 \end{bmatrix} = \begin{bmatrix} 0 & 0 & 0 \\ 0 & 0 & 0 \\ 0 & 0 & 0 \end{bmatrix}$$

† The cancellation law of (scalar) algebra says that for any scalars a, b, c, if $ab = ac$ and $a \neq 0$, then $b = c$.

‡ A *zero matrix*, written with a capital O, is a matrix whose elements are all zero.

To summarize the preceding discussion of matrix multiplication, if A is an $m \times n$ matrix and B is an $n \times p$ matrix, then

1. If $AB = D$, D is an $m \times p$ matrix.
2. BA is not defined, unless $m = p$.
3. If $m = p$, it is not necessarily true that $AB = BA$.
4. If $m = p$, and $AB = D$, $BA = E$, then D and E will be matrices of the same order only if $m = p = n$; in this case, A, B, D, and E will all be square matrices (of order n).
5. If A and B are square matrices, it is still not necessarily true that $AB = BA$.
6. If A is an $m \times n$ matrix, and B and C are $n \times p$ matrices, and if $AB = AC$, then it is not necessarily true that $B = C$.
7. If A is an $m \times n$ matrix and B is an $n \times p$ matrix, and if $AB = O$, it is not necessarily true that $A = O$ or $B = O$.

Despite these seemingly rather gloomy characteristics, matrix multiplication does satisfy the two most important laws of operations: the associative law of multiplication and the distributive law. Specifically, if

$$A = [a_{ij}]_{(m\times n)},\ B = [b_{ij}]_{(n\times p)},\ C = [c_{ij}]_{(p\times q)},\ D = [d_{ij}]_{(p\times q)},$$

and $F = [f_{ij}]_{(q\times r)}$, then

1. $(AB)C = A(BC)$ (Associative Law) **(2-21)**

2. $B(C + D) = BC + BD$ **(2-22a)**

2b. $(C + D)F = CF + DF$ **(2-22b)**

(Distributive Laws)

As we have seen, many of the laws of scalar algebra do not carry over to matrix algebra (and in particular, to matrix multiplication). Thus, we must be very careful to prove the validity of equations (2-21) and (2-22). Unfortunately, these proofs are somewhat complicated and involve manipulations of the summation symbol $\sum$. However, it is important that the reader becomes familiar with such manipulations.

Let us first establish the validity of equation (2-21). Let

$$AB = T = [t_{ik}]_{(m\times p)}$$

Thus, by definition,

$$t_{ik} = \sum_{l=1}^{n} a_{il}b_{lk} \tag{2-23}$$

Now, consider $TC = S = [s_{ij}]_{(m\times q)}$, and

$$s_{ij} = \sum_{k=1}^{p} t_{ik}c_{kj} \tag{2-24}$$

Upon substituting (2-23) into (2-24), we obtain

$$s_{ij} = \sum_{k=1}^{p} \left\{ \sum_{l=1}^{n} a_{il} b_{lk} \right\} c_{kj} \tag{2-25}$$

Now, if we multiply each element in the sum $\sum a_{il} b_{lk}$ by c_{kj}, we obtain

$$s_{ij} = \sum_{k=1}^{p} \left\{ \sum_{l=1}^{n} a_{il} b_{lk} c_{kj} \right\} \tag{2-26}$$

Thus, we have obtained $(AB)C = S$, where the elements of S are given by equation (2-26). Now, we follow the same type of procedure for the expression on the right-hand side of equation (2-21). Let $BC = V = [v_{lj}]_{(n \times q)}$; thus,

$$v_{lj} = \sum_{k=1}^{p} b_{lk} c_{kj} \tag{2-27}$$

Now, let $AV = W = [w_{ij}]_{(m \times q)}$, and

$$\begin{aligned} w_{ij} &= \sum_{l=1}^{n} a_{il} v_{lj} \\ &= \sum_{l=1}^{n} a_{il} \left\{ \sum_{k=1}^{p} b_{lk} c_{kj} \right\} \\ &= \sum_{l=1}^{n} \sum_{k=1}^{p} a_{il} b_{lk} c_{kj} \end{aligned} \tag{2-28}$$

The only difference between equation (2-26) and equation (2-28) is the *order of summation;* however, in the case of finite sums the order of summation is arbitrary; e.g., in equation (2-28), we may sum first over the index k and then over l, or vice versa. Thus, $s_{ij} = w_{ij}$ and S and W are matrices of the same order, $m \times q$, so that $S = W$ and equation (2-21) is valid for all matrices A, B, and C for which the products are defined.

Now, we demonstrate the validity of (2-22a). Letting $BC = T = [t_{ij}]_{(n \times q)}$ and $BD = V = [v_{ij}]_{(n \times q)}$ we have, by definition,

$$t_{ij} = \sum_{k=1}^{p} b_{ik} c_{kj} \tag{2-29a}$$

$$v_{ij} = \sum_{k=1}^{p} b_{ik} d_{kj} \tag{2-29b}$$

Thus, $BC + BD = T + V = [t_{ij} + v_{ij}]_{(n \times q)}$, and

$$t_{ij} + v_{ij} = \sum_{k=1}^{p} b_{ik} c_{kj} + \sum_{k=1}^{p} b_{ik} d_{kj} = \sum_{k=1}^{p} b_{ik} (c_{kj} + d_{kj}) \tag{2-30}$$

Also, letting $C + D = W = [w_{kj}]_{(p \times q)}$, we have

$$w_{kj} = c_{kj} + d_{kj} \tag{2-31}$$

Thus, $BW = S = [s_{ij}]_{(n \times q)}$, and

$$s_{ij} = \sum_{k=1}^{p} b_{ik} w_{kj} = \sum_{k=1}^{p} b_{ik}(c_{kj} + d_{kj}) \qquad \textbf{(2-32)}$$

Again, $t_{ij} + v_{ij} = s_{ij}$, and $(T + V)$ and S are matrices of the same order, $n \times q$, so that

$$(T + V) = (BC + BD) = S = BW = B(C + D),$$

and equation (2-21a) has been proved. Equation (2-21b) is proved in an analogous manner; the reader should work this proof out in detail to ensure his understanding of the procedure.

In proving the associative and distributive laws for matrix multiplication, we actually dealt only with the *elements* of the matrices involved, rather than with the matrices themselves. Such proofs are usually much more tedious than proofs which involve only manipulations with matrices. The latter type of proof will be employed whenever possible, both because it is more compact than proofs involving the elements of matrices, and because the reader should learn to take full advantage of the matrix properties and operations at his disposal. Following is an example of such a proof.

EXAMPLE

If A and B are $m \times n$ matrices, and C and D are $n \times p$ matrices, then

$$(A + B)(C + D) = AC + AD + BC + BD \qquad \textbf{(2-33)}$$

Proof: Let $C + D = E$, then

$$\begin{aligned} (A + B)E &= AE + BE \quad \text{(by equation (2-22b))} \\ &= A(C + D) + B(C + D) \\ &= (AC + AD) + (BC + BD) \quad \text{(by equation (2-22a))} \\ &= AC + AD + BC + BD. \end{aligned}$$

2.5 SPECIAL TYPES OF MATRICES

In this section we will define some special types of matrices, which are commonly encountered in numerous applications.

In the previous section, the *zero matrix* was introduced: a zero matrix is a matrix with all zero elements. If we wish to specify the order of the zero matrix, we shall write $O_{(m \times n)}$; otherwise, the capital letter O will always mean a zero matrix. Any zero matrix has the property that when it is pre- or postmultiplied by any other matrix (or scalar), the result is a zero matrix. If A is an $m \times n$ matrix, then

$$O_{(p\times m)}A = O_{(p\times n)} \tag{2-34a}$$

$$AO_{(n\times q)} = O_{(m\times q)} \tag{2-34b}$$

The validity of the equations (2-34) should be obvious from the definition of matrix multiplication.

For every $m \times n$ matrix $A = [a_{ij}]$, there is a corresponding $n \times m$ matrix which is obtained from A by interchanging its rows and columns and is called *the transpose of A*. The transpose of A (or A transpose) is denoted by A' or, in some texts, by A^T. The first row of A is the first column of A'; the second row of A is the second column of A'; and so on. Thus, if $A = [a_{ij}]$, then $A' = [a_{ji}]$.

EXAMPLES

(a) $A = \begin{bmatrix} a_{11} & a_{12} & a_{13} \\ a_{21} & a_{22} & a_{23} \end{bmatrix}$ $\qquad A' = \begin{bmatrix} a_{11} & a_{21} \\ a_{12} & a_{22} \\ a_{13} & a_{23} \end{bmatrix}$

(b) $A = \begin{bmatrix} 1 & -1 & 2 & 6 \\ 3 & 4 & -7 & 8 \\ 0 & -2 & 3 & 1 \end{bmatrix}$ $\qquad A' = \begin{bmatrix} 1 & 3 & 0 \\ -1 & 4 & -2 \\ 2 & -7 & 3 \\ 6 & 8 & 1 \end{bmatrix}$

(c) $\bar{x} = \begin{bmatrix} 3 \\ 2 \\ -1 \\ 0 \end{bmatrix}$ $\qquad \bar{x}' = [3 \quad 2 \quad -1 \quad 0]$

Example (c) indicates that a row vector can be considered as the transpose of a column vector. Thus, no special notation is required to distinguish row vectors from column vectors; we shall merely adopt the convention of thinking of all vectors as column vectors, unless the transpose symbol appears in the representation of the vector. Thus,

$$\bar{x} = \begin{bmatrix} x_1 \\ x_2 \\ \cdot \\ \cdot \\ \cdot \\ x_n \end{bmatrix} \quad \text{and} \quad \bar{x}' = [x_1\, x_2 \, \ldots \, x_n]$$

A special case of matrix multiplication of particular interest is the multiplication of a row vector by a column vector. If $\bar{x}$ and $\bar{y}$ are two n-component vectors, then

$$\bar{x}'\bar{y} = \sum_{j=1}^{n} x_j y_j \tag{2-35}$$

where

$$\bar{x}' = [x_1\ x_2\ \ldots\ x_n] \quad \text{and} \quad \bar{y} = \begin{bmatrix} y_1 \\ y_2 \\ \cdot \\ \cdot \\ \cdot \\ y_n \end{bmatrix}$$

The product (2-35) is called the *scalar product of* $\bar{x}$ *and* $\bar{y}$.

EXAMPLE

$$\bar{x} = \begin{bmatrix} 3 \\ 4 \\ 5 \\ -2 \end{bmatrix} \qquad \bar{y} = \begin{bmatrix} 1 \\ 0 \\ -2 \\ 3 \end{bmatrix}$$

$$\bar{x}'\bar{y} = [3 \quad 4 \quad 5 \quad -2] \begin{bmatrix} 1 \\ 0 \\ -2 \\ 3 \end{bmatrix}$$

$$= 3(1) + 4(0) + 5(-2) - 2(3) = -13$$

Note that $\bar{x}'\bar{y} = \bar{y}'\bar{x}$.

We shall now investigate several special types of *square* matrices. A square matrix which is equal to its own transpose is said to be a *symmetric matrix;* i.e., A is a symmetric matrix if $A = A'$.

The elements $a_{11}, a_{22}, \ldots, a_{nn}$ of a square matrix $A = [a_{ij}]_{(n\times n)}$ are called the *diagonal elements* of A (or, sometimes, the *main diagonal elements*); the other elements of A are called the *off-diagonal elements.* A square matrix whose off-diagonal elements are all zero is called a *diagonal matrix.* If the diagonal elements of a diagonal matrix are equal, the matrix is called a *scalar matrix.*

EXAMPLES

(a) diagonal matrices: $A = \begin{bmatrix} 3 & 0 & 0 \\ 0 & -1 & 0 \\ 0 & 0 & 4 \end{bmatrix}$ $\quad B = \begin{bmatrix} 0 & 0 & 0 \\ 0 & 2 & 0 \\ 0 & 0 & 0 \end{bmatrix}$

(b) scalar matrices: $D = \begin{bmatrix} 3 & 0 & 0 \\ 0 & 3 & 0 \\ 0 & 0 & 3 \end{bmatrix}$ $\quad E = \begin{bmatrix} \alpha & 0 & 0 \\ 0 & \alpha & 0 \\ 0 & 0 & \alpha \end{bmatrix}$

An *identity matrix* is a scalar matrix with all 1's on the diagonal. Identity matrices are denoted by I_n, where the subscript n denotes the order. For any $m \times n$ matrix A,

$$I_m A = A = A I_n \tag{2-36}$$

Note that every scalar matrix has the form

$$D_n = \begin{bmatrix} \alpha & 0 & \dots & 0 \\ 0 & \alpha & & 0 \\ \cdot & \cdot & & \cdot \\ \cdot & \cdot & & \cdot \\ \cdot & \cdot & & \cdot \\ 0 & 0 & \dots & \alpha \end{bmatrix}_{(n \times n)} \tag{2-37}$$

for some scalar α, and that

$$\alpha I_n = \alpha \begin{bmatrix} 1 & 0 & \dots & 0 \\ 0 & 1 & \dots & 0 \\ \cdot & & & \cdot \\ \cdot & & & \cdot \\ \cdot & & & \cdot \\ 0 & 0 & \dots & 1 \end{bmatrix} = \begin{bmatrix} \alpha & 0 & \dots & 0 \\ 0 & & \dots & 0 \\ \cdot & & & \\ \cdot & & & \\ \cdot & & & \\ 0 & 0 & \dots & \alpha \end{bmatrix} = D_n \tag{2-38}$$

Thus, a scalar matrix can be expressed as the product of a scalar times an identity matrix. Combining equations (2-36) and (2-38), we see that, for any $m \times n$ matrix A,

$$\begin{aligned} D_m A &= \alpha I_m A = \alpha A = \alpha(A I_n) \\ &= (\alpha A) I_n = (A \alpha) I_n = A(\alpha I_n) = A D_n \end{aligned} \tag{2-39}$$

Thus, multiplying A by a scalar matrix has the same effect as merely multiplying A by the scalar.

A square matrix B is called the *inverse* of a square matrix A (both of order n) if

$$AB = BA = I_n$$

B is usually denoted by A^{-1}. Not every square matrix has an inverse. However, if the matrix of coefficients of a system of n equations in n unknowns, $A\bar{x} = \bar{b}$, does possess an inverse, then we can write $A^{-1}(A\bar{x}) = (A^{-1}A)\bar{x} = I_n\bar{x} = \bar{x} = A^{-1}\bar{b}$.

2.6 PARTITIONED MATRICES

It is frequently desirable to divide the elements of a given matrix into groups of elements in such a way that each group is itself a matrix. This can be done by subdividing or partitioning the matrix into smaller rectangular arrays. Each rectangular array thus formed is called a *submatrix* of the original matrix. The partitioning itself is usually indicated by dashed lines. For example, consider the matrix A:

$$A = \begin{bmatrix} 1 & 3 & -6 & 5 & 4 \\ 3 & 2 & 0 & -1 & -3 \\ 1 & -1 & 2 & 3 & 0 \\ 0 & -3 & 4 & -2 & 1 \end{bmatrix}$$

One possible partitioned form is

$$A = \left[\begin{array}{ccc:cc} 1 & 3 & -6 & 5 & 4 \\ 3 & 2 & 0 & -1 & -3 \\ \hdashline 1 & -1 & 2 & 3 & 0 \\ 0 & -3 & 4 & -2 & 1 \end{array}\right]$$

The submatrices thus formed are

$$B_{11} = \begin{bmatrix} 1 & 3 & -6 \\ 3 & 2 & 0 \end{bmatrix} \qquad B_{12} = \begin{bmatrix} 5 & 4 \\ -1 & -3 \end{bmatrix}$$

$$B_{21} = \begin{bmatrix} 1 & -1 & 2 \\ 0 & -3 & 4 \end{bmatrix} \qquad B_{22} = \begin{bmatrix} 3 & 0 \\ -2 & 1 \end{bmatrix}$$

Thus, we could write

$$A = \begin{bmatrix} B_{11} & B_{12} \\ B_{21} & B_{22} \end{bmatrix}$$

where now the elements of A are not scalars, but submatrices. The subscripts assigned to these submatrices suggest that B_{11}, B_{12}, B_{21}, and B_{22} can be treated as if they were regular elements of A. Indeed, this is the case. However, one must be careful to partition the matrices so that, when operations (such as matrix addition or multiplication) are carried out on, for example, two partitioned matrices the *resulting* operations that must be carried out on the submatrices are defined. For example, if we wish to add two matrices of the same order, we must partition both matrices exactly the same way so that the corresponding submatrices may be added.

Let us see how two matrices that are to be multiplied together might be partitioned.

Let M be an $m \times n$ matrix and let N be an $n \times p$ matrix. Further, if we let

$$M = \begin{bmatrix} A_1 & B_1 \\ C_1 & D_1 \end{bmatrix} \qquad N = \begin{bmatrix} A_2 & B_2 \\ C_2 & D_2 \end{bmatrix}$$

then we want to write

$$\begin{aligned} MN &= \begin{bmatrix} A_1 & B_1 \\ C_1 & D_1 \end{bmatrix} \begin{bmatrix} A_2 & B_2 \\ C_2 & D_2 \end{bmatrix} \\ &= \begin{bmatrix} (A_1A_2 + B_1C_2) & (A_1B_2 + B_1D_2) \\ (C_1A_2 + D_1C_2) & (C_1B_2 + D_1D_2) \end{bmatrix} \end{aligned} \tag{2-40}$$

In order for (2-40) to be a valid expression, the matrix products A_1A_2, B_1C_2, A_1B_2, B_1D_2, and so forth, must be defined, and the pairs of matrices A_1A_2 and B_1C_2, A_1B_2 and B_1D_2, and so forth, must be of the same order. These conditions will be met if we partition M and N so that

$$\begin{array}{lll@{\qquad}lll} A_1 & \text{is} & r \times s & B_1 & \text{is} & r \times (n - s) \\ C_1 & \text{is} & (m - r) \times s & D_1 & \text{is} & (m - r) \times (n - s) \\ A_2 & \text{is} & s \times q & B_2 & \text{is} & s \times (p - q) \\ C_2 & \text{is} & (n - s) \times q & D_2 & \text{is} & (n - s) \times (p - q) \end{array}$$

EXAMPLE

Let

$$M = \left[\begin{array}{cc:c} 3 & -1 & 1 \\ 2 & 0 & 2 \\ 1 & 2 & 1 \\ \hdashline 0 & 3 & -1 \end{array}\right] \qquad N = \begin{bmatrix} 2 & 1 & 3 & 4 & 1 \\ 1 & -1 & 0 & -2 & 0 \\ 0 & 0 & 1 & 3 & 1 \end{bmatrix}$$

Thus, $m = 4$, $n = 3$, $p = 5$,

$$A_1 = \begin{bmatrix} 3 & -1 \\ 2 & 0 \\ 1 & 2 \end{bmatrix} \qquad B_1 = \begin{bmatrix} 1 \\ 2 \\ 1 \end{bmatrix} \qquad C_1 = [0 \quad 3] \qquad D_1 = [-1]$$

Hence, $r = 3$ and $s = 2$; and so we wish to partition N so that A_2 will have $s = 2$ rows. We can choose the number of columns, q, arbitrarily. For example, let $q = 3$. Then we have

$$N = \left[\begin{array}{ccc|cc} 2 & 1 & 3 & 4 & 1 \\ 1 & -1 & 0 & -2 & 0 \\ \hline 0 & 0 & 1 & 3 & 1 \end{array}\right] = \begin{bmatrix} A_2 & B_2 \\ C_2 & D_2 \end{bmatrix}$$

and, from equation (2-40),

$$MN =$$

$$\left[\begin{array}{c|c} \begin{bmatrix} 3 & -1 \\ 2 & 0 \\ 1 & 2 \end{bmatrix}\begin{bmatrix} 2 & 1 & 3 \\ 1 & -1 & 0 \end{bmatrix} + \begin{bmatrix} 1 \\ 2 \\ 1 \end{bmatrix}[0 \quad 0 \quad 1] & \begin{bmatrix} 3 & -1 \\ 2 & 0 \\ 1 & 2 \end{bmatrix}\begin{bmatrix} 4 & 1 \\ -2 & 0 \end{bmatrix} + \begin{bmatrix} 1 \\ 2 \\ 1 \end{bmatrix}[3 \quad 1] \\ \hline [0 \quad 3]\begin{bmatrix} 2 & 1 & 3 \\ 1 & -1 & 0 \end{bmatrix} + [-1][0 \quad 0 \quad 1] & [0 \quad 3]\begin{bmatrix} 4 & 1 \\ -2 & 0 \end{bmatrix} + [-1][3 \quad 1] \end{array}\right]$$

$$= \begin{bmatrix} \begin{bmatrix} 5 & 4 & 10 \\ 4 & 2 & 8 \\ 4 & -1 & 4 \end{bmatrix} & \begin{bmatrix} 17 & 4 \\ 14 & 4 \\ 3 & 2 \end{bmatrix} \\ [3 \quad -3 \quad -1] & [-9 \quad -1] \end{bmatrix}$$

$$= \begin{bmatrix} 5 & 4 & 10 & 17 & 4 \\ 4 & 2 & 8 & 14 & 4 \\ 4 & -1 & 4 & 3 & 2 \\ 3 & -3 & -1 & -9 & -1 \end{bmatrix} = \text{MN}$$

2.7 DETERMINANTS

Any system of two simultaneous equations in two unknowns can be represented by

$$a_{11}x_1 + a_{12}x_2 = b_1 \qquad \textbf{(2-41a)}$$

$$a_{21}x_1 + a_{22}x_2 = b_2 \qquad \textbf{(2-41b)}$$

The matrix

$$A = \begin{bmatrix} a_{11} & a_{12} \\ a_{21} & a_{22} \end{bmatrix} \qquad \textbf{(2-42)}$$

is called the *coefficient matrix* of the system of equations (2-41). The matrix

$$B = \begin{bmatrix} a_{11} & a_{12} & b_1 \\ a_{21} & a_{22} & b_2 \end{bmatrix}$$

is called the *augmented matrix* of the system. If $b_1 = b_2 = 0$, the system (2-41) is called a *homogeneous* system of equations; otherwise, the system is said to be *nonhomogeneous.*

Let us investigate the conditions under which (2-41) has a unique solution. If equation (2-41a) is multiplied by a_{22} and equation (2-41b) by a_{12}, we obtain two new equations:

$$(a_{11}a_{22})x_1 + (a_{12}a_{22})x_2 = a_{22}b_1 \qquad \textbf{(2-43a)}$$
$$(a_{12}a_{21})x_1 + (a_{12}a_{22})x_2 = a_{12}b_2 \qquad \textbf{(2-43b)}$$

If equation (2-43b) is then subtracted from equation (2-43a), the resulting equation is

$$(a_{11}a_{22} - a_{12}a_{21})x_1 = a_{22}b_1 - a_{12}b_1 \qquad \textbf{(2-44)}$$

If we now multiply equation (2-41a) by a_{21}, equation (2-41b) by a_{11}, and subtract, the resulting equation is

$$(a_{11}a_{22} - a_{12}a_{21})x_2 = a_{11}b_2 - a_{21}b_1 \qquad \textbf{(2-45)}$$

Equations (2-44) and (2-45) indicate that, in order for a unique solution of the system (2-41) to exist, it must be true that

$$a_{11}a_{22} - a_{12}a_{21} \neq 0 \qquad \textbf{(2-46)}$$

Furthermore, if $a_{11}a_{22} - a_{12}a_{21} \neq 0$, then

$$x_1 = \frac{a_{22}b_1 - a_{12}b_2}{a_{11}a_{22} - a_{12}a_{21}}$$
$$x_2 = \frac{a_{11}b_2 - a_{21}b_1}{a_{11}a_{22} - a_{12}a_{21}} \qquad \textbf{(2-47)}$$

The quantity $(a_{11}a_{22} - a_{12}a_{21})$ is called the *determinant* of the matrix A (defined by equation (2-42)), and is denoted by det A, or $|A|$. Thus,

$$\det A = a_{11}a_{22} - a_{12}a_{21} \qquad \textbf{(2-48)}$$

for any 2×2 matrix A. Applying definition (2-48), we can rewrite equation (2-47):

$$x_1 = \frac{\det C_1}{\det A}$$
$$x_2 = \frac{\det C_2}{\det A} \qquad \textbf{(2-49)}$$

where

$$C_1 = \begin{bmatrix} b_1 & a_{12} \\ b_2 & a_{22} \end{bmatrix} \qquad C_2 = \begin{bmatrix} a_{11} & b_1 \\ a_{21} & b_2 \end{bmatrix}$$

Equation (2-49) is known as *Cramer's rule.* It tells us that, to determine the solution to any system of equations of the form (2-41), one need only compute three determinants; namely, det A, det C_1, and det C_2. If det $A = 0$, then there is no unique solution to (2-41).

There may either be an infinite number of solutions or no solution at all. We shall investigate these cases in the next section.

The concept of determinants can be generalized from the 2×2 case: for every square matrix A (of any order), we can associate a scalar, uniquely defined,† called the determinant of A.

Some of the most important properties of determinants are the following:

1. Given a square matrix A, if a new matrix if formed by interchanging any two columns (or any two rows) of A, the determinant of the resulting matrix is $-\det A$.
2. For every square matrix A, $\det A = \det A'$. (Thus, every property that holds for columns also holds for rows.)
3. If a square matrix A has two identical columns (or rows), then $\det A = 0$.
4. If B is a square matrix which is identical with a matrix A, except that all elements of some column (or row) are some scalar multiple, α, of the elements of the corresponding column (or row) of A, then $\det B = \alpha \cdot \det A$.
5. Given a square matrix A, if a new matrix is formed by adding any multiple of one column (row) to a different column (row), the determinant of the resulting matrix is equal to the determinant of A.
6. If all the elements below the main diagonal of a square matrix A are zero,‡ then the determinant of A is equal to the product of the diagonal elements:

$$\det A = a_{11}a_{22} \cdot \ldots \cdot a_{nn} \tag{2-50}$$

7. If A and B are any two n^{th} order matrices, then

$$\det (AB) = (\det A)(\det B) \tag{2-51}$$

8. If a square matrix A has a column (or row) of all zeros, then $\det A = 0$.
9. If A_{ij} denotes the determinant of the matrix obtained from a square matrix A by deleting the ith row and the jth column of A (yielding a square matrix of order $n - 1$) then

$$\det A = \sum_{j=1}^{n} (-1)^{i+j} a_{ij} A_{ij}, \quad \text{for any } i \tag{2-52a}$$

$$\det A = \sum_{i=1}^{n} (-1)^{i+j} a_{ij} A_{ij}, \quad \text{for any } j \tag{2-52b}$$

---

† We shall not explicitly define determinants in this text, because the definition itself is of little use. The importance of determinants lies in their properties. Moreover, determinants are rarely calculated from their definition, and for the purposes of this text the method of calculating the determinant of any square matrix shall serve as the definition. For a detailed treatment of determinants, see Hohn, F. E.: *Elementary Matrix Algebra*, Second Edition. Macmillan Co., New York (1964).

‡ Such a matrix is called an *upper triangular matrix*.

The number $(-1)^{i+j}A_{ij}$ is called the *cofactor* of the element a_{ij}. Equation (2-52a) is called the cofactor expansion of det A by the ith row; equation (2-52b) is called the cofactor expansion of det A by the jth column.

Properties 1, 4, 5, and 6 suggest a method of computing the determinant of any square matrix. The operations described in properties 1, 4, and 5 are sometimes called *elementary row and column operations*. One procedure for calculating a determinant, then, is to perform elementary row and column operations on the matrix, until the matrix has been converted to an upper triangular matrix. The determinant is then computed from (2-50). To illustrate the procedure, let's consider the following matrix:

$$A = \begin{bmatrix} 2 & 0 & -2 & 4 \\ 1 & 3 & 5 & 8 \\ 2 & -1 & -1 & 3 \\ 6 & 6 & 3 & 2 \end{bmatrix}$$

Since we wish to convert A to upper triangular form, the first step will be to obtain all zero elements in the first column below $a_{11} = 2$. To obtain a zero in the second row, we employ property 5 and add $-\frac{1}{2}$ times row 1 to row 2, yielding

$$\det A = \begin{vmatrix} 2 & 0 & -2 & 4 \\ 1 & 3 & 5 & 8 \\ 2 & -1 & -1 & 3 \\ 6 & 6 & 3 & 2 \end{vmatrix} = \begin{vmatrix} 2 & 0 & -2 & 4 \\ 0 & 3 & 6 & 6 \\ 2 & -1 & -1 & 3 \\ 6 & 6 & 3 & 2 \end{vmatrix} \tag{2-53}$$

We can repeat this procedure to obtain zeros in the rest of column 1:

$$\begin{vmatrix} 2 & 0 & -2 & 4 \\ 0 & 3 & 6 & 6 \\ 2 & -1 & -1 & 3 \\ 6 & 6 & 3 & 2 \end{vmatrix} = \begin{vmatrix} 2 & 0 & -2 & 4 \\ 0 & 3 & 6 & 6 \\ 0 & -1 & 1 & -1 \\ 0 & 6 & 9 & -10 \end{vmatrix} \tag{2-54}$$

Each time we used some multiple of row 1 to obtain a zero for the first element of each remaining row. Next, we want to obtain zeros for each element in column 2 below $a_{22} = 3$. Thus, we now add appropriate multiples of row 2 to the remaining rows (3 and 4); in particular, adding $\frac{1}{3}$ times row 2 to row 3, and -2 times row 2 to row 4 will give us the desired result:

$$\begin{vmatrix} 2 & 0 & -2 & 4 \\ 0 & 3 & 6 & 6 \\ 0 & -1 & 1 & -1 \\ 0 & 6 & 9 & -10 \end{vmatrix} = \begin{vmatrix} 2 & 0 & -2 & 4 \\ 0 & 3 & 6 & 6 \\ 0 & 0 & 3 & 1 \\ 0 & 0 & -3 & -22 \end{vmatrix} \tag{2-55}$$

Now, adding 1 times row 3 to row 4 yields

$$\begin{vmatrix} 2 & 0 & -2 & 4 \\ 0 & 3 & 6 & 6 \\ 0 & 0 & 3 & 1 \\ 0 & 0 & -3 & -22 \end{vmatrix} = \begin{vmatrix} 2 & 0 & -2 & 4 \\ 0 & 3 & 6 & 6 \\ 0 & 0 & 3 & 1 \\ 0 & 0 & 0 & -21 \end{vmatrix}$$

Thus by equation (2-50)

$$\det A = (2)(3)(3)(-21) = -378$$

There is a slight modification to the foregoing procedure which sometimes simplifies the calculations. Before adding a multiple of the first row to each succeeding row, we could have first employed property 4 to obtain a 1 in the diagonal position of row 1 (a_{11}), by multiplying the first row by $\frac{1}{2}$ and the determinant of the resulting matrix by 2:

$$\det A = \begin{vmatrix} 2 & 0 & -2 & 4 \\ 1 & 3 & 5 & 8 \\ 2 & -1 & -1 & 3 \\ 6 & 6 & 3 & 2 \end{vmatrix} = 2 \begin{vmatrix} 1 & 0 & -1 & 2 \\ 1 & 3 & 5 & 8 \\ 2 & -1 & -1 & 3 \\ 6 & 6 & 3 & 2 \end{vmatrix}$$

Property 4 could again be employed in (2-54), to obtain a 1 in the diagonal position of row 2:

$$\det A = 2 \begin{vmatrix} 1 & 0 & -1 & 2 \\ 11 & 3 & 5 & 8 \\ 2 & -1 & -1 & 3 \\ 6 & 6 & 3 & 2 \end{vmatrix}$$

$$= 2 \begin{vmatrix} 1 & 0 & -1 & 2 \\ 0 & 3 & 6 & 6 \\ 0 & -1 & 1 & -1 \\ 0 & 6 & 9 & 10 \end{vmatrix} = (2)(3) \begin{vmatrix} 1 & 0 & -1 & 2 \\ 0 & 1 & 2 & 2 \\ 0 & -1 & 1 & -1 \\ 0 & 6 & 9 & 10 \end{vmatrix}$$

In general, it is computationally easier to first obtain a 1 in the diagonal position of the row of which multiples are to be added to succeeding rows.

EXAMPLES

(a)
$$A = \begin{bmatrix} a_{11} & a_{12} \\ a_{21} & a_{22} \end{bmatrix} \quad \text{(assume } a_{11} \neq 0\text{)}$$

$$\begin{aligned} \det A &= \begin{vmatrix} a_{11} & a_{12} \\ a_{21} & a_{22} \end{vmatrix} = a_{11} \begin{vmatrix} 1 & a_{12}/a_{11} \\ a_{21} & a_{22} \end{vmatrix} \\ &= a_{11} \begin{vmatrix} 1 & a_{12}/a_{11} \\ 0 & a_{22} - a_{21}(a_{12}/a_{11}) \end{vmatrix} \\ &= (a_{11})(1)(a_{22} - a_{21}a_{12}/a_{11}) \\ &= a_{11}a_{22} - a_{21}a_{12} \end{aligned}$$

which agrees with equation (2-48).

If $a_{11} = 0$, then we first interchange column 1 and column 2, remembering that this operation multiplies the determinant by (-1):

$$\begin{aligned} \det A &= \begin{vmatrix} a_{11} & a_{12} \\ a_{21} & a_{22} \end{vmatrix} = (-1) \begin{vmatrix} a_{12} & a_{11} \\ a_{22} & a_{21} \end{vmatrix} \\ &= (-1)(a_{12}) \begin{vmatrix} 1 & a_{11}/a_{12} \\ 0 & a_{21} - a_{22}(a_{11}/a_{12}) \end{vmatrix} \quad \text{(assuming } a_{12} \neq 0\text{)} \\ &= (-1)(a_{12})(1)[a_{21} - a_{22}(a_{11}/a_{12})] \\ &= (-1)(a_{12}a_{21} - a_{22}a_{11}) \\ &= a_{11}a_{22} - a_{12}a_{21} \end{aligned}$$

(If $a_{11} = 0$ and $a_{12} = 0$, then $\det A = 0$ by property 8.)

(b)
$$A = \begin{bmatrix} 0 & 1 & 3 \\ 1 & 2 & 6 \\ -5 & -4 & -12 \end{bmatrix}$$

Since $a_{11} = 0$, we must first interchange column 1 (or row 1) with some other column (or row). We interchange column 1 and column 2:

$$\begin{aligned} \det A &= \begin{vmatrix} 0 & 1 & 3 \\ 1 & 2 & 6 \\ -5 & -4 & -12 \end{vmatrix} = (-1) \begin{vmatrix} 1 & 0 & 3 \\ 2 & 1 & 6 \\ -4 & -5 & -12 \end{vmatrix} \\ &= (-1) \begin{vmatrix} 1 & 0 & 3 \\ 0 & 1 & 0 \\ 0 & -5 & 0 \end{vmatrix} \quad \text{(}-2 \text{ times row 1 added to row 2 and 4 times row 1 added to row 3)} \\ &= (-1) \begin{vmatrix} 1 & 0 & 3 \\ 0 & 1 & 0 \\ 0 & 0 & 0 \end{vmatrix} \quad \text{(5 times row 2 added to row 3)} \\ &= 0, \text{ by property 8.} \end{aligned}$$

A matrix whose determinant is zero is called a *singular matrix;* a matrix with a nonzero determinant is called *nonsingular.*

A system of n equations in n unknowns has a unique solution if and only if the matrix of coefficients is nonsingular.

A square matrix A has an inverse if and only if A is nonsingular.

Proofs of the above theorems may be found in Hohn.[4]

2.8 RANK

In the last section we saw how the concept of determinants could be used to determine whether or not a unique solution exists for any system of n equations in n unknowns. We would like to be able to extend this use of determinants to systems of equations in which the number of equations is not equal to the number of variables. In a system of m equations in n unknowns, $m \neq n$, the matrix of coefficients is not square and, hence, has no determinant associated with it. Thus, a more general concept than that of determinants must be introduced.

Note that by performing elementary row and column operations on a square matrix A, it is possible to obtain a new matrix B such that

$$B = \left[\begin{array}{c|c} I_r & O_{r\times(n-r)} \\ \hline O_{(n-r)\times r} & O_{(n-r)\times n(-r)} \end{array}\right] \qquad \textbf{(2-56)}$$

where n is the order of A.

Note also that B is nonsingular if and only if A is nonsingular, since performing elementary operations alters the value of the determinant in one of two ways:

1. Multiplies the determinant by (-1) (corresponding to the interchange of two parallel lines).

2. Multiplies the determinant by a nonzero constant (corresponding to the multiplication of a line by that constant).

Thus, if $\det A \neq 0$, then $\det B \neq 0$. If A is nonsingular, then in equation (2-56) $r = n$, and $B = I_n$.

Suppose, now, that elementary row and column operations are performed on a nonsquare matrix A of order $m \times n$. It is then possible to obtain a new $m \times n$ matrix B such that B will have one of the forms

$$B = \left[\begin{array}{c|c} I_r & O \\ \hline O & O \end{array}\right] \qquad \textbf{(2-57a)}$$

$$B = [I_r \mid O] \qquad \textbf{(2-57b)}$$

$$B = \left[\begin{array}{c} I_r \\ \hline O \end{array}\right] \qquad \textbf{(2-57c)}$$

In (2-57b), r must equal m; in (2-57c), r must equal n.

Regardless of the order in which the elementary operations are performed, only one of the forms defined by equations (2-57) will be obtained; moreover, r, the order of the identity matrix which appears in (2-57), is unique for any matrix A. The number r is called the *rank*† of the matrix A.

If a matrix B is obtained from a matrix A by means of one or more elementary operations, then A and B have the same rank. (This fact is a consequence of the definition of rank.) Two matrices which have the same order and rank are said to be *equivalent*. The symbol "$\sim$" between two matrices denotes the fact that they are equivalent.

EXAMPLE

$$A = \begin{bmatrix} 3 & 0 & 3 & 3 & 3 \\ 2 & -1 & 1 & 0 & 3 \\ 1 & 2 & 3 & 5 & -1 \end{bmatrix}$$

$$\sim \begin{bmatrix} 1 & 0 & 1 & 1 & 1 \\ 2 & -1 & 1 & 0 & 3 \\ 1 & 2 & 3 & 5 & -1 \end{bmatrix}$$

(multiplying row 1 by $\frac{1}{3}$: elementary operation 4)

$$\sim \begin{bmatrix} 1 & 0 & 1 & 1 & 1 \\ 0 & -1 & -1 & -2 & 1 \\ 0 & 2 & 2 & 4 & -2 \end{bmatrix}$$

(adding -2 times row 1 to row 2 and -1 times row 1 to row 3: elementary operation 5)

$$\sim \begin{bmatrix} 1 & 0 & 0 & 0 & 0 \\ 0 & -1 & -1 & -2 & 1 \\ 0 & 2 & 2 & 4 & -2 \end{bmatrix}$$

(adding -1 times column 1 to column 2, to column 3, to column 4, and to column 5: elementary operation 5)

$$\sim \begin{bmatrix} 1 & 0 & 0 & 0 & 0 \\ 0 & -1 & -1 & -2 & 1 \\ 0 & 0 & 0 & 0 & 0 \end{bmatrix}$$

(adding 2 times row 2 to row 3: elementary operation 5)

$$\sim \begin{bmatrix} 1 & 0 & 0 & 0 & 0 \\ 0 & 1 & 1 & 2 & -1 \\ 0 & 0 & 0 & 0 & 0 \end{bmatrix}$$

(multiplying row 2 by -1: elementary operation 4)

$$\sim \left[\begin{array}{cc:ccc} 1 & 0 & 0 & 0 & 0 \\ 0 & 1 & 0 & 0 & 0 \\ \hdashline 0 & 0 & 0 & 0 & 0 \end{array}\right]$$

(adding -1 times column 2 to column 3 and to column 5, and adding -2 times column 2 to column 4: elementary operation 4)

Thus, the rank of A is 2.

† Again, we have not actually defined the rank of a matrix, but have outlined a computational procedure for calculating it.

2.9 SYSTEMS OF *m* EQUATIONS IN *n* UNKNOWNS

We now wish to investigate the relationship between the existence of solutions of a system of m simultaneous equations in n unknowns and the rank of the system's coefficient matrix and augmented matrix.

We shall restrict our attention to the case $m < n$ since this is the case most often encountered in practice.

First, let us observe the following facts:

1. In any system of equations, if an equation is replaced by a new equation formed by multiplying the original equation by a nonzero scalar α, then the resulting system of equations is equivalent to the original system, in the sense that both systems will have the same solutions.

2. In any system of equations, if an equation is replaced by a new equation formed by adding k times any other equation to that equation, the resulting system of equations is equivalent to the original system.

3. In any system of equations, if two columns (for example column i and column j) of the coefficient matrix are interchanged, the resulting set of equations is equivalent to the original system, except that the vector $\bar{x}$ of unknowns now becomes

$$\bar{x} = \begin{bmatrix} x_1 \\ x_2 \\ \cdot \\ \cdot \\ \cdot \\ x_{i-1} \\ x_j \\ x_{i+1} \\ \cdot \\ \cdot \\ \cdot \\ x_{j-1} \\ x_i \\ x_{j+1} \\ \cdot \\ \cdot \\ \cdot \\ x_n \end{bmatrix} \tag{2-58}$$

That is, the components x_i and x_j are interchanged also.

Note that statements 1 and 2 represent *row* operations on the augmented matrix.

With these observations, we see that it is always possible to convert a system of m equations in n unknowns $(m < n)$ into an equivalent system of the form

$$\begin{bmatrix} 1 & \hat{a}_{12} & \hat{a}_{13} & \hat{a}_{14} & \dots & \hat{a}_{1r} & \dots & \hat{a}_{1n} \\ 0 & 1 & \hat{a}_{23} & \hat{a}_{24} & \dots & \hat{a}_{2r} & \dots & \hat{a}_{2n} \\ 0 & 0 & 1 & \hat{a}_{34} & \dots & \hat{a}_{3r} & \dots & \hat{a}_{3n} \\ \cdot & \cdot & \cdot & \cdot & & \cdot & & \cdot \\ \cdot & \cdot & \cdot & \cdot & & \cdot & & \cdot \\ \cdot & \cdot & \cdot & \cdot & & \cdot & & \cdot \\ 0 & 0 & 0 & 0 & \dots & 1 & \dots & \hat{a}_{rn} \\ 0 & 0 & 0 & 0 & \dots & 0 & \dots & 0 \\ \cdot & \cdot & \cdot & \cdot & & \cdot & & \cdot \\ \cdot & \cdot & \cdot & \cdot & & \cdot & & \cdot \\ \cdot & \cdot & \cdot & \cdot & & \cdot & & \cdot \\ 0 & 0 & 0 & 0 & \dots & 0 & \dots & 0 \end{bmatrix} \begin{bmatrix} x_1 \\ x_2 \\ x_3 \\ \cdot \\ \cdot \\ \cdot \\ \\ \\ \\ \\ \\ x_n \end{bmatrix} = \begin{bmatrix} \hat{b}_1 \\ \hat{b}_2 \\ \\ \cdot \\ \cdot \\ \cdot \\ \hat{b}_r \\ \hat{b}_{r+1} \\ \cdot \\ \cdot \\ \cdot \\ \hat{b}_m \end{bmatrix} \tag{2-59}$$

The caret ($\wedge$) above each element indicates that this is the new value of the element obtained by performing row operations on the augmented matrix of the original system. Again, the subscript r is the rank of the *coefficient matrix.*

Note that if $\hat{b}_{r+1} = \hat{b}_{r+2} = \dots = \hat{b}_m = 0$, then the last $m - r$ equations are now

$$0x_1 + 0x_2 + \dots + 0x_n = 0,$$

and can be eliminated from consideration. However, if any of these last $m - r$ components of the $\bar{b}$ vector are not 0, then at least one of the last $m - r$ equations is inconsistent; for example, if $\hat{b}_{r+1} \neq 0$, then the $(r + 1)^{\text{st}}$ equation becomes

$$0x_1 + 0x_2 + \dots + 0x_n = \hat{b}_{r+1}$$

or, $0 = \hat{b}_{r+1}$. Notice also, that the rank of the coefficient matrix equals the rank of the augmented matrix if and only if

$$\hat{b}_{r+1} = \hat{b}_{r+2} = \dots = \hat{b}_m = 0$$

Thus, *a system of m equations in n unknowns is consistent* (i.e., possesses solutions) *if and only if the rank of the coefficient matrix equals the rank of the augmented matrix.*

Moreover, if we continue to perform elementary row operations on the augmented matrix of (2-59), we can obtain an equivalent system of the form

$$\begin{bmatrix} I_r & P_{r\times(n-r)} \\ O_{(m-r)\times r} & O_{(m-r)\times(n-r)} \end{bmatrix} \begin{bmatrix} x_1 \\ x_2 \\ \cdot \\ \cdot \\ \cdot \\ x_r \\ x_{r+1} \\ \cdot \\ \cdot \\ \cdot \\ x_n \end{bmatrix} = \begin{bmatrix} Q_{r\times 1} \\ R_{(m-r)\times 1} \end{bmatrix} \tag{2-60}$$

The system is consistent only if $R = \bar{0}$, and it must be remembered that for every interchange of columns necessary to produce (2-60), the corresponding components of $\bar{x}$ must be interchanged. Writing $\bar{x}' = [\bar{x}_a', \bar{x}_b']$, with $\bar{x}_a$ being an r-component vector and $\bar{x}_b$ an $(n - r)$-component vector, we can rewrite (2-60) as

$$I_r\bar{x}_a + P\bar{x}_b = Q \tag{2-61}$$

or

$$\bar{x}_a = Q - P\bar{x}_b \tag{2-62}$$

Equation (2-62) expresses the first r components of $\bar{x}$ in terms of the remaining $n - r$ components; if the system is consistent, (2-62) represents the *general* solution of the system (2-58). The last $(n - r)$ components of $\bar{x}$ can be assigned values arbitrarily; any assignment of values to $x_{r+1}, \ldots, x_n$ yields a solution. If $r = n$, the solution is unique.

EXAMPLE

$$\left\{\begin{array}{rrrrr} 1x_1 & -2x_2 + & 2x_3 + 3x_4 & = & 2 \\ -1x_1 & 2x_2 & -1x_3 + x_4 & = & 4 \\ 2x_1 & -4x_2 + & 3x_3 + 2x_4 & = & -2 \end{array}\right\}$$

The augmented matrix is

$$A = \begin{bmatrix} 1 & -2 & 2 & 3 & 2 \\ -1 & 2 & -1 & 1 & 4 \\ 2 & -4 & 3 & 2 & -2 \end{bmatrix} \quad \text{and} \quad \bar{x} = \begin{bmatrix} x_1 \\ x_2 \\ x_3 \\ x_4 \end{bmatrix}$$

Adding appropriate multiples of row 1 to rows 2 and 3 yields

$$A \sim \begin{bmatrix} 1 & -2 & 2 & 3 & 2 \\ 0 & 0 & 1 & 4 & 6 \\ 0 & 0 & -1 & -4 & -6 \end{bmatrix}$$

Now, because the second diagonal element is 0, we must interchange either column 3 or column 4 with column 2 to yield a nonzero element in the diagonal position of row 2. (We can never interchange the *last* column with any other column, because the last column does not correspond to a variable, but rather to the $\bar{b}$ vector in the original system of equations $A\bar{x} = \bar{b}$).

We choose to interchange column 2 and column 3:

$$A \sim \begin{bmatrix} 1 & 2 & -2 & 3 & 2 \\ 0 & 1 & 0 & 4 & 6 \\ 0 & -1 & 0 & -4 & -6 \end{bmatrix}, \qquad \bar{x} = \begin{bmatrix} x_1 \\ x_3 \\ x_2 \\ x_4 \end{bmatrix}$$

Now, adding row 2 to row 3 yields

$$A \sim \begin{bmatrix} 1 & 2 & -2 & 3 & 2 \\ 0 & 1 & 0 & 4 & 6 \\ 0 & 0 & 0 & 0 & 0 \end{bmatrix} \tag{2-63}$$

Thus, we now have a set of equations of the form (2-59):

$$\begin{bmatrix} 1 & 2 & -2 & 3 \\ 0 & 1 & 0 & 4 \\ 0 & 0 & 0 & 0 \end{bmatrix} \begin{bmatrix} x_1 \\ x_3 \\ x_2 \\ x_4 \end{bmatrix} = \begin{bmatrix} 2 \\ 6 \\ 0 \end{bmatrix}$$

If we now add -2 times row 2 to row 1, in the matrix of (2-83) we obtain

$$A \sim \begin{bmatrix} 1 & 0 & -2 & -5 & -10 \\ 0 & 1 & 0 & 4 & 6 \\ 0 & 0 & 0 & 0 & 0 \end{bmatrix} \tag{2-64}$$

The matrix of (2-64) provides us with the equivalent system of equations of the form of equation (2-60):

$$\left[\begin{array}{cc|cc} 1 & 0 & -2 & -5 \\ 0 & 1 & 0 & 4 \\ \hline 0 & 0 & 0 & 0 \end{array}\right] \begin{bmatrix} x_1 \\ x_3 \\ x_2 \\ x_4 \end{bmatrix} = \left[\begin{array}{c} -10 \\ 6 \\ \hline 0 \end{array}\right]$$

or

$$\begin{bmatrix} I_2 & P_{2\times 2} \\ O_{1\times 2} & O_{1\times 2} \end{bmatrix} \begin{bmatrix} x_1 \\ x_3 \\ x_2 \\ x_4 \end{bmatrix} = \begin{bmatrix} Q_{2\times 1} \\ R_{1\times 1} \end{bmatrix}$$

where

$$P = \begin{bmatrix} -2 & -5 \\ 0 & 4 \end{bmatrix} \qquad Q = \begin{bmatrix} -10 \\ 0 \end{bmatrix} \qquad R = [0]$$

Since $R = \bar{0}$, the system is consistent, and its general solution, in the form of equation (2-62), is

$$\begin{bmatrix} x_1 \\ x_3 \end{bmatrix} = \begin{bmatrix} -10 \\ 6 \end{bmatrix} - \begin{bmatrix} -2 & -5 \\ 0 & 4 \end{bmatrix} \begin{bmatrix} x_2 \\ x_4 \end{bmatrix}$$

or

$$\left\{ \begin{aligned} x_1 &= -10 + 2x_2 + 5x_4 \\ x_3 &= 6 - 4x_4 \end{aligned} \right\}$$

2.10 SYSTEMS OF n EQUATIONS IN n UNKNOWNS

The discussion of the previous section is applicable to the special case in which the number of unknowns equals the number of equations. If we also assume that the rank of the coefficient matrix is equal to the number of equations (and to the number of unknowns) then the equivalent system (2-59) takes the form

$$\begin{bmatrix} 1 & \hat{a}_{12} & \hat{a}_{13} & \cdots & \hat{a}_{1n} \\ 0 & 1 & \hat{a}_{23} & \cdots & \hat{a}_{2n} \\ \cdot & \cdot & \cdot & \cdots & \cdot \\ \cdot & \cdot & \cdot & \cdots & \cdot \\ \cdot & \cdot & \cdot & \cdots & \cdot \\ 0 & 0 & 0 & \cdots & 1 \end{bmatrix} \begin{bmatrix} x_1 \\ x_2 \\ \cdot \\ \cdot \\ \cdot \\ x_n \end{bmatrix} = \begin{bmatrix} \hat{b}_1 \\ \hat{b}_2 \\ \cdot \\ \cdot \\ \cdot \\ \hat{b}_n \end{bmatrix} \tag{2-65}$$

Or, in equation form, (2-65) becomes

$$\left\{ \begin{aligned} x_1 + \hat{a}_{12}x_2 + \hat{a}_{13}x_3 + \ldots + \hat{a}_{1n}x_n &= \hat{b}_1 \\ x_2 + \hat{a}_{23}x_3 + \ldots + \hat{a}_{2n}x_n &= \hat{b}_2 \\ x_3 + \ldots + \hat{a}_{3n}x_n &= \hat{b}_3 \\ &\;\cdot \\ &\;\cdot \\ &\;\cdot \\ x_n &= \hat{b}_n \end{aligned} \right\} \tag{2-66}$$

From the nth equation of the system (2-66), we see that $x_n = \hat{b}_n$; if we substitute this value into the $(n-1)^{\text{st}}$ equation of (2-66), we can solve directly for x_{n-1}:

$$\begin{aligned} x_{n-1} &= \hat{b}_{n-1} - \hat{a}_{n-1,n} x_n \\ &= \hat{b}_{n-1} - \hat{a}_{n-1,n} \hat{b}_n \end{aligned} \tag{2-67}$$

Then, equation (2-67) can be substituted into the $(n-2)^{\text{nd}}$ equation of (2-66) and x_{n-2} can then be determined. Continuing in this manner, each of the unknowns can be solved in reverse order:

$$x_n, x_{n-1}, x_{n-2}, \ldots, x_2, x_1$$

The entire procedure of reducing a system of n equations in n unknowns and then solving successively for $x_n, x_{n-1}, \ldots, x_2, x_1$ is called the *method of reduction*, or *Gaussian Elimination*.† The latter part of the method (solving successively for $x_n, x_{n-1}, \ldots, x_2, x_1$) is called the *backward pass*.

EXAMPLE

$$\left\{\begin{aligned} x_1 - 2x_2 + x_3 &= 0 \\ 2x_1 - 2x_2 + 4x_3 &= 10 \\ x_1 + x_2 + 2x_3 &= 9 \end{aligned}\right\} \tag{2-68}$$

The augmented matrix of (2-68) is

$$A = \begin{bmatrix} 1 & -2 & 1 & 0 \\ 2 & -2 & 4 & 10 \\ 1 & 1 & 2 & 9 \end{bmatrix}$$

Adding -2 times row 1 to row 2 and -1 times row 1 to row 3 yields

$$A \sim \begin{bmatrix} 1 & -2 & 1 & 0 \\ 0 & 2 & 2 & 10 \\ 0 & 3 & 1 & 9 \end{bmatrix}$$

Multiplying row 2 by $\frac{1}{2}$ yields

$$A \sim \begin{bmatrix} 1 & -2 & 1 & 0 \\ 0 & 1 & 1 & 5 \\ 0 & 3 & 1 & 9 \end{bmatrix}$$

† For a more comprehensive treatment of Gaussian Elimination, see: Franklin, J. N.: *Matrix Theory*. Prentice-Hall, Inc., New Jersey (1968). Forsythe, G., and Moler, C. B.: *Computer Solution of Linear Algebraic Systems*. Prentice-Hall, Inc., New Jersey (1967).

Adding −3 times row 2 to row 3 yields

$$A \sim \begin{bmatrix} 1 & -2 & 1 & 0 \\ 0 & 1 & 1 & 5 \\ 0 & 0 & -2 & -6 \end{bmatrix}$$

Finally, multiplying row 3 by $-\frac{1}{2}$ yields

$$A \sim \begin{bmatrix} 1 & -2 & 1 & 0 \\ 0 & 1 & 1 & 5 \\ 0 & 0 & 1 & 3 \end{bmatrix}$$

Thus, the equivalent system of (2-68) in the form (2-65) is

$$\begin{bmatrix} 1 & -2 & 1 \\ 0 & 1 & 1 \\ 0 & 0 & 1 \end{bmatrix} \begin{bmatrix} x_1 \\ x_2 \\ x_3 \end{bmatrix} = \begin{bmatrix} 0 \\ 5 \\ 3 \end{bmatrix}$$

Performing the backward pass, we obtain

$$x_3 = 3$$
$$x_2 = 5 - 1x_3 = 5 - 1(3) = 2$$
$$x_1 = 0 - (-2)x_2 - 1x_3 = +2(+2) - 1(3) = 1$$

The solution to (2-68) is therefore

$$x_1 = 1 \qquad x_2 = 2 \qquad x_3 = 3$$

The reader should note that certain modifications must be made for Gaussian Elimination when the method is implemented on a digital computer or desk calculator. These modifications are necessary because some of the calculations when performed on a machine with a finite number of digits introduce round-off errors. These errors can become significantly large if certain precautions are not taken. These modifications primarily involve a reordering of the rows or columns of the augmented matrix in such a way that possible round-off effects tend to be minimized. For a thorough discussion of these procedures, the reader should consult Forsythe and Moler.[1]

2.11 LINEAR INDEPENDENCE; VECTOR SPACES

In this section we shall investigate some properties of sets of vectors. Let us begin by defining a "linear combination" of vectors:

If $\alpha_1, \alpha_2, \ldots, \alpha_k$ are scalars, then the vector

$$\bar{y} = \alpha_1\bar{x}_1 + \alpha_2\bar{x}_2 + \ldots + \alpha_k\bar{x}_k \qquad \textbf{(2-69)}$$

is called a *linear combination* of the vectors $\bar{x}_1, \bar{x}_2, \ldots, \bar{x}_k$.

Suppose, now, that we are given a set of vectors

$$X = \{\bar{x}_1, \bar{x}_2, \ldots, \bar{x}_k\}$$

These vectors are said to be *linearly independent* if the only set of scalars $\alpha_1, \alpha_2, \ldots, \alpha_k$ for which

$$\alpha_1\bar{x}_1 + \alpha_2\bar{x}_2 + \ldots + \alpha_k\bar{x}_k = \bar{0} \tag{2-70}$$

is the set $\alpha_1 = \alpha_2 = \ldots = \alpha_k = 0$. If there exists a set of scalars, not all zero, which satisfies equation (2-70), then the vectors $\bar{x}_1, \bar{x}_2, \ldots, \bar{x}_k$ are said to be *linearly dependent.*

If a set of vectors is linearly dependent, then at least one of the $\alpha_1, \alpha_2, \ldots, \alpha_k$ must be nonzero. If we assume that α_j is nonzero, then we can rewrite equation (2-70) as follows:

$$-\alpha_j\bar{x}_j = \alpha_1\bar{x}_1 + \alpha_2\bar{x}_2 + \ldots + \alpha_{j-1}\bar{x}_{j-1} + \alpha_{j+1}\bar{x}_{j+1} + \ldots + \alpha_k\bar{x}_k$$

or,

$$\bar{x}_j = \left(-\frac{\alpha_1}{\alpha_j}\right)\bar{x}_1 + \left(-\frac{\alpha_2}{\alpha_j}\right)\bar{x}_2 + \ldots + \left(-\frac{\alpha_{j-1}}{\alpha_j}\right)\bar{x}_{j-1} + \left(-\frac{\alpha_{j+1}}{\alpha_j}\right)\bar{x}_{j+1} + \ldots + \left(\frac{-\alpha_k}{\alpha_j}\right)\bar{x}_k \tag{2-71}$$

In equation (2-71) we have expressed $\bar{x}_j$ as a linear combination of $\bar{x}_1, \bar{x}_2, \ldots, \bar{x}_{j-1}, \bar{x}_{j+1}, \ldots, \bar{x}_k$. In general, we see that if a set of vectors is linearly dependent, then at least one of them can be expressed as a linear combination of the rest.

EXAMPLES

(a) $\bar{x}_1 = \begin{bmatrix} 1 \\ 3 \\ 1 \\ -1 \end{bmatrix} \quad \bar{x}_2 = \begin{bmatrix} 5 \\ 2 \\ 3 \\ 0 \end{bmatrix} \quad \bar{x}_3 = \begin{bmatrix} 7 \\ 8 \\ 5 \\ -2 \end{bmatrix}$

These vectors are linearly dependent, since

$$2\bar{x}_1 + \bar{x}_2 - \bar{x}_3 = \bar{0}$$

(b) $\bar{x}_1 = \begin{bmatrix} 1 \\ 3 \\ 1 \\ -1 \end{bmatrix} \quad \bar{x}_2 = \begin{bmatrix} 5 \\ 2 \\ 3 \\ 0 \end{bmatrix} \quad \bar{x}_3 = \begin{bmatrix} 7 \\ 8 \\ 5 \\ 1 \end{bmatrix}$

The vector equation

$$\alpha_1\bar{x}_1 + \alpha_2\bar{x}_2 + \alpha_3\bar{x}_3 = \bar{0} \tag{2-72}$$

becomes a set of simultaneous equations:

$$\left.\begin{cases} \alpha_1 + 5\alpha_2 + 7\alpha_3 = 0 & \text{(2-73a)} \\ 3\alpha_1 + 2\alpha_2 + 8\alpha_3 = 0 & \text{(2-73b)} \\ \alpha_1 + 3\alpha_2 + 5\alpha_3 = 0 & \text{(2-73c)} \\ -\alpha_1 \qquad + \alpha_3 = 0 & \text{(2-73d)} \end{cases}\right\}$$

From equation (2-73d), we find that $\alpha_3 = \alpha_1$. Equations (2-73a, b, c) then become

$$\left.\begin{cases} 8\alpha_1 + 5\alpha_2 = 0 & \text{(2-74a)} \\ 11\alpha_1 + 2\alpha_2 = 0 & \text{(2-74b)} \\ 6\alpha_1 + 3\alpha_2 = 0 & \text{(2-74c)} \end{cases}\right\}$$

From equation (2-74c), we find that $\alpha_2 = -2\alpha_1$. Substituting this result into equations (2-74a, b) yields

$$\left.\begin{cases} 8\alpha_1 + 5(-2\alpha_1) = -2\alpha_1 = 0 & \text{(2-75a)} \\ 11\alpha_1 + 2(-2\alpha_1) = 7\alpha_1 = 0 & \text{(2-75b)} \end{cases}\right\}$$

Equations (2-75) imply that $\alpha_1 = 0$; therefore, $\alpha_2 = -2\alpha_1 = 0$, $\alpha_3 = \alpha_1 = 0$. Thus, the only solution to equation (2-72) is the trivial solution $\alpha_1 = \alpha_2 = \alpha_3 = 0$. The vectors $\bar{x}_1$, $\bar{x}_2$, $\bar{x}_3$, $\bar{x}_4$ are therefore linearly independent.

Equation (2-70) may be considered as a set of n simultaneous equations in k unknowns:

Letting

$$\bar{x}_1 = \begin{bmatrix} x_{11} \\ x_{21} \\ x_{31} \\ \cdot \\ \cdot \\ \cdot \\ x_{n1} \end{bmatrix} \quad \bar{x}_2 = \begin{bmatrix} x_{12} \\ x_{22} \\ x_{32} \\ \cdot \\ \cdot \\ \cdot \\ x_{n2} \end{bmatrix} \quad \ldots \quad \bar{x}_k = \begin{bmatrix} x_{1k} \\ x_{2k} \\ x_{3k} \\ \cdot \\ \cdot \\ \cdot \\ x_{nk} \end{bmatrix} \quad \bar{\alpha} = \begin{bmatrix} \alpha_1 \\ \alpha_2 \\ \cdot \\ \cdot \\ \cdot \\ \alpha_k \end{bmatrix}$$

and $X_{(n\times k)} = [\bar{x}_1, \bar{x}_2, \ldots, \bar{x}_k]$, we obtain

$$X\bar{\alpha} = \bar{0} \tag{2-76}$$

Thus, the results of Section 2.9 indicate that there is a unique solution to equation (2-76) if and only if the rank of X is equal to k, the number of unknowns. In this case, obviously, the unique solution is $\bar{\alpha} = \bar{0}$. *The vectors* $\bar{x}_1, \bar{x}_2, \ldots, \bar{x}_k$ *are linearly independent, therefore, if and only if the rank of the matrix* $X = [\bar{x}_1, \bar{x}_2, \ldots, \bar{x}_k]$ *is* k.

It is clear that if $k > n$, then the rank of X cannot exceed n, and hence $\bar{x}_1, \bar{x}_2, \ldots, \bar{x}_k$ will always be linearly dependent. In other words, the maximum number of linearly independent n-component vectors is n. Any set of $n + 1$ vectors is linearly dependent.

Let us now show that given any set of n linearly independent (n-component) vectors, $\bar{x}_1, \bar{x}_2, \ldots, \bar{x}_n$ any other (n-component) vector $\bar{y}$ can be expressed as a unique linear combination of $\bar{x}_1, \bar{x}_2, \ldots, \bar{x}_n$. In other words, first we wish to show that there exists a set of scalars $\alpha_1, \alpha_2, \ldots, \alpha_n$ such that

$$\alpha_1\bar{x}_1 + \alpha_2\bar{x}_2 + \ldots + \alpha_n\bar{x}_n = \bar{y} \tag{2-77}$$

But equation (2-77) is nothing more than a system of n equations in n unknowns:

$$X\bar{\alpha} = \bar{y} \tag{2-78}$$

Since $\bar{x}_1, \bar{x}_2, \ldots, \bar{x}_n$ are linearly independent, the rank of the coefficient matrix is n, and the rank of the augmented matrix $[X, \bar{y}]$ is also n. (It cannot be $n + 1$, because the matrix only has n rows.) Thus, equation (2-78) always possesses solutions. Moreover, X is nonsingular; hence the solution is unique.

We turn now to a brief discussion of sets of vectors with certain special properties. First, we need a few definitions: A set of vectors is said to be *closed under addition if the sum of any two vectors in the set is also in the set. The set is said to be closed under scalar multiplication if the product of any vector in the set by any scalar yields a vector in the set.*

A set of vectors which is closed under addition and scalar multiplication is called a *vector space.* In the previous section, we noted that the maximum number of linearly independent vectors in the set of all n-component vectors is n. Thus, for every subset of this set, there exists some maximum number of linearly independent vectors. In particular, every vector space has a maximum number of linearly independent vectors. This number is called the *dimension* of the vector space. If a vector space has dimension k, then any set of k linearly independent vectors in the vector space is called a *basis* of the vector space. Any other vector in the vector space can be expressed as a unique linear combination of a given set of basis vectors.

EXAMPLE

The set $S = \left\{ \begin{bmatrix} x_1 \\ x_2 \\ x_3 \end{bmatrix} \middle| \, x_3 = 0 \right\}$ is a vector space.

To see this, let $\bar{x} = \begin{bmatrix} a \\ b \\ 0 \end{bmatrix}$ and $\bar{y} = \begin{bmatrix} c \\ d \\ 0 \end{bmatrix}$.

Then, $\bar{x} + \bar{y} = \begin{bmatrix} a + c \\ b + d \\ 0 \end{bmatrix} \in S$. (Since the scalars a, b, c, d are completely arbitrary, the vectors $\bar{x}$ and $\bar{y}$ denote any two vectors in S.) Also, for any scalar α, $\alpha\bar{x} = \begin{bmatrix} \alpha a \\ \alpha b \\ 0 \end{bmatrix} \in S$. Thus, S is closed under addition and scalar multiplication.

Let us now determine the dimension of S. Note that if X is a matrix whose columns are vectors in S, then X has three rows, and the last row contains only zeros. Thus, the rank of X must be less than or equal to two. Hence, the dimension of S is either one or two. To show that it is in fact, two, we need only find two linearly independent vectors. This is quite easy to do. Following are three such sets of two linearly independent vectors from S:

(i) $\begin{bmatrix} 1 \\ 0 \\ 0 \end{bmatrix}, \begin{bmatrix} 0 \\ 1 \\ 0 \end{bmatrix}$ (ii) $\begin{bmatrix} 2 \\ 1 \\ 0 \end{bmatrix}, \begin{bmatrix} 0 \\ 1 \\ 0 \end{bmatrix}$ (iii) $\begin{bmatrix} 1 \\ 1 \\ 0 \end{bmatrix}, \begin{bmatrix} 1 \\ -1 \\ 0 \end{bmatrix}$

Each of these three sets is a basis for S. Any vector in S can be expressed as a linear combination of each of these sets:

If $\bar{x} = \begin{bmatrix} a \\ b \\ 0 \end{bmatrix}$ is any element of S, then

$$\begin{bmatrix} a \\ b \\ 0 \end{bmatrix} = a\begin{bmatrix} 1 \\ 0 \\ 0 \end{bmatrix} + b\begin{bmatrix} 0 \\ 1 \\ 0 \end{bmatrix}$$

$$\begin{bmatrix} a \\ b \\ 0 \end{bmatrix} = (\tfrac{1}{2}a)\begin{bmatrix} 2 \\ 1 \\ 0 \end{bmatrix} + (b - \tfrac{1}{2}a)\begin{bmatrix} 0 \\ 1 \\ 0 \end{bmatrix}$$

$$\begin{bmatrix} a \\ b \\ 0 \end{bmatrix} = \left(\frac{a + b}{2}\right)\begin{bmatrix} 1 \\ 1 \\ 0 \end{bmatrix} + \left(\frac{a - b}{2}\right)\begin{bmatrix} 1 \\ -1 \\ 0 \end{bmatrix}$$

2.12 QUADRATIC FORMS AND DEFINITE MATRICES

We conclude this chapter with several definitions concerning a special type of matrix which we will encounter in later chapters. Consider the function $F(\bar{x}) \equiv F(x_1, x_2, \ldots, x_n)$ of the n variables $x_1, x_2, \ldots, x_n$:

$$
\begin{aligned}
F(\bar{x}) &= \sum_{i=1}^{n} \sum_{j=1}^{n} p_{ij}x_ix_j \\
&= (p_{11}x_1^2 + p_{12}x_1x_2 + \ldots + p_{1n}x_1x_n) \\
&\quad + (p_{21}x_2x_1 + p_{22}x_2^2 + \ldots + p_{2n}x_2x_n) \\
&\quad + \ldots + (p_{n1}x_nx_1 + \ldots + p_{nn}x_n^2)
\end{aligned} \quad \textbf{(2-79)}
$$

The function $F(\bar{x})$ is called a *quadratic form;* each term contains either the square of one variable or the product of two different variables. $F(\bar{x})$ can be expressed in matrix notation, as follows: Let $P = [p_{ij}]_{(n\times n)}$ and $\bar{x}' = [x_1, x_2, \ldots, x_n]$; then let $\bar{y} = P\bar{x}$. Thus, if $\bar{y}' = [y_1, y_2, \ldots, y_n]$, then

$$
y_i = \sum_{j=1}^{n} p_{ij}x_j \quad \textbf{(2-80)}
$$

Note that we can rewrite equation (2-79) as

$$
F(\bar{x}) = \sum_{i=1}^{n} x_i \sum_{j=1}^{n} p_{ij}x_j \quad \textbf{(2-81)}
$$

Substituting equation (2-80) into (2-81) yields

$$
F(\bar{x}) = \sum_{i=1}^{n} x_iy_i \quad \textbf{(2-82)}
$$

But, the right-hand side of equation (2-82) represents the scalar product of $\bar{x}$ with $\bar{y}$. Thus,

$$
F(\bar{x}) = \bar{x}'\bar{y} = \bar{x}'P\bar{x} \quad \textbf{(2-83)}
$$

The matrix P, then, completely specifies the quadratic form; P is called the matrix associated with the quadratic form $F(\bar{x})$.

Looking again at equation (2-79), we see that, except for the terms involving x_j^2, each product of the form x_ix_j $(i \neq j)$ appears twice; that is, equation (2-79) may be rewritten as

$$
\begin{aligned}
F(\bar{x}) & \\
&= (p_{11}x_1^2 + p_{22}x_2^2 + \ldots + p_{nn}^2x_n^2) \\
&\quad + [(p_{12} + p_{21})x_1x_2 + (p_{13} + p_{31})x_1x_3 + \ldots + (p_{1n} + p_{n1})x_1x_n] \\
&\quad + [(p_{23} + p_{32})x_2x_3 + (p_{24} + p_{42})x_2x_4 + \ldots + (p_{2n} + p_{n2})x_2x_n \\
&\quad + \ldots + [(p_{n-1,n} + p_{n,n-1})]x_{n-1}x_n
\end{aligned} \quad \textbf{(2-84)}
$$

Thus, for $i \neq j$, the coefficient of $x_i x_j$ is $(p_{ij} + p_{ji})$. Defining

$$A = [a_{ij}]_{(n \times n)}$$

where

$$a_{ij} = \frac{p_{ij} + p_{ji}}{2}, \qquad \text{all } i, j \tag{2-85}$$

we see that

$$a_{ij} + a_{ji} = p_{ij} + p_{ji}$$

and so

$$F(\bar{x}) = \bar{x}'P\bar{x} = \bar{x}'A\bar{x} \tag{2-86}$$

Hence, replacing P with A does not change the quadratic form. Observe that A is symmetric. Given any quadratic form $\bar{x}'P\bar{x}$, then, we can always replace P with a symmetric matrix. This argument also illustrates the fact that different matrices may be associated with the same quadratic form.

EXAMPLE

$$\begin{aligned} F(x_1, x_2, x_3) &= x_1^2 + 4x_1x_2 + 2x_1x_3 - 7x_2^2 - 6x_2x_3 + 5x_3^2 \\ &= [x_1, x_2, x_3]\begin{bmatrix} 1 & 4 & 2 \\ 0 & -7 & -6 \\ 0 & 0 & 5 \end{bmatrix}\begin{bmatrix} x_1 \\ x_2 \\ x_3 \end{bmatrix} \\ &= [x_1, x_2, x_3]\begin{bmatrix} 1 & 2 & 1 \\ 2 & -7 & -3 \\ 1 & -3 & 5 \end{bmatrix}\begin{bmatrix} x_1 \\ x_2 \\ x_3 \end{bmatrix} \end{aligned}$$

Quadratic forms may be either positive or negative or zero for any fixed $\bar{x}$; however, a quadratic form $F(\bar{x}) = \bar{x}'P\bar{x}$ may have the property that it is always positive (except for $F(\bar{0})$). Such a quadratic form is said to be *positive definite*. Similarly a quadratic form $\bar{x}'P\bar{x}$ is said to be *negative definite* if $\bar{x}'P\bar{x} < 0$ for all $\bar{x}$ except $\bar{x} = \bar{0}$. If a quadratic form $\bar{x}'P\bar{x}$ has the property that $\bar{x}'P\bar{x} \geq 0$ for all $\bar{x}$ and there exists at least one $\bar{x} \neq \bar{0}$ such that $\bar{x}'P\bar{x} = 0$, then $\bar{x}'P\bar{x}$ is said to be *positive semidefinite*. A similar definition for *negative semidefinite* quadratic forms is obtained by reversing the sense of the inequality. A quadratic form which is positive for some vectors $\bar{x}$ and negative for others is said to be *indefinite*. A symmetric matrix A is often referred to as a positive definite, positive semidefinite, negative definite, negative semidefinite, or indefinite matrix if the quadratic form associated with A is positive definite, positive semidefinite, negative definite, negative semidefinite, or indefinite, respectively.

EXAMPLE

(a) $A = \begin{bmatrix} 1 & 0 & 0 \\ 0 & 3 & 0 \\ 0 & 0 & 2 \end{bmatrix}$ is positive definite, since

$$\bar{x}'A\bar{x} = x_1^2 + 3x_2^2 + 2x_2^2 > 0$$

unless $x_1 = x_2 = x_3 = 0$

(b) $A = \begin{bmatrix} -1 & 0 & 0 \\ 0 & -1 & -1 \\ 0 & -1 & -1 \end{bmatrix}$ is negative semidefinite, since

$$\begin{aligned} \bar{x}'A\bar{x} &= -x_1^2 - x_2^2 + 2x_2x_3 - x_3^2 \\ &= -x_1^2 - (x_2 - x_3)^2 \leq 0 \end{aligned}$$

for all $\bar{x}$ and $\bar{x}'A\bar{x} = 0$ when $x_1 = 0$, $x_2 = x_3$ (e.g., $\bar{x}'A\bar{x} = 0$ when $\bar{x}' = [0, 1, 1]$)

EXERCISES

Operations with Matrices

1. Consider the following matrices:

$$A = \begin{bmatrix} 1 & 3 & -1 \\ 0 & 2 & 0 \\ 2 & 1 & 0 \\ 3 & 0 & 2 \end{bmatrix} \qquad B = \begin{bmatrix} 1 & -1 & 2 & 4 \\ 2 & 0 & -1 & 0 \\ 0 & -1 & 3 & 0 \end{bmatrix}$$

$$C = \begin{bmatrix} 0 & -1 & 0 & 2 \\ 2 & 0 & 2 & -1 \\ 1 & 0 & 3 & 1 \\ 0 & 1 & 4 & 0 \end{bmatrix} \qquad D = \begin{bmatrix} 1 & 0 & 0 & 0 \\ 0 & 2 & 0 & 0 \\ 0 & 0 & -1 & 0 \\ 0 & 0 & 0 & 0 \end{bmatrix}$$

$$E = \begin{bmatrix} 2 & 3 \\ 0 & 1 \\ 1 & 0 \\ 2 & -1 \end{bmatrix} \qquad \bar{x} = \begin{bmatrix} -1 \\ 2 \\ 0 \\ 1 \end{bmatrix} \qquad \bar{y} = \begin{bmatrix} 1 \\ -1 \\ 4 \\ 3 \end{bmatrix}$$

For each of the following operations state whether the operations are defined, and if so, perform the indicated operations:

(a) AB	(b) BA	(c) AC
(d) CA	(e) $A'C$	(f) $C'A$
(g) $(C'A)'$	(h) AD	(i) DA
(j) BE	(k) $E'B$	(l) $B\bar{x}$
(m) $\bar{x}'\bar{y}$	(n) $\bar{x}\bar{y}'$	(o) $\bar{x}\bar{y}$

2. Show that for any square matrix A and any scalar matrix D of the same order, $AD = DA$.

3. If D is an nth order diagonal matrix and A is any n^{th} order matrix, is it necessarily true that $AD = DA$? Illustrate with an example.

4. Show that $(AB)' = B'A'$. Be sure to state the orders of A and B required for the foregoing relations to be defined.

5. Show that, if A and B are both nonsingular square matrices, then $(AB)^{-1} = B^{-1}A^{-1}$. (Hint: Consider the equation

$$(AB)(AB)^{-1} = I.)$$

6. Use the result of exercise 4 to show that, for any matrix A, the matrices AA' and $A'A$ are symmetric.

7. The product of a square matrix A with itself is usually written $AA \equiv A^2$. Construct an example to show that, in general, $(A + B)(A - B) \neq A^2 - B^2$, and $(A + B)^2 \neq A^2 + 2AB + B^2$. Under what conditions do the preceding relations hold as equalities?

Determinants and Rank

8. Evaluate the following determinants:

$$\text{(a)} \begin{vmatrix} 1 & 0 & 2 \\ 1 & 1 & 3 \\ 2 & 1 & 6 \end{vmatrix} \qquad \text{(b)} \begin{vmatrix} 1 & 2 & 0 & 3 & 4 & 5 \\ 4 & 1 & 6 & 7 & 6 & 8 \\ -1 & 0 & 0 & 1 & 2 & 1 \\ 2 & 4 & 0 & 9 & 1 & 2 \\ 1 & 2 & 0 & 3 & 8 & 6 \\ 2 & 2 & 0 & 2 & 2 & 9 \end{vmatrix}$$

(c) $\begin{vmatrix} 0 & 1 & 2 & 3 \\ 0 & 2 & 4 & 5 \\ 1 & -2 & -3 & 1 \\ 1 & -2 & 1 & 1 \end{vmatrix}$ (d) $\begin{vmatrix} 0 & 0 & 0 & -4 \\ 0 & 0 & 3 & 1 \\ 0 & 2 & 1 & 1 \\ 1 & 2 & -1 & 3 \end{vmatrix}$

(e) $\begin{vmatrix} 0 & 0 & 0 & 0 & -4 \\ 0 & 0 & 0 & 3 & 1 \\ 0 & 0 & 2 & 1 & 1 \\ 0 & 1 & 2 & -1 & 3 \\ 1 & 0 & 0 & 0 & 0 \end{vmatrix}$ (f) $\begin{vmatrix} 1 & 3 & -2 & 1 \\ 2 & 0 & -4 & 3 \\ 0 & 1 & -1 & 0 \\ -1 & 5 & 0 & -2 \end{vmatrix}$

9. For what values of the scalar λ do each of the following determinants vanish?

(a) $\begin{vmatrix} 1-\lambda & 3 & 2 \\ 0 & 2-\lambda & 5 \\ 0 & 0 & 1-\lambda \end{vmatrix}$

(b) $\begin{vmatrix} 1-\lambda & 0 & 0 \\ 0 & 1-\lambda & 1 \\ 0 & 2 & 2-\lambda \end{vmatrix}$

10. If P is any nonsingular matrix and A is any square matrix (of the same order), show that $\det(A - \lambda I) = \det(P^{-1}AP - \lambda I)$ for any scalar λ.

11. Determine the ranks of the following matrices:

(a) $\begin{bmatrix} 0 & 1 & 2 & 3 \\ 0 & 2 & 4 & 5 \\ 1 & -2 & -3 & 1 \\ 1 & -2 & 1 & 1 \end{bmatrix}$

(b) $\begin{bmatrix} 0 & 1 & 2 & 3 & 1 & 0 \\ 0 & 2 & 4 & 5 & 0 & 1 \\ 1 & -2 & -3 & 1 & 0 & 0 \\ 1 & -2 & 1 & 1 & 0 & 0 \end{bmatrix}$

(c) $\begin{bmatrix} 1 & 3 & 1 & 2 & 0 \\ 1 & -1 & 2 & 0 & 1 \\ 2 & 2 & 3 & 2 & 1 \\ 3 & 1 & 5 & 2 & 2 \end{bmatrix}$

(d) $\begin{bmatrix} 1 & 0 & 1 & 0 \\ 0 & -1 & 2 & 3 \\ 1 & 2 & 0 & 1 \\ 2 & 1 & 3 & 4 \\ 2 & 2 & 1 & 1 \\ 1 & 1 & 2 & 4 \end{bmatrix}$

Systems of Equations

12. In a given system of m equations in n unknowns, $A\bar{x} = \bar{b}$, with $m \leq n$, show that if the rank of A is equal to m, then the system is consistent.

13. In a system of n equations in n unknowns, $A\bar{x} = \bar{b}$, if the rank of A is equal to the rank of $[A, \bar{b}]$, does it necessarily follow that A^{-1} exists? Illustrate with an example.

14. Obtain the general solution for each of the following systems of equations, or determine that the system is inconsistent:

(a) $$\begin{Bmatrix} x_1 + 3x_2 + x_3 + 2x_4 = 0 \\ x_1 - x_2 + 2x_3 = 1 \\ 2x_1 + 2x_2 + 3x_3 + 2x_4 = 1 \\ 3x_1 + x_2 + 5x_3 + 2x_4 = 2 \end{Bmatrix}$$

(b) $$\begin{Bmatrix} x_2 + 2x_3 + x_4 = 1 \\ x_1 - 2x_2 - 3x_3 + x_4 = 0 \\ x_1 - 2x_2 + x_3 + x_4 = 1 \end{Bmatrix}$$

(c) $$\begin{Bmatrix} x_2 + x_3 + x_4 + 2x_5 + 2x_6 = 7 \\ 2x_1 + x_2 + 5x_3 + 3x_4 + 4x_5 + 2x_6 = 17 \\ 3x_1 - x_2 + 5x_3 + 2x_4 + x_5 - 2x_6 = 8 \\ -x_1 + 2x_2 + x_4 + 3x_5 + 4x_6 = 9 \end{Bmatrix}$$

15. For each of the systems of equations of exercise 14, find an equivalent system of the form of equation (2-60).

16. Solve each of the following systems of equations by Gaussian Elimination:

$$\text{(a)}\quad \begin{Bmatrix} x_1 + 2x_2 + x_3 = 4 \\ x_1 - x_2 + 2x_3 = 2 \\ x_2 + x_3 = 2 \end{Bmatrix}$$

$$\text{(b)}\quad \begin{Bmatrix} 4x_2 - x_3 = 3 \\ x_1 + x_3 = 2 \\ x_1 - 3x_2 = -2 \end{Bmatrix}$$

$$\text{(c)}\quad \begin{Bmatrix} x_1 + 3x_2 + x_3 = 1 \\ 5x_2 - 6x_3 = 0 \\ x_1 - 2x_2 + 4x_3 = 1 \end{Bmatrix}$$

Linear Independence and Vector Spaces

17. Determine which of the following sets of vectors is linearly independent and which is linearly dependent:

$$\text{(a)}\quad \begin{bmatrix} 1 \\ 2 \\ 3 \end{bmatrix}, \begin{bmatrix} 3 \\ 4 \\ -1 \end{bmatrix}, \begin{bmatrix} 3 \\ 2 \\ 0 \end{bmatrix}, \begin{bmatrix} 6 \\ 0 \\ 1 \end{bmatrix}$$

$$\text{(b)}\quad \begin{bmatrix} 1 \\ 2 \\ 3 \\ 4 \\ 5 \end{bmatrix}, \begin{bmatrix} 2 \\ -1 \\ 0 \\ -1 \\ 1 \end{bmatrix}, \begin{bmatrix} 5 \\ 0 \\ 3 \\ 2 \\ 7 \end{bmatrix}$$

$$\text{(c)}\quad \begin{bmatrix} 1 \\ 2 \\ 3 \\ 4 \\ 5 \end{bmatrix}, \begin{bmatrix} 2 \\ -1 \\ 0 \\ -1 \\ 1 \end{bmatrix}, \begin{bmatrix} 5 \\ 0 \\ 3 \\ 2 \\ 7 \end{bmatrix}, \begin{bmatrix} 5 \\ 1 \\ -1 \\ 3 \\ 2 \end{bmatrix}$$

18. Show that if some subset of a set of vectors is linearly dependent, then the entire set is linearly dependent.

19. Show that if a set of vectors is linearly independent, then every subset of this set is a linearly independent set of vectors.

20. Show that every set of k n-component vectors is linearly dependent, if $k > n$.

21. Determine whether each of the following sets is a vector space:

(a) $S_1 = \{[x_1, x_2] \mid x_1 = 0\}$
(b) $S_2 = \{[x_1, x_2] \mid x_1 = 1\}$
(c) $S_3 = \{[x_1, x_2, x_3] \mid 2x_1 - 3x_3 = 0\}$
(d) $S_4 = \{[x_1, x_2, x_3] \mid x_1 + x_2 = 3\}$
(e) $S_5 = \{[x_1, x_2, x_3, x_4, x_5] \mid x_1 = 0;\ x_2 = x_3\}$
(f) $S_6 = \{[x_1, x_2, x_3, x_4, x_5] \mid x_1 = 0;\ x_2 = x_3;\ x_4 = 3x_5\}$
(g) $S_7 = \{[x_1, x_2, x_3, x_4, x_5] \mid x_1 = 0;\ x_2 = x_3 = x_4\}$

22. For each of the sets in exercise 21 which is a vector space, determine the dimension of the vector space and find a basis for it.

Quadratic Forms and Definite Matrices

23. For each of the following quadratic forms, determine an identical quadratic form with a symmetric matrix associated with it:

(a) $f(\bar{x}) = x_1^2 + x_2^2$
(b) $f(\bar{x}) = x_1^2 - 2x_1x_2 + x_2^2$
(c) $f(\bar{x}) = 2x_1^2 - x_1x_2 - 3x_2^2$

24. For each of the quadratic forms of exercise 23, determine whether the associated symmetric matrix is positive or negative definite or semidefinite or whether it is indefinite.

25. Show that, if a matrix A is positive definite, then each of its diagonal elements must be strictly positive.

REFERENCES

1. Forsythe, G. E., and Moler, C. B.: *Computer Solution of Linear Algebraic Systems.* Prentice-Hall, Inc., Englewood Cliffs, N.J. (1967).
2. Franklin, J. N.: *Matrix Theory.* Prentice-Hall, Inc., Englewood Cliffs, N.J. (1968).
3. Hadley, G.: *Linear Algebra.* Addison-Wesley, Reading, Massachusetts (1961).
4. Hohn, Franz E.: *Elementary Matrix Algebra.* Second Edition, Macmillan Co., New York (1964).

Chapter 3

n-DIMENSIONAL GEOMETRY AND CONVEX SETS

3.1 INTRODUCTION: HYPERPLANES

Much insight into the nature of solutions to constrained optimization problems can be obtained by considering such problems from a geometric point of view. For example, consider the following problem:

Find the values of x_1 and x_2 which maximize the linear function

$$z = 3x_1 + 2x_2 \tag{3-1}$$

subject to the constraints that x_1 and x_2 be nonnegative and also that

$$\begin{Bmatrix} x_1 + 2x_2 \leq 3 \\ 3x_1 + 1x_2 \leq 4 \end{Bmatrix} \tag{3-2}$$

The shaded region in Figure 3-1 represents the set of all points (x_1, x_2) which satisfy the constraints (3-2) and the additional constraint that x_1 and x_2 be nonnegative. Thus, from this infinite collection of points, we wish to find a point which yields the largest value of z in equation (3-1).

In this example, the point which yields the largest value of z is the point (1, 1) which is one of the "corners" of the region. We shall see later in the chapter that the optimal point, or solution, to a linear programming problem is always at a "corner" of the region.

The set of all points (x_1, x_2) is called a *two-dimensional euclidean space*,† and is denoted by E^2. Note that a point in E^2 can also be

† Note: The set of all pairs of numbers (x_1, x_2) is not a euclidean space unless the concept of distance between any two such pairs is defined.

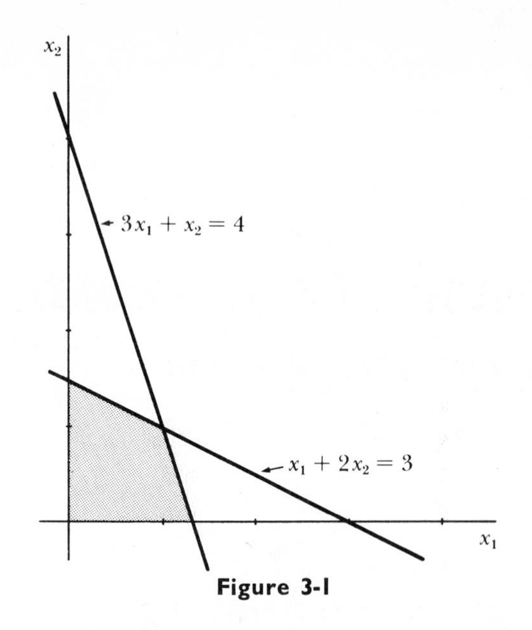

Figure 3-1

considered as a two-component vector. Thus, the two terms "point" and "vector" can be used interchangeably.

The distance between two points in E^2 can be found by use of Pythagoras' Theorem: The square of the hypotenuse of a right triangle is equal to the sum of the squares of the other two sides. If (a_1, b_1) and (a_2, b_2) are any two points in E^2, then from Figure 3-2

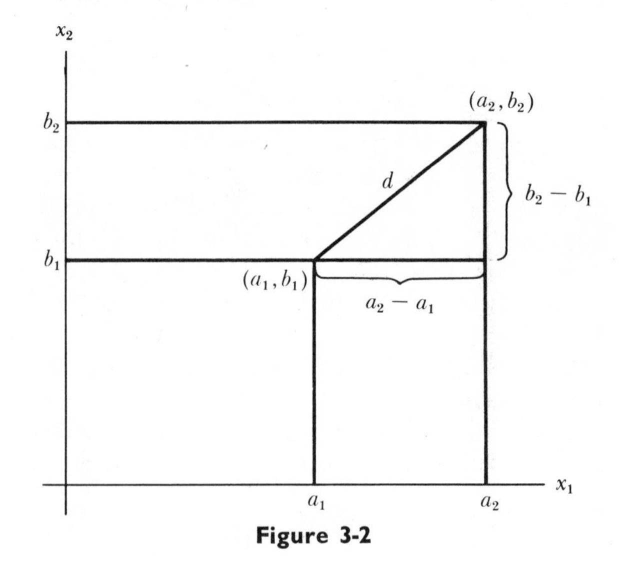

Figure 3-2

we see that if d is the distance between these two points then

$$d^2 = (a_2 - a_1)^2 + (b_2 - b_1)^2 \tag{3-3}$$

Now, let's consider the two points (a_1, b_1) and (a_2, b_2) as vectors:

$$\bar{x}_1 = \begin{bmatrix} a_1 \\ b_1 \end{bmatrix} \qquad \bar{x}_2 = \begin{bmatrix} a_2 \\ b_2 \end{bmatrix}$$

Notice that the scalar product of $(\bar{x}_2 - \bar{x}_1)$ with itself is

$$(\bar{x}_2 - \bar{x}_1)'(\bar{x}_2 - \bar{x}_1) = (a_2 - a_1)^2 + (b_2 - b_1)^2 = d^2$$

Thus, the distance between any two vectors $\bar{x}_1$, $\bar{x}_2$ in E^2 is given by

$$d = [(\bar{x}_2 - \bar{x}_1)'(\bar{x}_2 - \bar{x}_1)]^{\frac{1}{2}} \tag{3-4}$$

This distance will be denoted by $|\bar{x}_2 - \bar{x}_1|$.

In a like manner, we can consider three-component vectors as points in a three-dimensional euclidean space, E^3. The distance between any two points $\bar{x}_1$, $\bar{x}_2$ in E^3 is given by

$$\begin{aligned} d &= [(a_2 - a_1)^2 + (b_2 - b_1)^2 + (c_2 - c_1)^2]^{\frac{1}{2}} \\ &= [(\bar{x}_2 - \bar{x}_1)'(\bar{x}_2 - \bar{x}_1)]^{\frac{1}{2}} \end{aligned}$$

where $\bar{x}_1 = (a_1, b_1, c_1)'$ and $\bar{x}_2 = (a_2, b_2, c_2)'$.

Let us extend this analogy between points and vectors to n-component vectors, by first defining the set of all n-component vectors as an n-dimensional euclidean space, E^n. Thus, a point in E^n is an n-component vector, or *n-tuple*. Extending our definition of distance to points in E^n, the distance between two points $\bar{x}$ and $\bar{y}$ in E^n is given by

$$d = [(\bar{x} - \bar{y})'(\bar{x} - \bar{y})]^{\frac{1}{2}} = \left[\sum_{j=1}^{n} (x_j - y_j)^2 \right]^{\frac{1}{2}} \tag{3-5}$$

where $\bar{x} = (x_1, x_2, \ldots, x_n)'$ and $\bar{y} = (y_1, y_2, \ldots, y_n)'$.

Now, we would like to be able to generalize other geometric shapes (e.g., line, plane, circle, sphere, polygon, and the like) to E^n. Let us consider the equation of a line in E^2. If $\bar{x} = (x_1, x_2)$ is a point in E^2 and a_1, a_2, b are constants, then the equation

$$a_1x_1 + a_2x_2 = b \tag{3-6a}$$

represents a line in E^2. Equation (3-6a) could also be written as

$$[a_1, a_2]\begin{bmatrix} x_1 \\ x_2 \end{bmatrix} = b \tag{3-6b}$$

Moreover, the line represented by equation (3-6b) could be described as the set of all points (x_1, x_2) satisfying equation (3-6b); or, using set notation, the set

$$L = \left\{ \bar{x} = (x_1, x_2)' \;\middle|\; [a_1, a_2]\begin{bmatrix} x_1 \\ x_2 \end{bmatrix} = b \right\} \tag{3-7}$$

is the set of all points in E^2 which lie on the line represented by equation (3-6b).

Similarly, a plane in E^3 may be represented by the equation

$$a_1x_1 + a_2x_2 + a_3x_3 = b \tag{3-8a}$$

or

$$[a_1, a_2, a_3]\begin{bmatrix} x_1 \\ x_2 \\ x_3 \end{bmatrix} = b \tag{3-8b}$$

where a_1, a_2, a_3, and b are constants. Again, using the notation, the set of all points in E^3 which lie in the plane of equation (3-8b) is the set

$$P = \left\{ \bar{x} = (x_1, x_2, x_3)' \;\middle|\; [a_1, a_2, a_3]\begin{bmatrix} x_1 \\ x_2 \\ x_3 \end{bmatrix} = b \right\} \tag{3-9}$$

The natural extension of the sets L and P in E^n is the set

$$H = \left\{ \bar{x} = (x_1, x_2, \ldots, x_n)' \;\middle|\; [a_1, a_2, \ldots, a_n]\begin{bmatrix} x_1 \\ x_2 \\ \cdot \\ \cdot \\ \cdot \\ x_n \end{bmatrix} = b \right\} \tag{3-10}$$

where a_1, a_2, . . . , a_n, and b are constants. The set H is called a *hyperplane*. Thus, the equation

$$a_1x_1 + a_2x_2 + \ldots + a_nx_n = b \tag{3-11a}$$

or, equivalently,

$$[a_1, a_2, \ldots, a_n]\begin{bmatrix} x_1 \\ x_2 \\ \cdot \\ \cdot \\ \cdot \\ x_n \end{bmatrix} = b \tag{3-11b}$$

is the equation of a hyperplane in E^n. Note that for $n = 2$ a hyperplane is merely a line; for $n = 3$ it becomes a plane.

From the previous section we see that a linear equation in n unknowns is merely the equation of a hyperplane in E^n. Hence, a system of m equations in n unknowns describes the set of points in E^n which lie in each of the m hyperplanes; that is, which lie in the intersection of the m hyperplanes.

Just as in E^2 or E^3, a *line* in E^n is determined by two points. In E^2 the line passing through the two points $\bar{x}_1$ and $\bar{x}_2$ may be expressed as the set of all points $\bar{x}$ satisfying

$$\begin{aligned} \bar{x} &= \bar{x}_1 + \lambda(\bar{x}_2 - \bar{x}_1) \\ &= \lambda\bar{x}_2 + (1 - \lambda)\bar{x}_1 \end{aligned} \tag{3-12}$$

for all values of the real scalar λ.

If $\bar{x}_1$ and $\bar{x}_2$ are considered as points in E^n, then, again, the line passing through $\bar{x}_1$ and $\bar{x}_2$ is the set of all points in the set

$$L_1 = \{\bar{x} \mid \bar{x} = \lambda\bar{x}_2 + (1 - \lambda)\bar{x}_1, \quad \text{all real } \lambda\} \tag{3-13}$$

Notice that for $\lambda = 0$, $\bar{x} = \bar{x}_1$ and for $\lambda = 1$, $\bar{x} = \bar{x}_2$, so that for λ between 0 and 1, $\bar{x}$ will be a point on that portion of the line L_1 which is between $\bar{x}_1$ and $\bar{x}_2$. Thus, we define the *line segment* between $\bar{x}_1$ and $\bar{x}_2$ to be the set of all points in

$$L_2 = \{\bar{x} \mid \bar{x} = \lambda\bar{x}_2 + (1 - \lambda)\bar{x}_1, 0 \leq \lambda \leq 1\} \tag{3-14}$$

EXAMPLE

Consider the line in E^2 which passes through the two points $\bar{x}_1 = (2, 1)$ and $\bar{x}_2 = (1, 3)$. The equation of this line (see Figure 3-3) is easily found to be

$$2x + y = 5 \tag{3-15}$$

that is, any point $\bar{x} = (x, y)$ satisfying equation (3-15) lies on the line which passes through the points (2, 1) and (1, 3). From equation (3-12), we see that any such point $\bar{x}$ may be expressed as

$$\bar{x} = \lambda\bar{x}_2 + (1 - \lambda)\bar{x}_1$$

or

$$\begin{aligned} \begin{bmatrix} x \\ y \end{bmatrix} &= \lambda\begin{bmatrix} 1 \\ 3 \end{bmatrix} + (1 - \lambda)\begin{bmatrix} 2 \\ 1 \end{bmatrix} \\ &= \begin{bmatrix} 2 - \lambda \\ 1 + 2\lambda \end{bmatrix} \end{aligned}$$

For $\lambda = \frac{1}{2}$, we obtain $(x, y) = (1\frac{1}{2}, 2)$, which lies on the line between (2, 1) and 1, 3). For $\lambda = 2$, we obtain $(x, y) = (0, 5)$, which lies on the line, but not between (2, 1) and (1, 3).

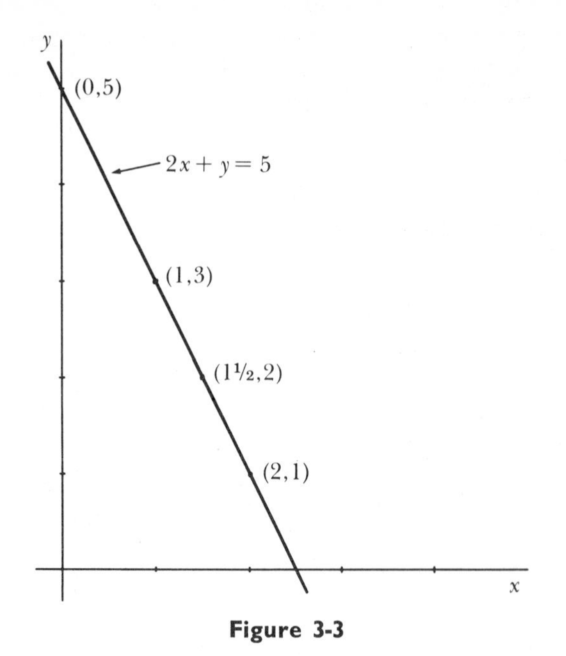

Figure 3-3

From the preceding discussion we have seen that linear equality constraints in optimization problems (involving variables) may be interpreted as hyperplanes in E^n. Since many optimization problems contain linear inequality constraints as well, let us now examine the geometric properties in E^n of linear inequalities. Consider the following inequality:

$$a_1x_1 + a_2x_2 + \ldots + a_nx_n \leq b \qquad \textbf{(3-16a)}$$

This expression may also be written as

$$[a_1, a_2, \ldots, a_n]\begin{bmatrix} x_1 \\ x_2 \\ \cdot \\ \cdot \\ \cdot \\ x_n \end{bmatrix} \leq b \qquad \textbf{(3-16b)}$$

or

$$\bar{a}'\bar{x} \leq b \qquad \textbf{(3-16c)}$$

The inequality (3-16a) divides all of E^n into two "halves" or sets:

$$S_1 = \{\bar{x} \mid \bar{a}'\bar{x} \leq b\} \qquad \textbf{(3-17a)}$$
$$S_2 = \{\bar{x} \mid \bar{a}'\bar{x} > b\} \qquad \textbf{(3-17b)}$$

Each point $\bar{x}$ in E^n is an element of one and only one of the two sets S_1 and S_2. The sets S_1 and S_2 are called *half-spaces* of E^n. One special type of half-space which occurs in practically all optimization problems is the nonnegativity requirement imposed on a variable, say x_j. That is,

$$x_j \geq 0 \tag{3-18}$$

defines a half-space of E^n. Thus, if the constraints of an optimization problem are linear equalities and linear inequalities, then the set of solutions (points) satisfying these constraints is the intersections of hyperplanes (the equalities) and half-spaces (the inequalities). We shall investigate the properties of this set of solutions (called the set of *feasible solutions*) in Section 3.4.

In a linear programming problem, the set of feasible solutions is the intersection of the hyperplanes and half-spaces defined by the constraints. In addition, the objective function,

$$z = \bar{c}'\bar{x} = \sum_{j=1}^{n} c_j x_j$$

defines a hyperplane, for a fixed value of z. Thus, in solving a linear programming problem, we are seeking the maximum value of z such that at least one point in the hyperplane $\bar{c}'\bar{x} = z$ lies in the intersection of the half-spaces and hyperplanes formed by the constraints. The question then naturally arises: given a point $\bar{x}_0$ in the set of feasible solutions, with a corresponding value of z, $z_0 = \bar{c}'\bar{x}_0$ (that is, the hyperplane† $\bar{c}'\bar{x} = z_0$ passes through the point $\bar{x}_0$) in which direction can we move in order to obtain a new point $\bar{x}_1$ for which $z_1 = \bar{c}'\bar{x}_1$ is greater than z_0?

In order to answer this question, let us first introduce the concept of *parallel hyperplanes:* Two hyperplanes $\bar{a}'\bar{x} = b_1$ and $\bar{a}'\bar{x} = b_2$ are said to be parallel hyperplanes. This definition extends to E^n the concept of parallel lines in E^2 (and E^3) and of parallel planes in E^3.

For example, consider the lines $x_1 + 2x_2 = -4$, $x_1 + 2x_2 = 0$, and $x_1 + 2x_2 = 2$, in E^2. These lines are shown in Figure 3-4. The direction of increasing values of z (with $z_0 = -4$, $z_1 = 0$, $z_2 = 2$) is obvious. Suppose now, that we wish to move from the point $(-2, -1)$ on the line $x_1 + 2x_2 = -4$ to some point on the line $x_1 + 2x_2 = 0$. The shortest distance between $(-2, -1)$ and the line $x_1 + 2x_2 = 0$ is given by the perpendicular line from $(-2, -1)$ to $x_1 + 2x_2 = 0$, and the equation of this perpendicular line is $2x_1 - x_2 = -3$. It intersects the line $x_1 + 2x_2 = 0$ at the point $(-\frac{6}{5}, \frac{3}{5})$. Any point (x_1, x_2) on this perpendicular line may be

† Note that if $\bar{c}$ and z_0 are given constants, then $\bar{c}'\bar{x} = z_0$ defines the hyperplane consisting of all points $\bar{x}$ satisfying $\bar{c}'\bar{x} = z_0$; on the other hand, given a vector $\bar{c}$ and a point $\bar{x}_0$, then a hyperplane passing through $\bar{x}_0$ is found by first calculating z_0 from $z_0 = \bar{c}'\bar{x}_0$.

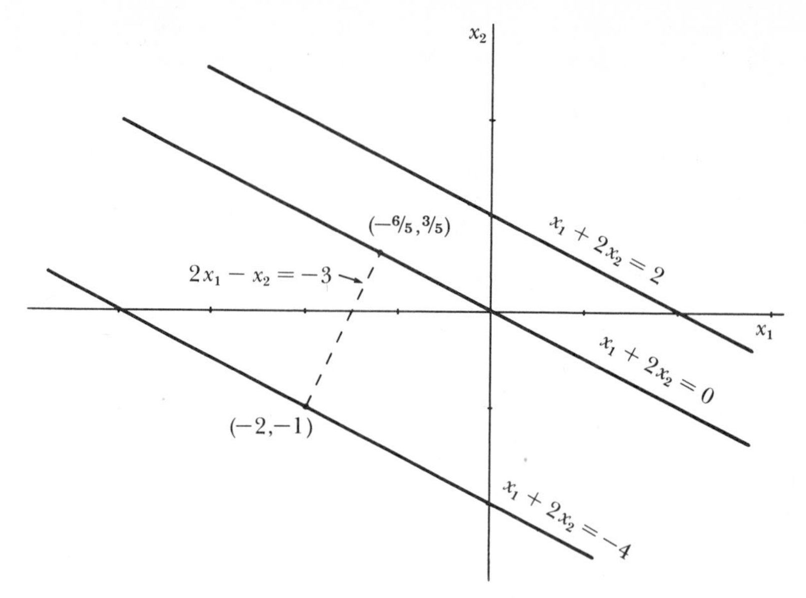

Figure 3-4

expressed parametrically by

$$\begin{bmatrix} x_1 \\ x_2 \end{bmatrix} = \lambda \begin{bmatrix} -\frac{6}{5} \\ \frac{3}{5} \end{bmatrix} + (1 - \lambda) \begin{bmatrix} -2 \\ -1 \end{bmatrix} \tag{3-19}$$

$$= \begin{bmatrix} -2 \\ -1 \end{bmatrix} + \lambda \begin{bmatrix} \frac{4}{5} \\ \frac{8}{5} \end{bmatrix}$$

$$= \begin{bmatrix} -2 \\ -1 \end{bmatrix} + \lambda(\tfrac{4}{5}) \begin{bmatrix} 1 \\ 2 \end{bmatrix} \tag{3-20}$$

Or, letting $\alpha = \frac{4}{5}\lambda$, equation (3-20) becomes

$$\begin{bmatrix} x_1 \\ x_2 \end{bmatrix} = \begin{bmatrix} -2 \\ -1 \end{bmatrix} + \alpha \begin{bmatrix} 1 \\ 2 \end{bmatrix} \tag{3-21}$$

Note that if the line $x_1 + 2x_2 = -4$ is written using hyperplane notation, then it becomes

$$[1, 2] \begin{bmatrix} x_1 \\ x_2 \end{bmatrix} = -4$$

Letting $\bar{c}' = [1, 2]$, $\bar{x}_0' = [-2, -1]$, equation (3-21) becomes

$$\bar{x}_1 \equiv \begin{bmatrix} x_1 \\ x_2 \end{bmatrix} = \bar{x}_0 + \alpha \bar{c} \tag{3-22}$$

Equation (3-22) tells us that moving from the point $\bar{x}_0$ along the line perpendicular to $x_1 + 2x_2 = -4$ is the same as moving from the point $\bar{x}_0$ *in the direction of* $\bar{c}$.

Moreover, if $\alpha > 0$, and $\bar{c}'\bar{x} = z_0$ and $\bar{c}'\bar{x} = z_1$ are parallel hyperplanes passing through the points $\bar{x}_0$ and $\bar{x}_1$, respectively, then $z_1 > z_0$. To see this, observe that from equation (3-22)

$$\begin{aligned} z_1 = \bar{c}'\bar{x}_1 &= \bar{c}'[\bar{x}_0 + \alpha\bar{c}] \\ &= \bar{c}'\bar{x}_0 + \bar{c}'(\alpha\bar{c}) \\ &= z_0 + \alpha\bar{c}'\bar{c} \end{aligned} \tag{3-23}$$

Now, since $\bar{c}'\bar{c} = \sum_{j=1}^{n} c_j^2 > 0$, if $\bar{c} \neq \bar{0}$, and $\alpha > 0$, $\alpha\bar{c}'\bar{c} > 0$, and $z_1 > z_0$.

Since equations (3-22) and (3-23) contain nothing which would restrict their validity to E^2, we can generalize the foregoing discussion to E^n. To summarize, then, if a hyperplane $\bar{c}'\bar{x} = z_0$ is moved parallel to itself in the direction of $\bar{c}$, to yield a new (parallel) hyperplane $\bar{c}'\bar{x} = z_1$, then $z_1 > z_0$.

3.2 HYPERSPHERES; OPEN, CLOSED, AND BOUNDED SETS

Let us now consider the n-dimensional extension of circles and spheres. In E^2, the equation of a circle whose center is at the point (a_1, a_2) and whose radius is r is

$$(x_1 - a_1)^2 + (x_2 - a_2)^2 = r^2 \tag{3-24}$$

Equation (3-24) states that any point $\bar{x}' = (x_1, x_2)$ lies on this circle if (and only if) the distance between $\bar{x}$ and the center of the circle $\bar{a}' = (a_1, a_2)$ is r. From equation (3-4), then,

$$[(\bar{x} - \bar{a})'(\bar{x} - \bar{a})]^{\frac{1}{2}} = r \tag{3-25a}$$

or,

$$|\bar{x} - \bar{a}| = r \tag{3-25b}$$

Using set notation, we may state that a circle with radius r and center $\bar{a}$ is the set of points

$$C = \{\bar{x} \mid \bar{x} \in E^2 \quad \text{and} \quad |\bar{x} - \bar{a}| = r\} \tag{3-26}$$

Also, the *interior of the circle* C is the set

$$I_c = \{\bar{x} \mid \bar{x} \in E^2 \quad \text{and} \quad |\bar{x} - \bar{a}| < r\} \tag{3-27}$$

Similarly, in E^3, the equation of a sphere with radius r and center at the point $\bar{a}' = (a_1, a_2, a_3)$ is

$$(x_1 - a_1)^2 + (x_2 - a_2)^2 + (x_3 - a_3)^2 = r^2 \tag{3-28}$$

Again, this equation may be rewritten

$$[(\bar{x} - \bar{a})'(\bar{x} - \bar{a})]^{\frac{1}{2}} = r \tag{3-29a}$$

or,

$$|\bar{x} - \bar{a}| = r \tag{3-29b}$$

This sphere, then, is the set of points

$$S = \{\bar{x} \mid \bar{x} \in E^3 \quad \text{and} \quad |\bar{x} - \bar{a}| = r\} \tag{3-30}$$

The interior of S is the set

$$I_s = \{\bar{x} \mid \bar{x} \in E^3 \quad \text{and} \quad |\bar{x} - \bar{a}| < r\} \tag{3-31}$$

which represents the set of all points "inside" the sphere.

Upon comparing equations (3-26) and (3-30), we see that the natural extension to points E^n is

$$H = \{\bar{x} \mid \bar{x} \in E^n \quad \text{and} \quad |\bar{x} - \bar{a}| = r\} \tag{3-32}$$

The set H defines a *hypersphere* in E^n with center at

$$\bar{a}' = (a_1, a_2, \ldots, a_n)$$

and radius r. The equation of the hypersphere is thus

$$\begin{aligned} r &= |\bar{x} - \bar{a}| \\ &= \left[\sum_{i=1}^{n} (x_i - a_i)^2\right]^{\frac{1}{2}} \end{aligned} \tag{3-33}$$

The interior of the hypersphere H is the set of points

$$I_H = \{\bar{x} \mid \bar{x} \in E^n \quad \text{and} \quad |\bar{x} - \bar{a}| < r\} \tag{3-34}$$

The interior of a hypersphere with center at $\bar{a}$ and radius ε is also called an *ε-neighborhood of $\bar{a}$* (since ε represents a radius, we restrict ε to be strictly positive: $\varepsilon > 0$).

In E^2, the notion of a "boundary" of a set of points is intuitively clear. For example, in Figure 3-5, we can immediately determine that the points $\bar{a}$ and $\bar{b}$ are boundary points of the set S_1, the point $\bar{c}$ lies in the interior of S_1, the points $\bar{d}$ and $\bar{e}$ are boundary points if the set S_2 and the points $\bar{f}$ and $\bar{g}$ are interior points of S_2. However, we need to translate this intuition into definitions of boundary and interior point. Observe that if we draw a circle around each interior point, with radius ε, that by choosing ε small enough, all points lying on the interior of the circle will be points in the set; on the other hand, if we attempt to draw a circle around a boundary point (with the boundary point as its center) no matter how small we choose ε, some of the points lying in the interior of this circle will lie outside the set. This is illustrated in Figure 3-6, in which $\bar{a}$ is an interior point and $\bar{b}$ is a boundary point.

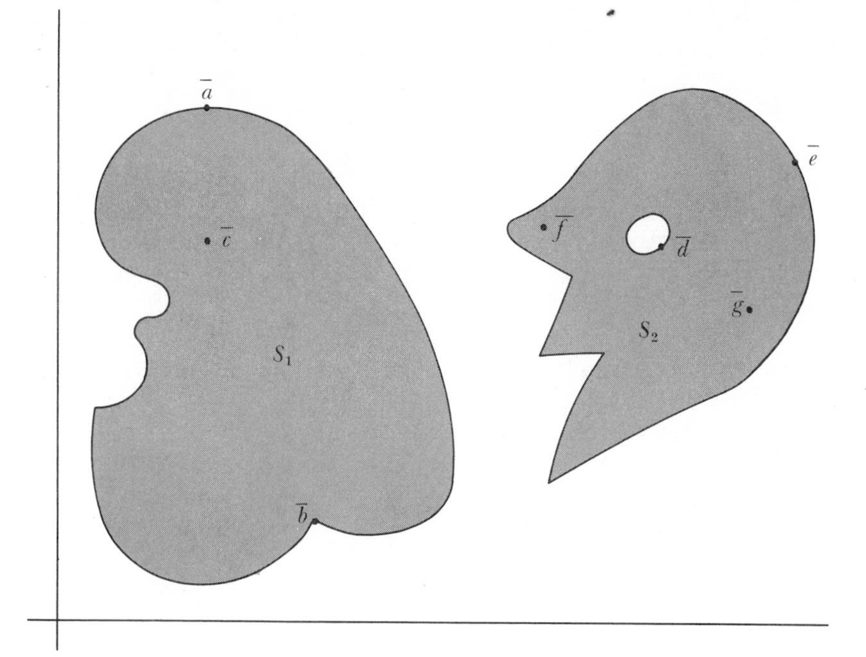

Figure 3-5

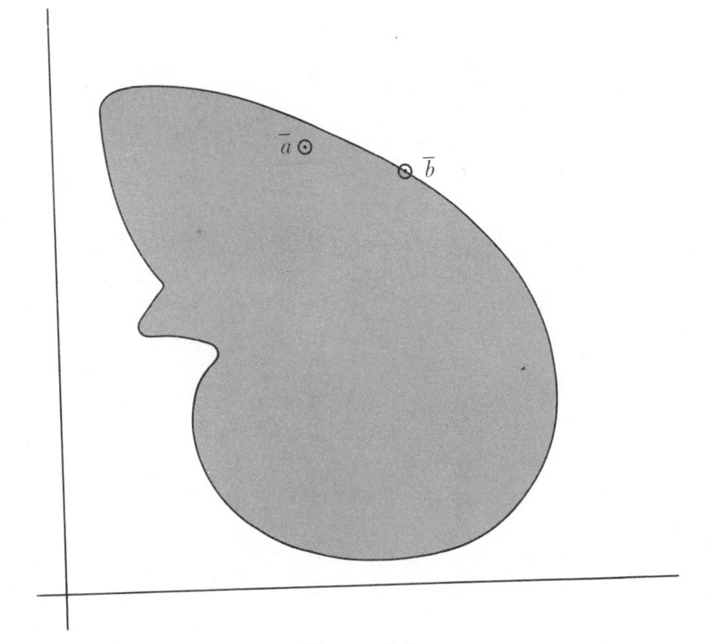

Figure 3-6

Thus far we have seen that our intuitive notions of boundary and interior points in E^2 have led us to discover a precise distinction between the two. Since the interiors of the circles described in the preceding discussion are really ε-neighborhoods in E^2, we are led to the following definitions in E^n of an interior point and a boundary point.

A point $\bar{a}$ is an interior point of the set S if there exists an ε-neighborhood of $\bar{a}$ which contains only points of the set S. In other words, $\bar{a}$ is an interior point of S if we can find an ε so that the resulting ε-neighborhood of $\bar{a}$ contains only points in S.

A point $\bar{a}$ is a boundary point of the set S if every ε-neighborhood of $\bar{a}$ (regardless of how small $\varepsilon > 0$ may be) *contains points which are in the set S and points which are not in the set S.*

With these definitions we can determine, for any set S and any point $\bar{x}$, whether $\bar{x}$ is an interior point of S, a boundary point of S, or a point not in S. Note, however, that a boundary point of a set S need not be a member of S.

EXAMPLES

(a) $S_1 = \{(x_1, x_2) \mid x_1^2 + x_2^2 < 1\}$

(b) $S_2 = \{(x_1, x_2) \mid x_1^2 + x_2^2 \leq 1\}$

(c) $S_3 = \{(x_1, x_2) \mid x_1^2 + x_2^2 = 1\}$

(d) $S_4 = \{(x_1, x_2) \mid x_1^2 + x_2^2 < 1 \text{ if } x_2 \geq 0 \text{ and } x_1^2 + x_2^2 < 1 \text{ for } x_2 < 0\}$

Set S_1 consists of all points *inside* the unit circle $x_1^2 + x_2^2 = 1$, but not points *on* the circle; the boundary of S_1 consists of all points on the circle. Thus, every point in S_1 is an interior point. Set S_2, on the other hand, consists of the points in S_1 plus the points on the circle $x_1^2 + x_2^2 = 1$; its boundary consists of all the points on the circle, and so all the boundary points of S_2 are elements of S_2. Similarly, all the boundary points of S_3 are elements of S_3; in fact, every element of S_3 is a boundary point. Set S_4 has the same boundary as the other three sets; however, some points of this boundary are elements of S_4, while others are not. For example, the point $(0, 1)$ is a boundary point of S_4 and is also an element of S_4; the point $(0, -1)$, on the other hand, is a boundary point but not an element of S_4.

The following definitions will be useful:

A set S is an open set if it contains only interior points (e.g., set S_1 in (a)); an open set, then, contains none of its boundary points.

A set S is a closed set if it contains all of its boundary points (e.g., sets S_2 and S_3 in (b) and (c)). A set S is neither open nor closed if it contains some but not all of its boundary points (e.g., set S_4 in (d)).

Now since the sets that we will be dealing with primarily are sets of solutions to optimization problems, it will often be desirable to require that each point in the set have only finite components. In E^2, we see that if we can enclose a set S in a circle with center at the origin, then all the points in S must have finite components. Similarly, in E^n, if a set S can be enclosed by a hypersphere with center at the origin, then every point in S must have only finite components. Such a set is said to be *strictly bounded.*

Another type of restriction which occurs frequently in optimization problems is that of requiring upper or lower bounds (or both) on the variables—the components of the points $\bar{x}$. A set S is said to be *bounded from above* if there is a finite upper bound for each component of the points in S (or, equivalently, if there exists an $\bar{r}$, with each component finite, such that for each point $\bar{x}$ in S, $\bar{x} \leq \bar{r}$).† A set S is said to be *bounded from below* if there is a finite lower bound for each component of the points in S.

EXAMPLES

(a) The set defined by the shaded region in Figure 3-1 is strictly bounded.
(b) Lines, planes, and hyperplanes are unbounded.
(c) The set $N = \{\bar{x} \mid x_j \geq 0, j = 1, 2, \ldots, n\}$, called the *nonnegative orthant of* E^n, is bounded from below.

Let us now prove some results which will be very useful in the next section:

Theorem 3.1. The half-space S_2 defined by equation (3-17b) is an open set.

Proof:

$$S_2 = \{\bar{x} \mid \bar{a}'\bar{x} > b\} \tag{3-17b}$$

Let $\bar{x}_0$ be any point in S_2, and let d be the distance from $\bar{x}_0$ to the hyperplane $\bar{a}'\bar{x} = b$. If we then define ε to be, for example, $d/3$, then all points in this ε-neighborhood of $\bar{x}_0$ will be in S_2.

Theorem 3.2. Every point on a hyperplane is a boundary point.

Proof: Consider the hyperplane

$$H = \{\bar{x} \mid \bar{a}'\bar{x} = b\} \tag{3-35}$$

† The vector inequality $\bar{x} \leq r$ is defined as the set of inequalities $x_j \leq r_j, j = 1, 2, \ldots, n$, where x_j and r_j are respectively the jth components of $\bar{x}$ and $\bar{r}$.

Let $\bar{x}_0$ be any point in H; we wish to show that every ε-neighborhood of $\bar{x}_0$ contains points in H and points not in H. Consider the point

$$\bar{x}_1 = \bar{x}_0 + \lambda \bar{c} \tag{3-36}$$

where λ is a positive scalar. Then

$$\bar{x}_1 - \bar{x}_0 = \lambda \bar{c}$$
$$|\bar{x}_1 - \bar{x}_0| = \lambda \, |\bar{c}|$$

Now, if we let $\lambda = \dfrac{\varepsilon}{2\,|\bar{c}|}$, then

$$|\bar{x}_1 - \bar{x}_0| = \frac{\varepsilon}{2} < \varepsilon$$

Or, for all positive ε, the point $\bar{x}_1$ is in an ε-neighborhood of $\bar{x}_0$. Now, let us show that $\bar{x}_1$ is not in H

$$\begin{aligned} \bar{a}'\bar{x}_1 &= \bar{a}'\left[\bar{x}_0 + \frac{\varepsilon}{2\,|\bar{c}|}\,\bar{c}\right] \\ &= \bar{a}'\bar{x}_0 + \frac{\varepsilon}{2\,|\bar{c}|}\,(\bar{c}'\,\bar{c}) \\ &= b + \frac{\varepsilon}{2} \end{aligned}$$

Thus, for all positive ε, $\bar{a}'\bar{x}_1 > b$, and hence $\bar{x}_1$ does not lie on the hyperplane H. However, since $\bar{x}_0$ is in H, we have shown that every ε-neighborhood of $\bar{x}_0$ contains points in H (e.g., $\bar{x}_0$) and points not in H (e.g., $\bar{x}_1$).

Theorem 3.3. A hyperplane is a closed set.

Proof: We wish to show that if $\bar{x}_0$ is a boundary point of the hyperplane H (defined by equation (3-35)), then $\bar{x}_0$ must be in H. However, if $\bar{x}_0$ is not in H, then it must be in one of the two half-spaces

$$S_1 = \{\bar{x} \mid \bar{a}'\bar{x} > b\} \tag{3-37a}$$
$$S_2 = \{\bar{x} \mid \bar{a}'x < b\} \tag{3-37b}$$

However, using the argument of the proof of Theorem 3.1 leads us to conclude that if $\bar{x}_0$ were in S_1 (or S_2), then there exists an ε-neighborhood of $\bar{x}_0$ containing only points in S_1 (or S_2). Thus, it must be true that $\bar{x}_0 \in H$.

We can use analogous methods to prove:

Theorem 3.4. The half-space

$$S_1 = \{\bar{x} \mid \bar{a}'\bar{x} \leq b\} \tag{3-17a}$$

is a closed set; its boundary consists of the hyperplane (3-35).

Thus, the half-space of equation (3-17a) is called a *closed half-space*, and the half-space of equation (3-17b) an *open half-space.*

3.3 CONVEX SETS

In a constrained optimization problem, the geometric shape of the set of solutions in E^n is perhaps the most crucial characteristic of the problem, with respect to the degree of difficulty we are likely to encounter in attempting to solve the problem. There is only one special type of set—a convex set—for which any appreciable amount of theory has been developed. In this section we shall investigate some of the properties of convex sets as well as the types of constraints which lead to convex sets of solutions.

We shall begin by defining a convex set: *A set C is convex if for any two points*† *in C, the line segment joining the points is in the set.* Thus, for every pair of points $\bar{x}_1$, $\bar{x}_2$ in a convex set C, the set

$$L = \{\bar{x} \mid \bar{x} = \lambda\bar{x}_1 + (1 - \lambda)\bar{x}_2, \quad 0 \leq \lambda \leq 1\} \qquad \textbf{(3-38)}$$

is a subset of C: $L \subseteq C$

The expression $\lambda\bar{x}_1 + (1 - \lambda)\bar{x}_2$, $0 \leq \lambda \leq 1$, is called a *convex combination* of the points $\bar{x}_1$, $\bar{x}_2$. For fixed λ, a convex combination of $\bar{x}_1$, $\bar{x}_2$ yields a point on the line segment L (of equation (3-38)).

Intuitively, then, a convex set must be "solid" (no "holes"), and its boundaries must not curve "into" the set. The two sets C_1, C_2 shown in Figure 3-7 are not convex, since the line segments joining the pairs of points $(\bar{x}_1, \bar{x}_2)$, $(\bar{x}_3, \bar{x}_4)$ do not lie entirely within C_1 and C_2, respectively.

It is obvious that a plane in E^3 is a convex set, since the line segment connecting any two points in a plane lies within that plane. We shall now show:

Theorem 3.5. A hyperplane is a convex set.

Proof: Consider the hyperplane $\bar{a}'\bar{x} = b$

Let $\bar{x}_1$, $\bar{x}_2$ lie on the hyperplane. Form the convex combination $\bar{x}_3 = \lambda\bar{x}_1 + (1 - \lambda)\bar{x}_2$, $0 \leq \lambda \leq 1$. We must show, then, that $\bar{x}_3$ lies on the hyperplane (i.e., that $\bar{a}'\bar{x}_3 = b$). But,

$$\begin{aligned} \bar{a}'\bar{x}_3 &= \bar{a}'[\lambda\bar{x}_1 + (1 - \lambda)\bar{x}_2] \\ &= \lambda\bar{a}'\bar{x}_1 + (1 - \lambda)\bar{a}'\bar{x}_2 \\ &= \lambda b + (1 - \lambda)b \qquad (\text{since } \bar{a}'\bar{x}_1 = b,\ \bar{a}'\bar{x}_2 = b) \\ &= b \end{aligned}$$

† By convention, sets containing no points (the empty set) and sets containing only one point are also said to be convex.

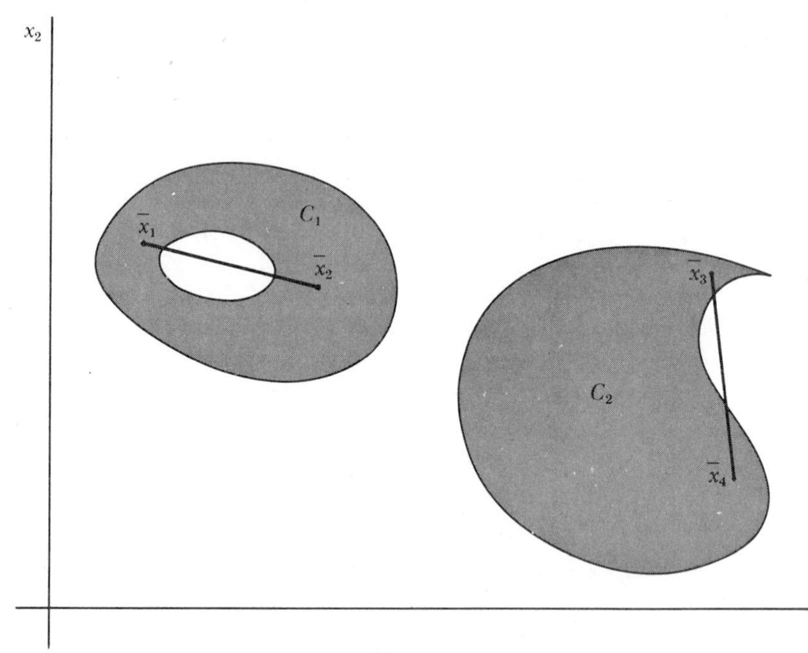

Figure 3-7

In a completely analogous manner, we can prove (and we leave it as an exercise for the reader):

Theorem 3.6. Closed and open half-spaces are convex sets.

Theorems 3.5 and 3.6 tell us that if the constraints of an optimization problem consist of linear equalities or linear inequalities, then each of these constraints forms a convex set. However, it is the *intersection* of these sets which we are interested in. We shall now show that this intersection is also convex.

Theorem 3.7. The intersection of a finite number of convex sets is also a convex set.

Proof: Let $C_1, C_2, \ldots, C_r$ be convex sets and let

$$D = \bigcap_{i=1}^{r} C_i \tag{3-39}$$

We wish to show, then, that given any two points $\bar{x}_1$, $\bar{x}_2$ in D, that $\lambda\bar{x}_1 + (1 - \lambda)\bar{x}_2$ is also in D for all λ, $0 \leq \lambda \leq 1$. But, if $\bar{x}_1$, $\bar{x}_2$ are elements of D then $\bar{x}_1$, $\bar{x}_2$ are also elements of each of the sets C_1, $C_2, \ldots, C_r$. Moreover, since $C_1, C_2, \ldots, C_r$ are convex, then $\lambda\bar{x}_1 + (1 - \lambda)\bar{x}_2$, $0 \leq \lambda \leq 1$, is in each of the sets $C_1, C_2, \ldots, C_r$. Therefore, $\lambda\bar{x}_1 + (1 - \lambda)\bar{x}_2$, $0 \leq \lambda \leq 1$, is in the intersection of $C_1, C_2, \ldots, C_r$.

Observe now that if C_1 and C_2 are two closed sets, and if $D = C_1 \cap C_2$, then D is also closed. To see this, note that any boundary point of D must be a boundary point of either C_1 or C_2 (or both), and that C_1 and C_2 contain all their boundary points. (See Figure 3-8.) We can generalize the preceding discussion and state that the intersection of a finite number of closed sets is also a closed set. In particular, combining the results of Theorems 3.3, 3.4, 3.5, 3.6, and 3.7, we see that the intersection of hyperplanes or closed half-spaces is a closed convex set. We shall make use of this fact later in the chapter.

We have observed previously that the optimal solution to the linear programming problem of Section 3.1 occurred at a "corner" of the set of solutions. Note also that this set of solutions is a convex set and that, except for the corner points, any point in the set lies on some line segment connecting two other points in the set. That is, we can find a line segment which lies entirely within the set and which passes through any given point in the set (except for the corner points). The "corner" points of a convex set are called *extreme points*. More precisely, *a point $\bar{x}$ is an extreme point of a convex set if there do not exist points $\bar{x}_1$, $\bar{x}_2$ ($\bar{x}_1 \neq \bar{x}_2$) in the set such that for some λ, $0 < \lambda < 1$, $\bar{x} = \lambda\bar{x}_1 + (1 - \lambda)\bar{x}_2$.* (The requirement that λ be between 0 and 1 ensures that $\bar{x}$ lies on a line segment between $\bar{x}_1$ and $\bar{x}_2$.)

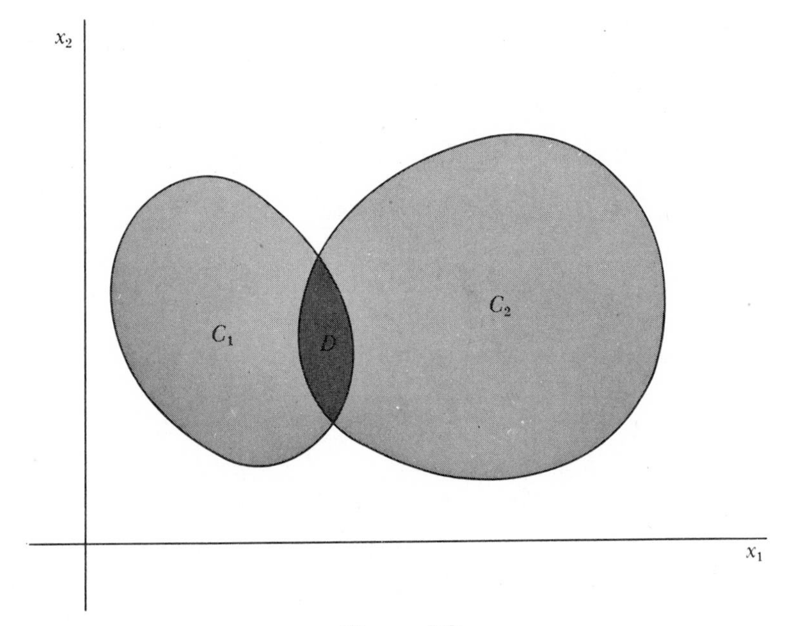

Figure 3-8

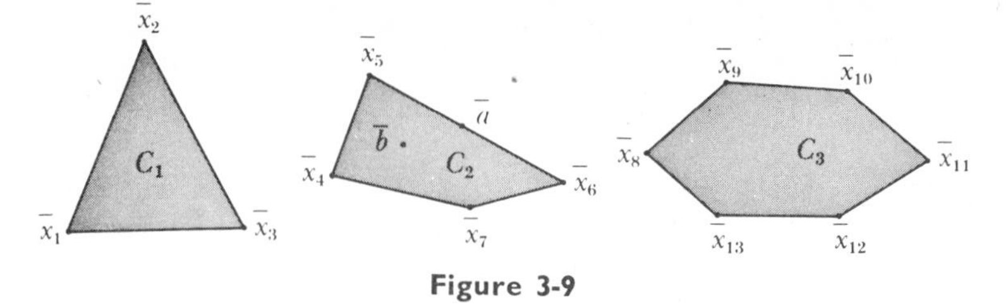

Figure 3-9

The convex sets C_1, C_2, and C_3 shown in Figure 3-9 contain 3, 4, and 6 extreme points, respectively, denoted by $\bar{x}_1, \bar{x}_2, \ldots, \bar{x}_{13}$. The points $\bar{a}$ and $\bar{b}$ in C_2 are obviously not extreme points. Notice that all the extreme points are boundary points, but that not all the boundary points are extreme points (e.g., the point $\bar{a}$).

It is also fairly clear and we shall now prove that no interior point of a convex set can be an extreme point.

Theorem 3.8. An extreme point of a convex set is a boundary point of the set.

Proof: Let $\bar{x}_0$ be any interior point of a convex set C. Then, there is an ε-neighborhood of $\bar{x}_0$ containing only points in C. Let $\bar{x}_1$ be a point in this ε-neighborhood: $|\bar{x}_0 - \bar{x}_1| < \varepsilon$. Now, define $\bar{x}_2 = 2\bar{x}_0 - \bar{x}_1$. Then, $\bar{x}_2 - \bar{x}_0 = \bar{x}_0 - \bar{x}_1$, and so

$$|\bar{x}_2 - \bar{x}_0| = |\bar{x}_0 - \bar{x}_1| < \varepsilon.$$

Thus, $\bar{x}_2$ is in the ε-neighborhood of $\bar{x}_0$, and hence is a point in C. But, note that $\bar{x}_0 = \frac{1}{2}\bar{x}_1 + \frac{1}{2}\bar{x}_2$. Thus, with $\lambda = \frac{1}{2}$, we have represented $\bar{x}_0$ as a convex combination of $\bar{x}_1, \bar{x}_2$. Thus, $\bar{x}_0$ cannot be an extreme point.

A convex combination of two points defines a point on the line segment between the two points. Given any two points, the set of all convex combinations of the two points (i.e., the line segment joining them) is clearly a convex set. Let us now generalize the concept of a convex combination: *A convex combination* of a (finite) set of points $\bar{x}_1, \bar{x}_2, \ldots, \bar{x}_r$ is defined as a point

$$\bar{x} = \sum_{i=1}^{r} \mu_i \bar{x}_i; \quad \mu_i \geq 0, i = 1, 2, \ldots, r; \quad \sum_{i=1}^{r} \mu_i = 1 \qquad \textbf{(3-40)}$$

For $r = 2$, we have

$$\bar{x} = \mu_1\bar{x}_1 + \mu_2\bar{x}_2; \quad \mu_1 \geq 0, \mu_2 \geq 0; \quad \mu_1 + \mu_2 = 1$$

or, letting $\lambda = \mu_1$, $\mu_2 = 1 - \mu_1 = 1 - \lambda$, $\bar{x} = \lambda\bar{x}_1 + (1 - \lambda)\bar{x}_2$. Thus, the generalized definition is consistent with the previous definition. Let us now investigate the properties of the set of all convex combinations of a set of r points $\bar{x}_1, \bar{x}_2, \ldots, \bar{x}_r$.

Theorem 3.9. The set C of all convex combinations of a finite number of points $\bar{x}_1, \bar{x}_2, \ldots, \bar{x}_r$ is a convex set.

Proof:

$$C = \left\{ \bar{x} \mid \bar{x} = \sum_{i=1}^{r} \mu_i \bar{x}_i; \quad \mu_i \geq 0, i = 1, 2, \ldots, r; \quad \sum_{i=1}^{r} \mu_i = 1 \right\} \tag{3-41}$$

We must show that, for any two points $\bar{y}$, $\bar{z}$ in C the convex combination $\bar{w} = \lambda \bar{y} + (1 - \lambda)\bar{z}$ is in C for all λ, $0 \leq \lambda \leq 1$. Let

$$\bar{y} = \sum_{i=1}^{r} \alpha_i \bar{x}_i, \; \alpha_i \geq 0, \; i = 1, 2, \ldots, r \quad \sum_{i=1}^{r} \alpha_i = 1 \tag{3-42a}$$

$$\bar{z} = \sum_{i=1}^{r} \beta_i \bar{x}_i, \; \beta_i \geq 0, \; i = 1, 2, \ldots, r \quad \sum_{i=1}^{r} \beta_i = 1 \tag{34-2b}$$

Then

$$\begin{aligned} \bar{w} &= \lambda \bar{y} + (1 - \lambda)\bar{z} \\ &= \lambda \left[\sum_{i=1}^{r} \alpha_i \bar{x}_i \right] + (1 - \lambda) \left[\sum_{i=1}^{r} \beta_i \bar{x}_i \right] \\ &= \sum_{i=1}^{r} [\lambda \alpha_i + (1 - \lambda)\beta_i] \bar{x}_i \end{aligned} \tag{3-43}$$

Now, $\bar{w}$ will be in C if the right-hand side of equation (3-43) represents a convex combination of $\bar{x}_1, \bar{x}_2, \ldots, \bar{x}_r$. Let

$$\gamma_i = \lambda \alpha_i + (1 - \lambda)\beta_i, \; i = 1, 2, \ldots, r$$

We must show, then, that $\gamma_i \geq 0$, $i = 1, 2, \ldots, r$, and $\sum_{i=1}^{r} \gamma_i = 1$. But,

$$\begin{aligned} \sum_{i=1}^{r} \gamma_i &= \sum_{i=1}^{r} [\lambda \alpha_i + (1 - \lambda)\beta_i)] \\ &= \lambda \sum_{i=1}^{r} \alpha_i + (1 - \lambda) \sum_{i=1}^{r} \beta_i \\ &= \lambda + (1 - \lambda) = 1 \end{aligned}$$

Also, since $\lambda \geq 0$, $(1 - \lambda) \geq 0$, all $\alpha_i \geq 0$, all $\beta_i \geq 0$, it is obvious that $\gamma_i \geq 0$, $i = 1, 2, \ldots, r$. Thus C is convex.

In Figure 3-10, the set C of all convex combinations of the seven points $\bar{x}_1, \bar{x}_2, \ldots, \bar{x}_7$ is represented by the region marked C. Note that C is a polygon, and that the only extreme points of C are $\bar{x}_1$, $\bar{x}_2$, $\bar{x}_3$, $\bar{x}_4$, $\bar{x}_5$. Clearly, in E^2 the set of all convex combinations of a finite set of points $\bar{x}_1, \bar{x}_2, \ldots, \bar{x}_r$ is a polyhedron; moreover, the only possible extreme points are elements of $\{\bar{x}_1, \bar{x}_2, \ldots, \bar{x}_r\}$.

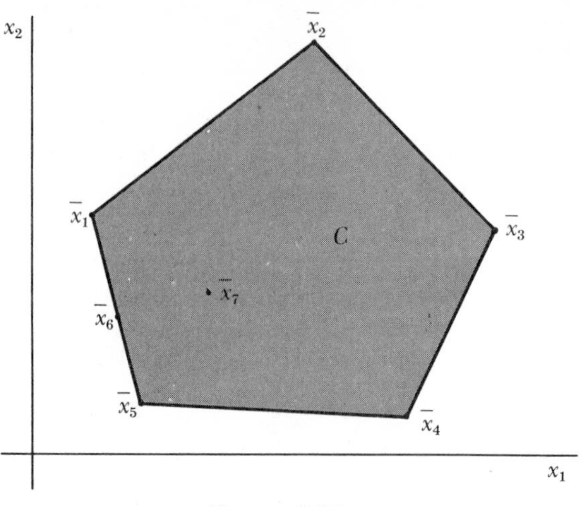

Figure 3-10

3.4 CONVEX HULLS AND CONVEX POLYHEDRA

For any set S which is not convex, we may wish to find a convex set C which contains S as a subset. Since E^n itself is a convex set and any set S is a subset of E^n, it is always possible to find such a convex set C. In fact, an infinite number of such sets exist. The intersection of all such convex sets which contain a set S as a subset is called the *convex hull of S*. Being the intersection of convex sets, the convex hull is a convex set. Moreover, it is the "smallest" convex set which contains S, in the sense that it contains all other convex sets of which S is a subset (by the definition of "intersection"). Intuitively, the convex hull "straightens out" the boundaries of a set and fills in any "holes." Thus, the convex hulls of the two sets C_1, C_2 of Figure 3-7 are shown in Figure 3-11, denoted by H_1, H_2, respectively. (The dashed lines represent that portion of the boundaries of C_1 and C_2 which are not on the boundaries of H_1 and H_2.) The convex hull of the set of points $\bar{x}_1, \bar{x}_2, \ldots, \bar{x}_7$ of Figure 3-10 is the polyhedron C (of Figure 3-10) which, remember, was formed by taking all convex combinations of $\bar{x}_1, \bar{x}_2, \ldots, \bar{x}_7$. This last observation leads us to:

Theorem 3.10. The convex hull of a finite set of points $\bar{x}_1, \bar{x}_2, \ldots, \bar{x}_r$ is the set of all convex combinations of $\bar{x}_1, \bar{x}_2, \ldots, \bar{x}_r$. (Thus, the convex hull of the set $S = \{\bar{x}_1, \bar{x}_2, \ldots, \bar{x}_r\}$ is the set C defined by equation (3-41).)

Proof: Theorem 3.9 states that C is convex. In order to show that C is the convex hull of S, we must show that every convex

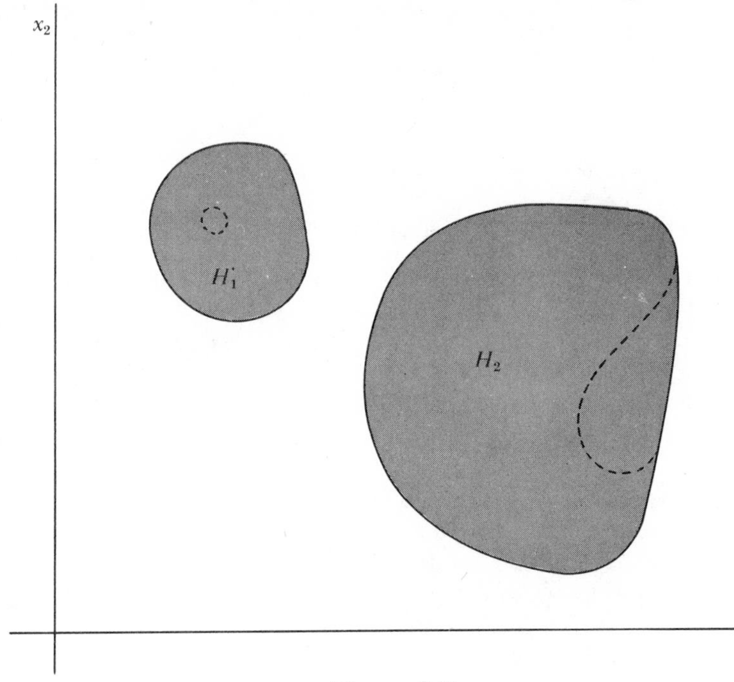

Figure 3-11

set which contains S also contains C (and hence C is in the intersection of such sets). We shall proceed by induction on r, the number of points in S. For $r = 1$, S contains only one point $\bar{x}_1$ and clearly $C = \{\bar{x}_1\}$. Now, assume the theorem is true for $r = p - 1$. Thus, the convex hull of $S = \{\bar{x}_1, \bar{x}_2, \ldots, \bar{x}_{p-1}\}$ is the set

$$\hat{C} = \left\{\bar{x} \mid \bar{x} = \sum_{i=1}^{p-1} \alpha_i \bar{x}_i; \quad \alpha_i \geq 0, i = 1, 2, \ldots, p - 1; \quad \sum_{i=1}^{p-1} \alpha_i = 1\right\} \tag{3-44}$$

Now, let us construct the convex hull C of $\bar{x}_1, \bar{x}_2, \ldots, \bar{x}_p$. First of all, C must contain $\bar{x}_p$ and it also must contain $\hat{C}$. Moreover, C must contain all convex combinations of $\bar{x}_p$ with any point in C (since C is convex). Hence, all points of the form

$$\bar{x} = \lambda\left(\sum_{i=1}^{p-1} \alpha_i \bar{x}_i\right) + (1 - \lambda)\bar{x}_p, \quad 0 \leq \lambda \leq 1 \tag{3-45}$$

must be in C. But, letting $\mu_i = \lambda\alpha_i$, $i = 1, 2, \ldots, p - 1$, and $\mu_p = 1 - \lambda$, equation (3-45) becomes

$$\bar{x} = \sum_{i=1}^{p-1} \mu_i \bar{x}_i + \mu_p \bar{x}_p = \sum_{i=1}^{p} \mu\, \bar{x}_i \tag{3-46}$$

and

$$\sum_{i=1}^{p} \mu_i = \sum_{i=1}^{p-1} \lambda\alpha_i + (1 - \lambda) = \lambda \sum_{i=1}^{p-1} \alpha_i + (1 - \lambda)$$
$$= \lambda 1 + (1 - \lambda) = 1$$

Also, since $\lambda \geq 0$, $\alpha_i \geq 0$ $(i = 1, 2, \ldots, p - 1)$,. $(1 - \lambda) \geq 0$, all $\mu_i \geq 0$, $i = 1, 2, \ldots, p$. Moreover, since each α_i and λ can have any value between 0 and 1, it must be true that each μ_i can have any value between 0 and 1 (although, of course, the relation $\sum_{i=1}^{p} \mu_i = 1$ must be satisfied).

Thus, equation (3-46) and the preceding discussion tell us that all convex combinations of $\bar{x}_1, \bar{x}_2, \ldots, \bar{x}_p$ must be in any convex set which contains $\bar{x}_1, \bar{x}_2, \ldots, \bar{x}_p$. Therefore, the convex hull of $\bar{x}_1, \bar{x}_2, \ldots, \bar{x}_p$ is indeed the set of all convex combinations of $\bar{x}_1, \bar{x}_2, \ldots, \bar{x}_p$.

We have already noted that in E^2 the convex hull of a set of points $\bar{x}_1, \bar{x}_2, \ldots, \bar{x}_r$ is a polyhedron. In E^n, we shall call the convex hull of a finite set of points the *convex polyhedron* of those points. Convex polyhedra are also sometimes called convex polytopes. In E^n, the boundary of a convex polyhedron is formed by planes. That is, each face of the polyhedron is a plane; in Figure 3-12, for example, the boundary of the polyhedron consists of seven faces; there are seven extreme points.

In E^n, each face of a convex polyhedron is a hyperplane. Clearly, the convex polyhedron of r points $\bar{x}_1, \bar{x}_2, \ldots, \bar{x}_r$ cannot have more than r extreme points; moreover, only the points $\bar{x}_1, \bar{x}_2, \ldots, \bar{x}_r$ may be extreme points, since all other points in the convex polyhedron

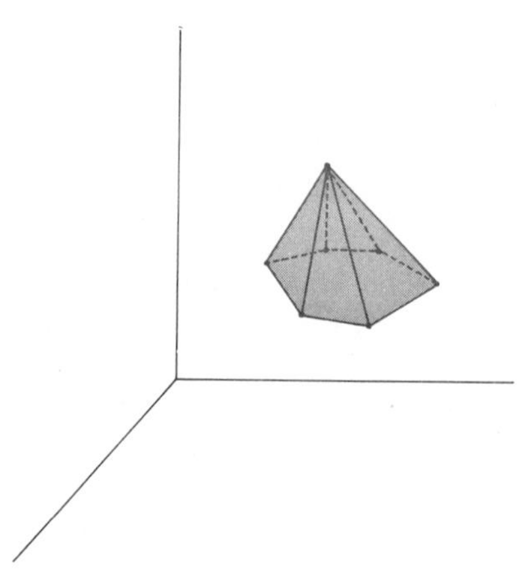

Figure 3-12

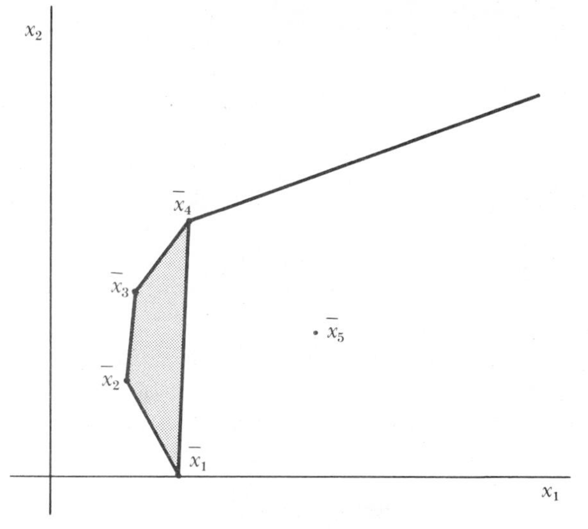

Figure 3-13

are convex combinations of $\bar{x}_1, \bar{x}_2, \ldots, \bar{x}_r$. However, some of the points $\bar{x}_1, \bar{x}_2, \ldots, \bar{x}_r$ may be interior points or boundary (but not extreme) points of the convex polyhedron. The extreme points $\{\bar{x}_1^*, \bar{x}_2^*, \ldots, \bar{x}_p^*\}$ of the convex polyhedron C of a set of points $S = \{\bar{x}_1, \bar{x}_2, \ldots, \bar{x}_r\}$ thus are a subset of S; moreover, the convex polyhedron of $\bar{x}_1^*, \bar{x}_2^*, \ldots, \bar{x}_p^*$ is also C. Thus, C is the set of all convex combinations of its extreme points. Intuitively, at least, the foregoing argument leads us to:

Theorem 3.11. Any point in a convex polyhedron can be represented as a convex combination of the extreme points of the polyhedron.

Although a convex polyhedron is determined by (and has) a finite number of extreme points, it is not true that every convex set which has a finite number of extreme points has the property that every point in the set can be expressed as a convex combination of the extreme points. The convex set S shown in Figure 3-13 has only four extreme points $\bar{x}_1, \bar{x}_2, \bar{x}_3, \bar{x}_4$, but the point $\bar{x}_5$ cannot be expressed as a convex combination of $\bar{x}_1, \bar{x}_2, \bar{x}_3, \bar{x}_4$, since it is not in the convex hull of $\bar{x}_1, \bar{x}_2, \bar{x}_3, \bar{x}_4$ (indicated by the darker region). If the set S were bounded, however, we could state:

Theorem 3.12. Any strictly bounded closed convex set S with a finite number of extreme points is the convex hull of the extreme points (and hence any point in S can be expressed as a convex combination of the extreme points).

The truth of Theorem 3.12 is intuitively obvious. Consider, for example, the set of Figure 3-13. In order to make it strictly bounded, at least one additional boundary line would have to be added. One such line, for example, is the line segment connecting $\bar{x}_1$ and $\bar{x}_4$, thus forming the convex hull of $\bar{x}_1$, $\bar{x}_2$, $\bar{x}_3$, $\bar{x}_4$. Any other boundary line which is added to make the set bounded would introduce new extreme points, but will still form a convex polyhedron, and hence is the convex hull of the extreme points. In Figure 3-14, we have added a boundary line to the set of Figure 3-13 to produce a strictly bounded set. Note that the resulting set is a convex polyhedron, with extreme points $\bar{x}_1$, $\bar{x}_2$, $\bar{x}_3$, $\bar{x}_4$, $\bar{x}_6$, $\bar{x}_7$.

The preceding discussion was limited to convex sets with a finite number of extreme points. In terms of constraints to optimization problems, such convex sets are the intersection of linear inequalities (half-spaces) and linear equalities (hyperplanes). Any convex set whose boundary consists in whole or in part of a curved surface (e.g., a hypersphere) contains an infinite number of extreme points. In Figure 3-15, for example, the extreme points of the convex set are $\bar{x}_1$, $\bar{x}_2$, $\bar{x}_3$, as well as any point on the curved boundary between $\bar{x}_1$ and $\bar{x}_3$, such as $\bar{x}_4$ (no line segment passing through $\bar{x}_4$ lies entirely within the set for which $\bar{x}_4$ is not an end point of that line segment).

Although the intersection of a finite number of half-spaces and hyperplanes does produce a convex set with a finite number of extreme points, it is clear that the actual number of extreme points may be rather large. The intersection of k linear equalities in n

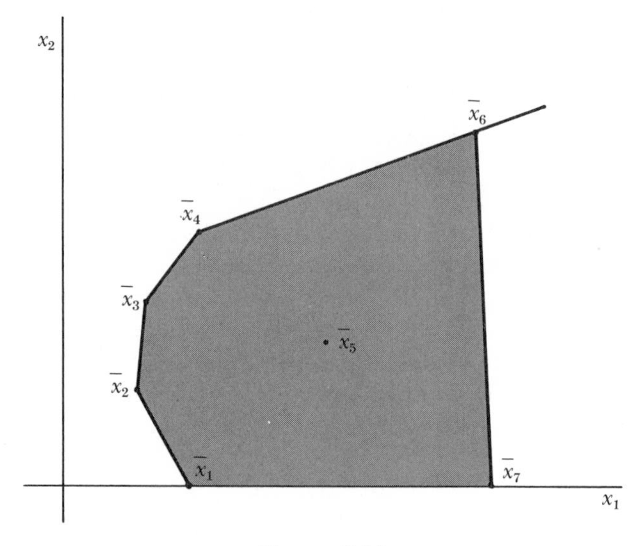

Figure 3-14

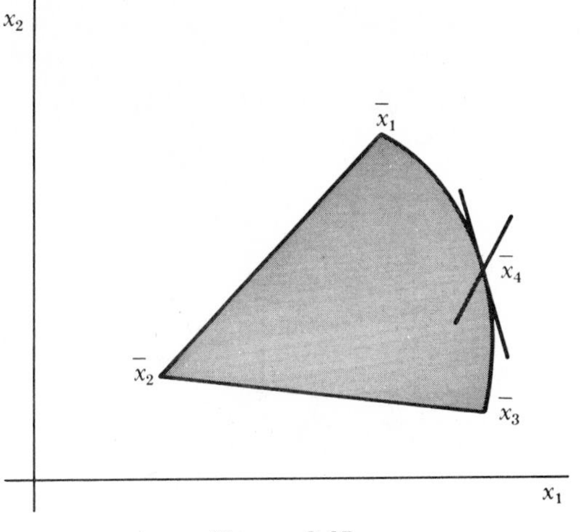

Figure 3-15

unknowns could form a convex set in E^n with as many as $\binom{n}{k} = \dfrac{n!}{(n-k)!\,k!}$ extreme points.†

Thus, for example, a system of five equations in ten unknowns could form a convex set with $\binom{10}{5} = 252$ extreme points; a system of ten equations in twenty unknowns could form a convex set with 184,756 extreme points. Thus, even in optimization problems with linear constraints, for which it is known that the optimal solution is an extreme point, total enumeration of all the extreme points becomes impractical for all but relatively small problems. It is for this reason that special techniques for solving such problems have been devised in which only a small subset of the extreme points need be examined. Some of these techniques will be discussed in Chapters 6 and 7.

We shall close this section with the definition of a special type of convex polyhedron with relatively few extreme points: *A simplex in E^n is the convex hull of any set of $n + 1$ points which do not all lie on one hyperplane in E^n.*

A simplex in E^2, then, is formed by any three points which do not all lie on a line. Thus, in E^3 a simplex is formed by four points which do not all lie in the same plane. Since three (noncolinear) points determine a plane in E^3, the fourth point must lie outside this

† The actual number of extreme points possible is currently unknown. Current research (see Grünbaum[1]) indicates that it may be somewhat less than $\binom{n}{k}$. However, for large n, the number of extreme points will tend to be quite large. The number $\binom{n}{k}$ yields an upper bound on the total number of extreme points.

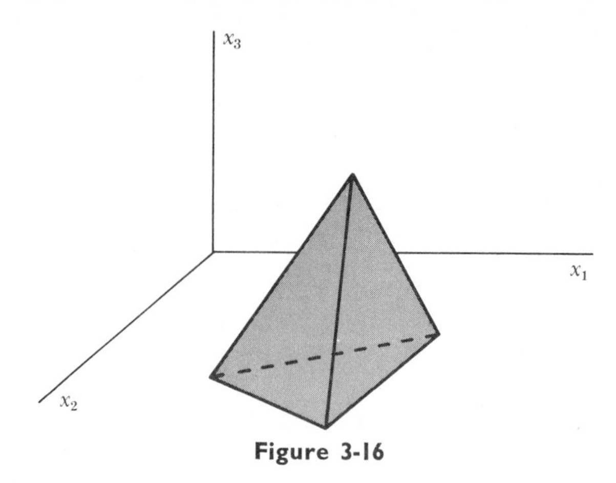

Figure 3-16

plane, and the resulting simplex is a pyramid, or tetrahedron. An example of a simplex in E^3 is shown in Figure 3-16.

3.5 SEPARATING AND SUPPORTING HYPERPLANES

We shall conclude this chapter with a brief presentation and discussion of several theorems which are very important in the theory of linear programming. Our discussion will be on an intuitive level, from a geometric point of view. We shall not prove the theorems rigorously, since the proofs are quite lengthy and are not particularly illuminating; they may be found in Hadley.[2] The theorems combine to give us the very important result that if a linear programming problem has an optimal solution, then at least one extreme point is optimal. Let us proceed.

Theorem 3.13. For any closed convex set C in E^n and any point $\bar{y}$ in E^n but not in C, there exists a hyperplane which passes through $\bar{y}$ and which has the property that all of C is contained in one of the two open half-spaces produced by the hyperplane.

Theorem 3.13 states that, for any point $\bar{y} \notin C$, we can find a hyperplane $\bar{a}'\bar{x} = b$ with the following properties:

1. $\bar{a}'\bar{y} = b$ ($\bar{y}$ is on the hyperplane)
2. If $H_1 = \{\bar{x} \mid \bar{a}'\bar{x} < b\}$

   $H_2 = \{\bar{x} \mid \bar{a}'\bar{x} > b\}$

   then either $C \subseteq H_1$ or $C \subseteq H_2$

Intuitively, this result is obvious (at least in E^2 and E^3); it merely says that a convex set C can be contained in some open half-space.

Note that C need not be bounded for Theorem 3.13 to hold. If C is all of E^n, however, then there is no point $\bar{y}$ not in C, and the theorem is obviously not valid. Thus, for all convex sets C which are proper subsets of E^n, Theorem 3.13 is valid. The hypothesis of Theorem 3.13 does not impose any restrictions on the point $\bar{y}$ other than it not be in C. Thus, the theorem is valid for $\bar{y}$ arbitrarily close to C. In the limiting case, then, when $\bar{y}$ is a boundary point of C, we might expect that a slightly modified version of Theorem 3.13 would also be valid. This is in fact so, and the result can be stated as:

Theorem 3.14. If $\bar{y}$ is a boundary point of a closed convex set C, there exists a hyperplane which passes through $\bar{y}$ and which has the property that all of C is contained in one of the two *closed* half-spaces produced by the hyperplane.

Such a hyperplane is called a *supporting hyperplane of C at $\bar{y}$*. If $\bar{a}'\bar{x} = b$ is a supporting hyperplane of C at $\bar{y}$, then all of C lies either in the closed half-space $\bar{a}'\bar{x} \geq b$ or the closed half-space $\bar{a}'\bar{x} \leq b$. Figures 3-17a and 3-17b illustrate separating hyperplanes for two convex sets C_1 and C_2 (note that C_2 is not bounded). Figures 3-18a and 3-18b illustrate supporting hyperplanes for C_1 and C_2. Figure 3-19 illustrates some supporting hyperplanes for a convex polyhedron in E^3. In Figure 3-19a, the boundary point $\bar{y}_1$ lies on the bottom face (plane) of the polyhedron, and the supporting hyperplane at $\bar{y}_1$ is thus the same plane; that is, its infinite extensions. It is the only supporting hyperplane passing through $\bar{y}_1$. In Figure 3-19b, the boundary point $\bar{y}_2$ lies along an "edge" of the polyhedron (that is, along the intersection of two faces—planes). A supporting hyperplane at $\bar{y}_2$ must contain all points on this edge of the polyhedron. There are an infinite number of such hyperplanes. In Figure 3-19c, the boundary point $\bar{y}_3$ is an extreme point of the polyhedron. A supporting hyperplane at $\bar{y}_3$ need only contain $\bar{y}_3$; again, there are an infinite number of such hyperplanes.

Notice that in all three cases, the supporting hyperplanes contain extreme points of the polyhedron. Recall that each face of a convex polyhedron in E^n lies on a hyperplane; the intersection of two such faces produces an "edge" of the polyhedron. Notice that the edges of the convex polyhedron of Figure 3-19 are line segments, each of which is determined by two extreme points of the polyhedron. Notice also that not every pair of extreme points produces an edge.

In order to extend these ideas to E^n, we define an *edge of a convex set C* as *the line segment joining any two extreme points of C if it is the intersection of C with a supporting hyperplane of C*. (See Figure 3-19b.) Moreover, two extreme points of C are called *adjacent extreme points* if the line segment joining them is an edge. (If C is unbounded, it is also possible for C to have an edge containing only one extreme

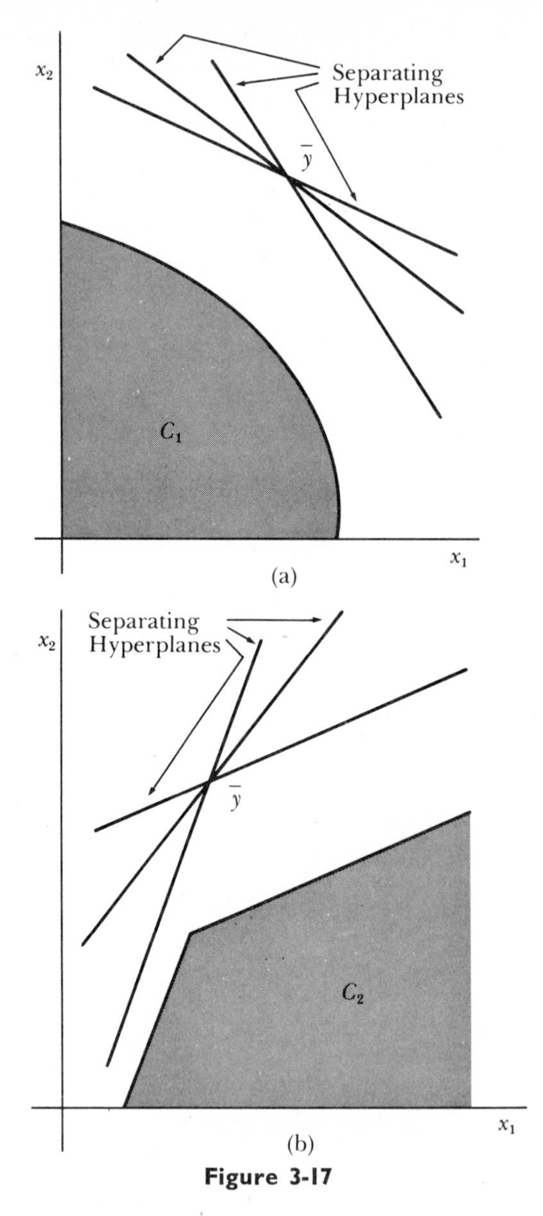

Figure 3-17

point; in Figure 3-18b, for example, the x_1-axis is such an edge. The above definition of "edge" is easily extended to include this case.)

Thus, in E^n, we may also consider a boundary point of a convex polyhedron as being a point on one face (hyperplane), on an edge, or as an extreme point. In each of the three cases, as in E^3, a supporting hyperplane must contain at least one extreme point of the polyhedron. More generally, we state:

Theorem 3.15. If C is a closed convex set which is bounded

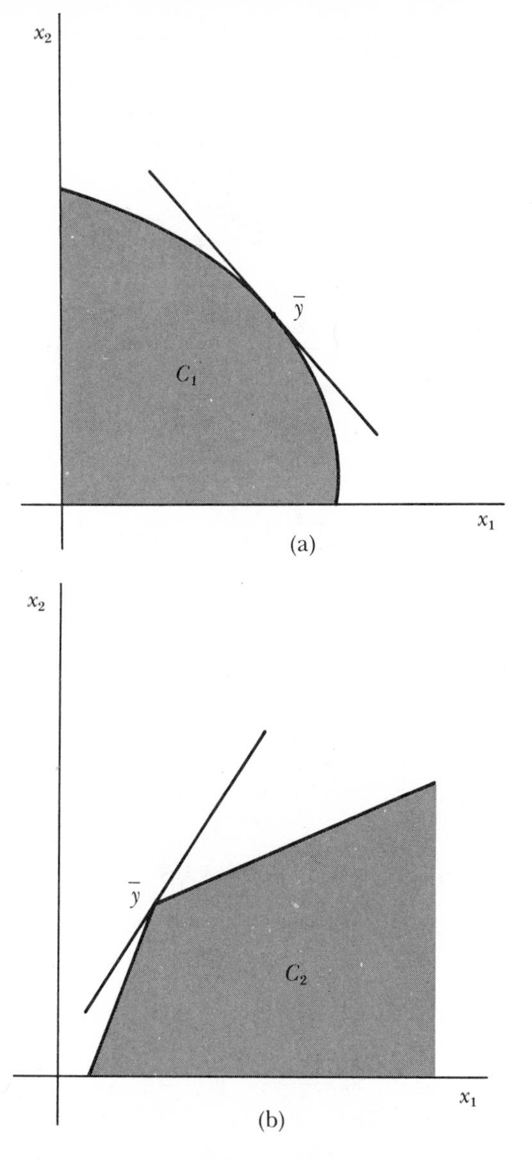

Figure 3-18

from below,† then every supporting hyperplane of C contains at least one extreme point of C.

In particular, in a linear programming problem, the optimal value of the objective function defines a hyperplane; we shall prove that this hyperplane is a supporting hyperplane. Thus, since all points on this hyperplane (which are also in the convex set of

† The fact that C is bounded from below implies that C has at least one extreme point.

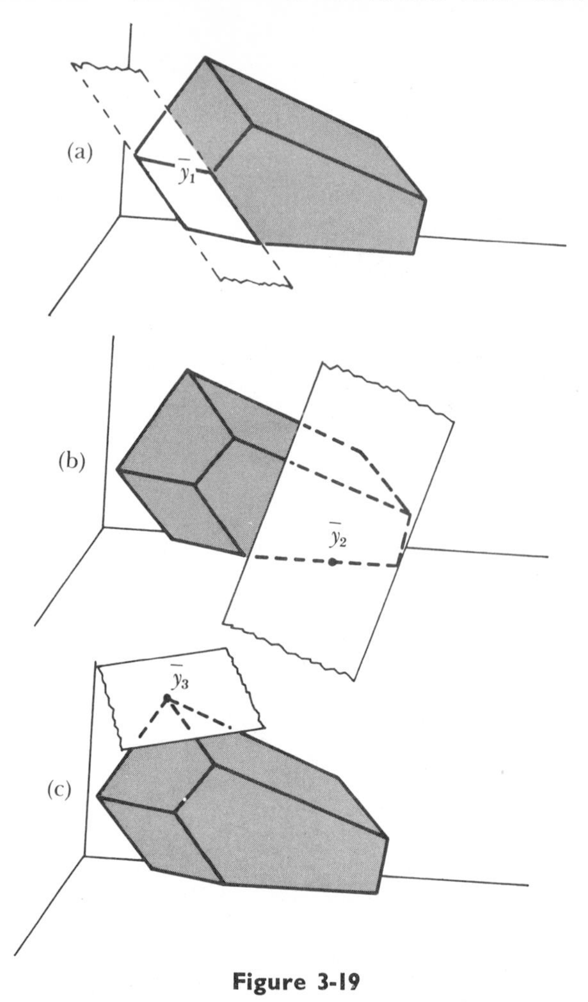

Figure 3-19

solutions) are optimal solutions, *at least one extreme point of the convex set of solutions to a linear programming problem is optimal*, provided that the objective function is not unbounded. It is this fact which is the basis of the various methods for solving linear programming problems which will be discussed in Chapters 6 and 7.

Theorem 3.16. If z^* is the optimal value of the objective function $z = \bar{c}'\bar{x}$ and $\bar{x}^*$ is an optimal solution (i.e., $z^* = \bar{c}'\bar{x}^*$) to a linear programming problem, then $z^* = \bar{c}'\bar{x}$ is a supporting hyperplane of the convex set of solutions C.

Proof: First, we must show that $\bar{x}^*$ is a boundary point of C. We can do this by showing that no optimal solution can be an interior point: If $\bar{x}^*$ were an interior point, then we could find an

ε-neighborhood of $\bar{x}^*$ which contained only points in C. Define

$$\bar{y} = \bar{x}^* + \left(\frac{\varepsilon}{2\,|\bar{c}|}\right)\bar{c}$$

Then

$$|\bar{x} - \bar{y}| = \left|\left(\frac{\varepsilon}{2\,|\bar{c}|}\right)\bar{c}\right| = \frac{\varepsilon}{2} < \varepsilon$$

Thus, $\bar{y}$ is in the ε-neighborhood of $\bar{x}^*$. But,

$$\bar{c}'\bar{y} = \bar{c}'\left[\bar{x}^* + \left(\frac{\varepsilon}{2\,|\bar{c}|}\right)\cdot\bar{c}\right] = \bar{c}'\bar{x}^* + \frac{\varepsilon}{2}\,|\bar{c}| = z^* + \frac{\varepsilon}{2}\,|\bar{c}| > z^*$$

This contradicts the assumption that z^* was the largest value attained by the objective function over points in C. Hence, we have shown that all optimal solutions are boundary points.

Now, let us consider a supporting hyperplane to C at $\bar{x}^*$. We wish to show that $z^* = \bar{c}'\bar{x}$ is such a hyperplane. If there is any point $\bar{y} \in C$ such that $\bar{c}'\bar{y} > z^*$, then $\bar{x}^*$ would not be optimal. Thus, for all points $\bar{x}$ in C, it must be true that $\bar{c}'\bar{x} \leq z^*$. Moreover, since $\bar{c}'\bar{x}^* = z^*$, at least one point in C lies on the hyperplane. Thus, $z^* = \bar{c}'\bar{x}$ is a supporting hyperplane of C.

EXERCISES

n-Dimensional Sets

1. (a) Determine the hyperplane in E^3 which contains the vectors $\bar{x}_1' = (1, 0, 1)$, $\bar{x}_2' = (2, 2, -1)$, $\bar{x}_3' = (0, 0, 0)$.

   (b) Show that the hyperplane of part (a) is a vector space. What is its dimension?

2. Show that any hyperplane through the origin is a vector space. If the hyperplane is in E^n, what is its dimension?

3. Determine the equation for the line segment in E^3 between the points $(0, 1, 0, 1)$ and $(-1, 2, 0, 3)$.

4. Consider the following sets in E^2:

$$S_1 = \{[x_1, x_2] \mid x_1^2 + x_2^2 = 4\}$$
$$S_2 = \{[x_1, x_2] \mid x_1^2 + x_2^2 < 4\}$$
$$S_3 = \{[x_1, x_2] \mid x_1^2 + x_2^2 \geq 4\}$$
$$S_4 = \{[x_1, x_2] \mid x_1 - x_2 = 1\}$$

$$S_5 = \{[x_1, x_2] \mid x_1 - x_2 \leq 1\}$$
$$S_6 = \{[x_1, x_2] \mid x_1 - x_2 > 1\}$$
$$S_7 = \{[x_1, x_2] \mid 3x_1 + 2x_2 \leq 6\}$$
$$S_8 = \{[x_1, x_2] \mid x_1 \geq 0\}$$
$$S_9 = \{[x_1, x_2] \mid x_2 \geq 0\}$$

Sketch each of the following sets:

(a) $S_1 \cup S_4$
(b) $S_1 \cap S_4$
(c) $S_2 \cap S_5$
(d) $S_3 \cap S_5$
(e) $S_2 \cap S_6$
(f) $S_5 \cap S_7$
(g) $S_6 \cap S_7 \cap S_8$
(h) $S_6 \cap S_7 \cap S_8 \cap S_9$
(i) $S_5 \cap S_7 \cap S_8 \cap S_9$
(j) $S_6 \cap S_7 \cap S_8 \cap S_9 \cap S_2$

5. For each of the sets sketched in exercise 4 state whether the set is open, closed, or neither, and whether it is bounded or unbounded.

Convex Sets

6. For each of the sets sketched in exercise 4 determine whether the set is convex.

7. Determine whether each of the following sets is convex:

(a) $S_1 = \{[x_1, x_2, x_3] \mid x_1 = 1, x_2 = x_3\}$
(b) $S_2 = \{[x_1, x_2, x_3] \mid x_1^2 + 2x_2^2 + 6x_3^2 \leq 9\}$
(c) $S_3 = \{[x_1, x_2, x_3] \mid x_1 - x_2^2 \leq 4\}$
(d) $S_4 = \{[x_1, x_2, x_3] \mid x_1 - x_2^2 \geq 4\}$
(e) $S_5 = \{[x_1, x_2, x_3] \mid x_1 + 2x_2 = 3\}$

8. Prove Theorem 3.6: Closed and open half-spaces are convex sets.

9. Determine the extreme points of the set S, where S is the set of all solutions to:

$$\begin{Bmatrix} x_1 + x_2 \leq 4 \\ 2x_1 + x_2 \leq 6 \\ x_2 \leq 3 \\ x_1, x_2 \geq 0 \end{Bmatrix}$$

10. Express the point $\bar{a} = [1, 1, 1, 1]$ as a convex combination of the points

$$\bar{x}_1 = [1, 2, 0, 3], \ \bar{x}_2 = [2, 0, 2, -1],$$

and $\bar{x}_3 = [0, 0, 2, -1]$.

Convex Hulls and Polyhedra

11. Sketch the convex polyhedra generated by the following sets of points:

    (a) $(0, 0), (1, 2), (1, -1), (3, 4)$
    (b) $(3, 2), (2, 1), (-3, 4)$
    (c) $(-1, 0), (1, 0), (2, 0)$
    (d) $(1, 0, 0), (0, 1, 0), (0, 0, 1), (0, 0, 0)$

12. Sketch the convex polyhedra formed by the intersection of the nonnegative orthant with each of the polyhedra of exercise 11.

13. Determine the extreme points for each of the polyhedra of exercise 12.

14. Sketch the convex hull of each of the following sets:

    (a) $S_1 = \{[x_1, x_2] \mid x_1^2 + x_2^2 \geq 1\}$
    (b) $S_2 = \{[x_1, x_2] \mid x_1^2 + x_2^2 \leq 9, x_1 \geq 0, x_2^2 \geq 4\}$
    (c) $S_3 = \{[x_1, x_2] \mid x_1^2 + x_2^2 \leq 9, 2x_1^2 + 3x_2^2 \geq 6\}$

15. (a) Sketch the convex polyhedron generated by the following set of points:

    $$(1, 0), (3, 2), (4, 3), (-1, 2), (-3, -2)$$

    (b) Sketch a supporting hyperplane to the polyhedron of part (a) at the point $(4, 3)$. Is this hyperplane unique?

16. For the polyhedron of exercise 15 sketch a supporting hyperplane at $(3, 2)$. Is this hyperplane unique?

17. For the polyhedron of exercise 15, sketch a supporting hyperplane at $(4, 3)$ which also passes through the point $(-1, 2)$. Is this hyperplane unique?

18. For the polyhedron of exercise 15, is there a supporting hyperplane at the point $(4, 3)$ which also passes through the point $(-3, -2)$? Explain.

19. For the convex set $S = \{[x_1, x_2] \mid x_1^2 + x_2^2 \leq 5\}$ determine the equation of a supporting hyperplane to S at the point (2, 1). Is this hyperplane unique? Explain.

20. If x_1 and x_2 are adjacent extreme points of a convex polyhedron, does a supporting hyperplane to the polyhedron at $\bar{x}_1$ always exist which also passes through $\bar{x}_2$? Explain.

REFERENCES

1. Grünbaum, B.: *Convex Polytopes*. Interscience Publishers, New York (1967).
2. Hadley, G.: *Linear Algebra* (Chapter 6). Addison-Wesley, Reading, Massachusetts (1961).
3. Valentine, F. A.: *Convex Sets*. McGraw-Hill, New York (1964).

Chapter 4

CLASSICAL OPTIMIZATION

4.1 INTRODUCTION

We shall be primarily concerned, in this chapter, with stating and describing the so-called classical optimization procedures. These derive, after suitable embellishments to make them useful, from the application of the calculus to the basic problem of finding the maximum or minimum of a continuous function. These techniques have an intrinsic utility in that they can sometimes be used to solve problems that are not too complex and do not involve more than a few variables. More important, however, is the theoretical significance of the results of classical optimization theory. This theory is of importance in the development of solution methods to nonlinear programming problems. The theory underlies any discussion of the development of algorithms for solving any complex optimization problem.

4.2 FUNCTIONS OF ONE VARIABLE: MATHEMATICAL BACKGROUND

We shall here summarize some definitions and results that will be required in subsequent sections.

A function $f(x)$ takes on its *absolute maximum* at a point x^* if $f(x) \leq f(x^*)$ for all x over which the function $f(x)$ is defined. The absolute maximum is often called the *global maximum*. The definition of *global minimum* can be obtained from the preceding definition by reversing the inequality between $f(x)$ and $f(x^*)$.

Let $f(x)$ be defined in some neighborhood δ about a point x^0. Then the function $f(x)$ is said to have a *strong relative maximum* (or *proper relative maximum*) at x^0 if there exists an ε, $0 < \varepsilon < \delta$, such that for all x satisfying $0 < |x - x^0| < \varepsilon$, it is the case that $f(x) < f(x^0)$.

Intuitively, what this says is merely that if a function $f(x)$ has a strong relative maximum at some point x^0, then there is an interval including x^0, no matter how small, such that for all x in this interval, $f(x)$ is *strictly less* than $f(x)$. It is the "strictly less" that makes this a *strong* relative maximum.

We can also define a *weak relative maximum* (or *improper relative maximum*) as follows. Again, let $f(x)$ be defined in some neighborhood δ about a point x^0. Then the function $f(x)$ is said to have a weak relative maximum at x^0 if there exists an ε, $0 < \varepsilon < \delta$, such that for all x satisfying $0 < |x - x^0| < \varepsilon$, it is the case that $f(x) \leq f(x^0)$ and there is at least one point x in the interval $[x^0 - \varepsilon, x^0 + \varepsilon]$ such that $f(x) = f(x^0)$.

We will usually not distinguish between strong and weak relative maxima but merely call them relative maxima. Relative maxima are often referred to as *local maxima* as well.

The definitions of *strong and weak relative minima* are analogous to the preceding definitions given for maxima except that in the former case, "$f(x) > f(x^0)$" is substituted for "$f(x) < f(x^0)$" and in the latter, "$f(x) \geq f(x^0)$" is substituted for "$f(x) \leq f(x^0)$."

In Figure 4-1 some examples of local and global optima are shown. In Figure 4-1a we see several examples of relative maxima and minima as well as a global maximum designated. In Figure 4-1b, the points designated A and B are examples of weak relative maxima. The point A cannot be a strong relative maximum, since the point B yields the same value of $f(x)$.

There are three other terms, used quite often in the literature of optimization, that we shall define here. A *stationary point* is any point at which

$$f'(x) = \frac{df(x)}{dx} = 0.$$

Most writers use the term stationary point and *critical point* interchangeably. However, there are departures from this. Some authors consider critical points as any points which may be candidates for global optima. As long as we do not restrict the interval over which x is considered, this causes no difficulty, since, as we shall see later, any local optimum may be a candidate for the global optimum. However, if a finite interval for x is considered, which is usually the case in practical problems, a global optimum may occur at a point which is not a local optimum. This will be discussed in more detail later.

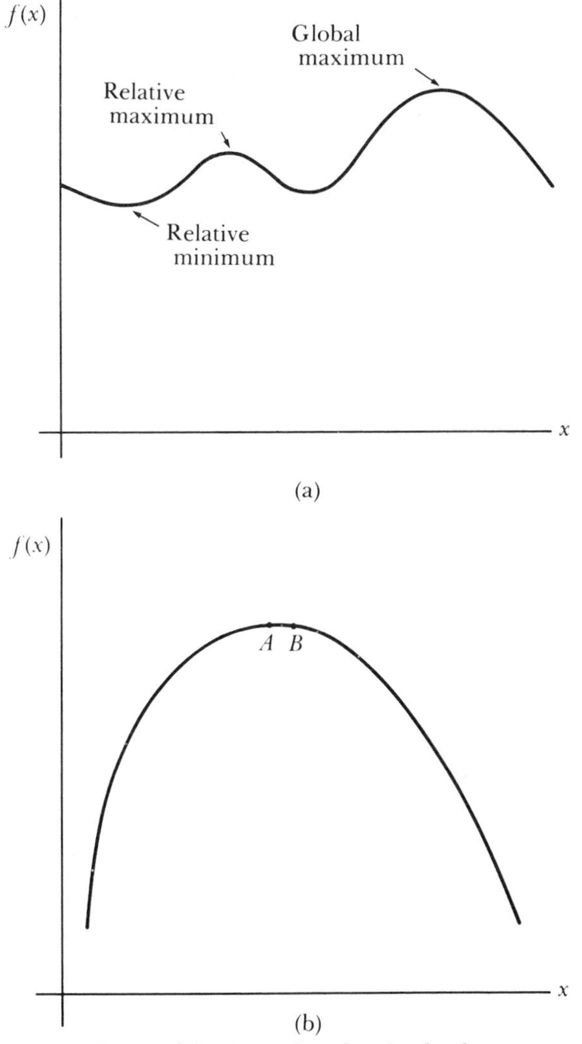

Figure 4-1 Examples of optimal values.

As we have noted, a point at which a function $f(x)$ has an extreme value is called a relative optimum. However, we may consider the function derived from $f(x)$ by differentiation, i.e., $f'(x)$. $f'(x)$ gives the slope of the curve $f(x)$. A point at which $f'(x)$ has a relative extreme value is called a *point of inflection.* In Figure 4-2 we illustrate these ideas. In Figure 4-2, the function $f(x) = x^3$ is shown. The origin, $x = 0$ is a stationary point, since $f'(x) = 3x^2 = 3(0) = 0$. Note, however, that $x = 0$ is not a relative optimum. However, $x = 0$ is a point of inflection, since as the plot of $f'(x)$ shows (the dashed line), $x = 0$ is a point at which

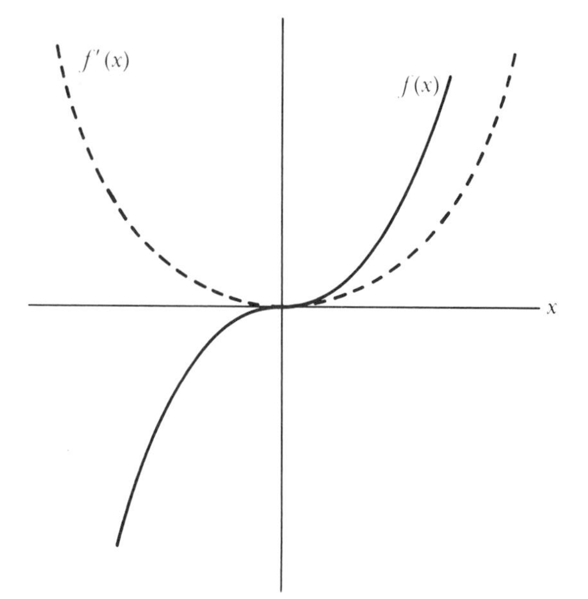

Figure 4-2 Point of inflection.

the slope of $f'(x)$ changes, i.e., it is a relative optimum point of $f'(x)$.

There is an important characteristic of certain special functions of a single variable which is related to the existence of minima and maxima. This property is known as *convexity*, and its opposite *concavity*.

Formally, a function $f(x)$ is said to be *convex* over some interval in x, if for any two points x_1, x_2 in the interval and for all λ, $0 \le \lambda \le 1$,

$$f[\lambda x_2 + (1 - \lambda)x_1] \le \lambda f(x_2) + (1 - \lambda)f(x_1) \qquad \textbf{(4-1)}$$

Similarly, a function $f(x)$ is said to be *concave* over some interval in x, if for any two points x_1, x_2 in the interval and for all λ, $0 \le \lambda \le 1$,

$$f[\lambda x_2 + (1 - \lambda)x_1] \ge \lambda f(x_2) + (1 - \lambda)f(x_1) \qquad \textbf{(4-2)}$$

It is clear from the definitions that if $f(x)$ is convex, then $-f(x)$ is concave and vice versa.

Intuitively, a function is convex if a line segment drawn between any two points on its graph falls entirely on or above the graph. A function is concave if a line segment drawn between any two points on its graph falls entirely on or below its graph. If the line segment falls entirely above (below) the graph the function is said to be *strictly convex* (*strictly concave*). In Figure 4-3 these concepts are illustrated. Figure 4-3a shows a convex function and Figure 4-3b shows a concave function. These are strictly convex and concave

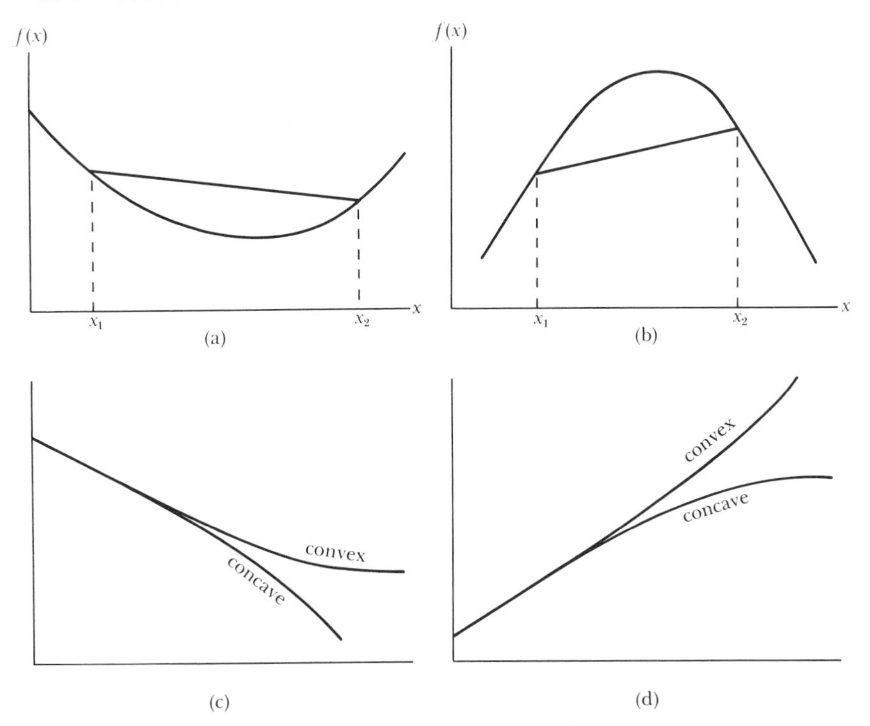

Figure 4-3 Convex and concave functions.

respectively. It will be noted that one has a minimum and the other has a maximum. These relationships will be explored further in the next section. In Figures 4-3c and 4-3d some other possible examples of convex and concave functions are given.

It should be obvious that not every function must be convex or concave. In fact, very often a function is neither convex nor concave. For example, with reference to Figure 4-2, with $f(x) = x^3$, it can be seen that $f(x)$ is neither convex nor concave. However, in the interval, $(-\infty, 0)$, $f(x)$ is concave and in the interval, $(0, \infty)$, $f(x)$ is convex.

An important theorem which will be recalled from calculus is Taylor's theorem. It may be stated in a variety of ways. We will state it as follows. If $f(x)$ is continuous and has a continuous first derivative over some interval in x, then for any two points x_1 and x_2 in this interval, where $x_2 = x_1 + h$, there exists a θ, $0 \leq \theta \leq 1$ such that

$$f(x_2) = f(x_1) + hf'[\theta x_1 + (1 - \theta)x_2] \quad \textbf{(4-3)}$$

It can also be shown that if, in addition, $f(x)$ has a continuous second derivative, then there exists a θ, $0 \leq \theta \leq 1$ such that

$$f(x_2) = f(x_1) + hf'(x_1) + \frac{h^2}{2}f''[\theta x_1 + (1 - \theta)x_2] \quad \textbf{(4-4)}$$

Taylor's theorem can be extended to any order if $f(x)$ has continuous derivatives of the requisite order. The general statement of Taylor's theorem then becomes that there exists a θ, $0 \leq \theta \leq 1$ such that

$$f(x_2) = f(x_1) + hf'(x_1) + \frac{h^2}{2!}f''(x_1) + \dots + \frac{h^n}{n!} f^{(n)}[\theta x_1 + (1 - \theta)x_2] \qquad \textbf{(4-5)}$$

4.3 OPTIMIZATION OF FUNCTIONS OF A SINGLE VARIABLE

We shall consider in this section the problem of how to find relative maxima and minima of functions of a single variable. It is well to first consider functions of only one variable to illustrate some of the ideas which become much more complex in the context of optimization of functions of many variables.

In the chapter on search methods, Chapter 5, we shall consider methods of minimization and maximization which directly carry out these processes, i.e., in one way or another, we calculate the value of $f(x)$ for some value $x = x_1$ and then use some process of reasoning to lead us to another point, $x = x_2$ such that, if we are minimizing, $f(x_2) < f(x_1)$ and so on. This is quite different, both in philosophy and methodology, from the methods we are about to consider. The approach of classical mathematics, as exemplified in the calculus, is to construct a set of necessary and sufficient conditions which the optimal value of x must satisfy. These conditions are derived analytically from $f(x)$. The practical problem will then remain, unfortunately as we shall discover, of finding the value of x which satisfies these conditions. First, however, we shall attend to the derivation and discussion of these conditions.

If $f(x)$ has a relative minimum at a point x^0, then we know from the definition of relative minimum that there exists an $\varepsilon > 0$ such that for some interval or neighborhood, about x^0, $f(x) \geq f(x^0)$. In particular, consider points in this neighborhood about x^0 of the form $x = x^0 + h$, $0 < |h| < \varepsilon$. Then we may write

$$f(x^0 + h) - f(x^0) \geq 0 \qquad \textbf{(4-6)}$$

for all h, $0 < |h| < \varepsilon$. This may then be divided by h to obtain

$$\frac{f(x^0 + h) - f(x^0)}{h} \geq 0 \qquad h > 0 \qquad \textbf{(4-7)}$$

$$\frac{f(x^0 + h) - f(x^0)}{h} \leq 0 \qquad h < 0 \qquad \textbf{(4-8)}$$

If we take the limit in (4-7) as $h \to 0$ we obtain from (4-7) that $\frac{df(x^0)}{dx} \geq 0$ and similarly from (4-8) that $\frac{df(x^0)}{d} \leq 0$. These two results together inevitably lead to the conclusion that

$$\frac{df(x^0)}{dx} = 0 \tag{4-9}$$

An analogous argument, with inequalities reversed, can be made for maximization and it too yields equation (4-9). Therefore equation (4-9) tells us that if x^0 is a point where $f(x)$ takes on a relative maximum or minimum, the first derivative of $f(x)$ is equal to zero.

It will be recalled from the previous section, that any point at which the first derivative is zero is called a stationary point. Therefore, what we have shown is that a *necessary* condition for x^0 to be a relative maximum or minimum is that x^0 be a stationary point. For example, consider the function

$$f(x) = 3x^3 - 36x \tag{4-10}$$

This function has two stationary points, $x = \pm 2$. These may or may not be maxima or minima. It can be seen that the preceding necessary condition does not provide much information, even when we can solve the equation, $f'(x) = 0$, since at a stationary point a function (assuming, of course, that it is differentiable) may have either a maximum, a minimum, or neither. Consider some obvious examples:

$f_1(x) = 3x^2$ has a minimum
$f_2(x) = -4x^2$ has a maximum
$f_3(x) = 10x^3$ has neither a maximum nor a minimum

4.4 SUFFICIENT CONDITIONS FOR THE EXISTENCE OF AN OPTIMUM

It is clear that if the necessary condition we have derived, i.e., $f'(x) = 0$, is to be of use to us, we shall have to try to find a set of *sufficient* conditions to identify the various cases that may arise at stationary points. Perhaps the simplest way to derive these sufficient conditions is to examine the Taylor series expansion derived from the application of Taylor's theorem.

Suppose that $f(x)$ and its first two derivatives are continuous at x^0. Using Taylor's theorem we can write

$$f(x^0 + h) = f(x^0) + hf'(x^0) + \frac{h^2}{2} f''[\theta x^0 + (1 - \theta)(x^0 + h)] \qquad 0 \leq \theta \leq 1 \tag{4-11}$$

If $f(x)$ has a relative minimum at x^0, then we know from the necessary condition, that $f'(x^0) = 0$. Therefore we may write (4-11) as

$$f(x^0 + h) - f(x^0) = \frac{h^2}{2} f''[\theta x^0 + (1 - \theta)(x^0 + h)] \quad \textbf{(4-12)}$$

$$0 \leq \theta \leq 1$$

If $f(x^0)$ is to be a minimum, then it follows that

$$f(x^0 + h) - f(x^0) = \frac{h^2}{2} f''[\theta x^0 + (1 - \theta)(x^0 + h)] > 0 \quad \textbf{(4-13)}$$

$$0 \leq \theta \leq 1$$

Suppose $f''(x^0) < 0$. Then it follows from continuity that $f''[\theta x^0 + (1 - \theta)(x^0 + h)] < 0$ and therefore

$$f(x^0 + h) - f(x^0) = \frac{h^2}{2} f''[\theta x^0 + (1 - \theta)(x^0 + h)] < 0 \quad \textbf{(4-14)}$$

and therefore x^0 cannot be a minimum. Conversely, if $f''(x^0) > 0$ at $f'(x^0) = 0$, $f(x^0)$ is a minimum.

What we have shown is that if $f'(x^0) = 0$, a *sufficient* condition for $f(x)$ to be a minimum at x^0 is that

$$f''(x^0) > 0 \quad \textbf{(4-15)}$$

By an analogous argument, if $f'(x^0) = 0$, a *sufficient* condition for $f(x)$ to be a maximum at x^0 is that

$$f''(x^0) < 0 \quad \textbf{(4-16)}$$

Unfortunately, equations (4-15) and (4-16) do not quite dispose of the matter. It is perfectly possible that at a point x^0, both the first and second derivatives vanish. In this case we must examine higher order derivatives. We now prove the following theorem to determine sufficient conditions for any case.

Theorem 4.1. Assume that $f(x)$ and its first n derivatives are continuous. Then $f(x)$ has a relative maximum or minimum at x^0 if and only if n is even, where n is the order of the first nonvanishing derivative at x^0. The function $f(x)$ has a maximum at x^0 if $f^{(n)}(x^0) < 0$ and a minimum if $f^{(n)}(x^0) > 0$.

Proof: We assume that the first $n - 1$ derivatives of $f(x)$ vanish at x^0, i.e.,

$$f'(x^0) = f''(x^0) = \ldots = f^{(n-1)}(x^0) = 0 \quad \textbf{(4-17)}$$

We also assume that $f^{(n)}(x^0) \neq 0$ and that n is an even number. Using Taylor's theorem (equation (4-5)) and equation (4-17), and letting $x_1 = x^0$ and $x_2 = x^0 + h$ in equation (4-5), we have

$$f(x^0 + h) - f(x^0) = \frac{h^n}{n!} f^{(n)}[\theta x^0 + (1 - \theta)(x^0 + h)] \quad \textbf{(4-18)}$$

$$0 \leq \theta \leq 1$$

The continuity of $f^{(n)}(x)$, which was assumed, assures us that $f^{(n)}[\theta x^0 + (1 - \theta)(x^0 + h)]$ will have the same sign as $f^{(n)}(x^0)$. This then leads to the conclusion that $f(x^0 + h) - f(x^0)$ will have the same sign as $f^{(n)}(x^0)$, which follows from (4-18) and the fact that n is even. Therefore, we see that $f(x^0 + h) - f(x^0)$ will have the same sign as $f^{(n)}(x^0)$. This means that $f(x^0 + h) - f(x^0)$ will be positive whenever $f^{(n)}(x^0)$ is positive and negative whenever $f^{(n)}(x^0)$ is negative. Therefore x^0 will be a relative maximum or minimum, and whether or not it will be a maximum or minimum depends on the sign of $f^{(n)}(x^0)$.

The foregoing argument proves the "if" part of the theorem. We shall prove the "only if" part by contradiction. Let us assume the contradiction of our hypothesis, i.e., assume that the order of the first nonvanishing derivative is odd. Let us assume further that $f^{(n)}(x^0) > 0$, in order to be definite. Using Taylor's theorem we again write

$$f(x^0 + h) - f(x^0) = \frac{h^n}{n!} f^{(n)}[\theta x^0 + (1 - \theta)(x^0 + h)] \qquad \textbf{(4-18)}$$

$$0 \leq \theta \leq 1$$

As we know, if x^0 is to be an extremal point (maximum or minimum), then the quantity shown on either side of (4-18) has to be either nonpositive or nonnegative, depending on whether it is a maximum or minimum. We may note, however, that when n is odd, which we assumed, the quantity h^n is positive or negative as h is positive or negative. Because of the continuity of $f(x)$, as we noted, the sign of $f^{(n)}[\theta x^0 + (1 - \theta)(x^0 + h)]$ does not change as h goes from positive to negative. Therefore, the term on the right hand side of (4-18)

$$\frac{h^n}{n!} f^{(n)}[\theta x^0 + (1 - \theta)(x^0 + h)]$$

changes its sign as h^n changes its sign. Therefore, from (4-18), we see that $f(x^0 + h) - f(x^0)$ will have different signs depending upon whether h is positive or negative. This leads us to a contradiction, since if this is the case, x^0 cannot be an extremal point, this result being a direct contradiction of the definition of a relative maximum or minimum. This concludes the proof of the theorem.

As an example of the meaning of the foregoing result, consider the function

$$z = f(x) = x^4 - 4x^3 + 6x^2 - 4x + 1 = (x - 1)^4$$

$$\frac{dz}{dx} = 4(x - 1)^3 \qquad \frac{d^2z}{dx^2} = 12(x - 1)^2$$

$$\frac{d^3z}{dx^3} = 24(x - 1) \qquad \frac{d^4z}{dx^4} = 24$$

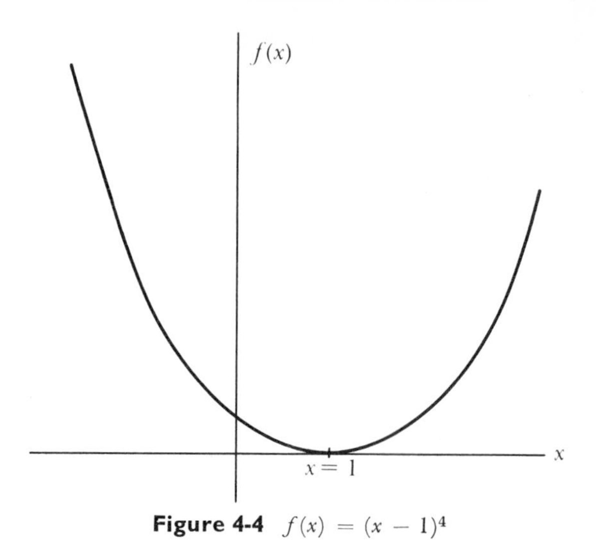

Figure 4-4 $f(x) = (x - 1)^4$

It can be seen that the first three derivatives are zero when $x = 1$. The first nonvanishing derivative is $\dfrac{d^4z}{dx^4}$. Therefore $f(x)$ has a minimum at $x = 1$ since $\dfrac{d^4z}{dx^4}$, the first nonvanishing derivative, is even and positive. A graph of the function $f(x) = (x - 1)^4$ is shown in Figure 4-4. If, however, we consider the function

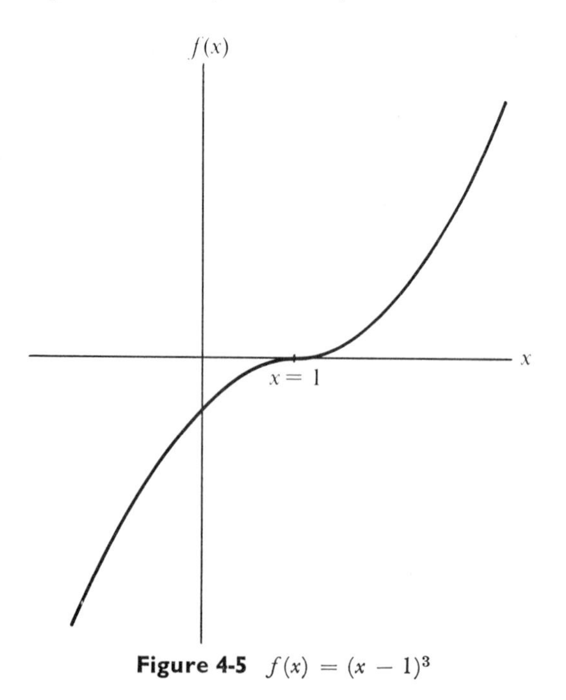

Figure 4-5 $f(x) = (x - 1)^3$

$z = f(x) = x^3 - 3x^2 + 3x - 1 = (x - 1)^3$, we now have

$$\frac{dz}{dx} = 3(x - 1)^2 \qquad \frac{d^2z}{dx^2} = 6(x - 1) \qquad \frac{d^3z}{dx^3} = 6$$

The first nonvanishing derivative is the third derivative. Even though the necessary condition is satisfied at $x = 1$, since $\frac{dz}{dx} = 0$, we have neither a maximum nor a minimum at $x = 1$, but a stationary point since $f'(x)$ changes sign as it passes through $x = 1$. The order of the first nonvanishing derivative is not even in this case. The graph of $f(x) = (x - 1)^3$ is shown in Figure 4-5.

4.5 GLOBAL EXTREMA: ONE VARIABLE

What we have shown in the previous section is how to determine whether or not there is a maximum or minimum of some continuous and differentiable function, $f(x)$, in some small interval. For this reason, these extrema, as we have already indicated, are called relative or local extrema. It is clear, however, in any practical optimization problem, that one is usually most concerned with the absolute or global extrema. For example, suppose the profit associated with the sales of x units of some manufactured item is of the form

$$P = \frac{Ax - B}{x^2 - C} + kx \tag{4-19}$$

What we are interested in is the value of x which makes P as large as possible, i.e., its global maximum. All the relative optima of P are not usually required. There may be as many as four relative optima in this problem, for calculating $\frac{dP}{dx} = 0$ leads to

$$x^4 - \left(\frac{A + 2Ck}{k}\right) x^2 + \frac{2B}{k} x + \left(\frac{kC^4 - AC}{k}\right) = 0$$

This equation must now be solved for its roots to determine the stationary points of $P(x)$. In some instances, we may wish to ascertain all the stationary points. The information of prime interest, however, in this problem is the global maximum.

We therefore have to consider how to find the global extrema of $f(x)$ for $a \leq x \leq b$, since in most practical problems there are practical bounds on what values x may be allowed to assume. In Figure 4-6, a flow chart is given for finding the global maximum of a continuous function $f(x)$ in the interval $a \leq x \leq b$. It will be noted that it is not necessary to use the sufficient conditions to sort out the

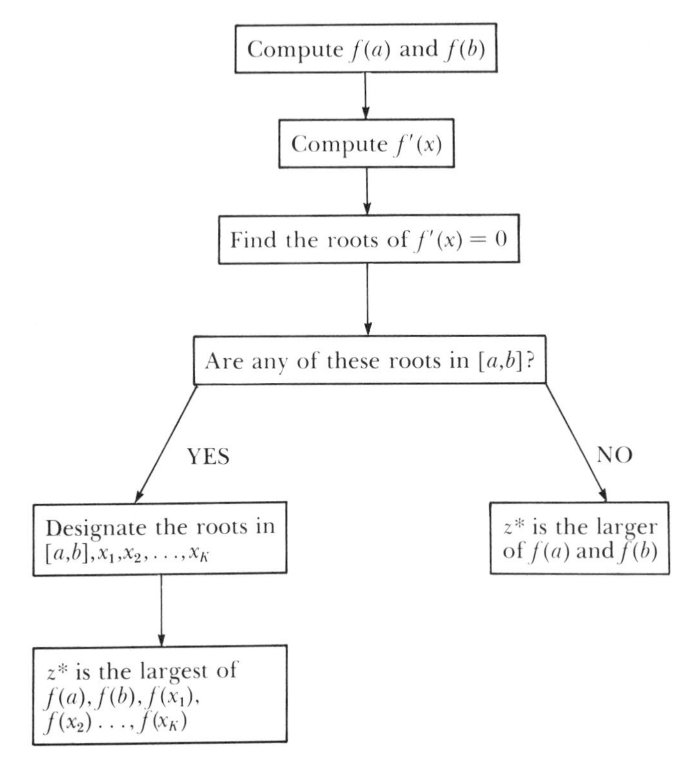

Figure 4-6 Procedure for finding global maximum of $z = f(x)$ in $a \leq x \leq b$.

relative maxima and minima, since this usually involves more computational labor than simply evaluating the function at all relative optima in the interval $a \leq x \leq b$. The procedure for finding the global minimum is similar, in an obvious way.

An example of the procedure shown in Figure 4-6 is as follows:

EXAMPLE

Find the global maximum of

$$f(x) = x^3 - x^2 - x + 3 \quad \text{for} \quad 0 \leq x \leq 2$$

First we compute $f(0) = 3$ and $f(2) = 5$.

$$f'(x) = 3x^2 - 2x - 1$$

Solving $f'(x) = 3x^2 - 2x - 1 = 0$, we find that

$$x_1 = -\tfrac{1}{3} \qquad x_2 = 1$$

We see that x_1 is not in $[0, 2]$ so we can ignore it. However x_2 is in

[0, 2], so we evaluate $f(1) = 2$. We see therefore that

$$z^* = \text{Max}\,[f(0) = 3, f(2) = 5, f(1) = 2] = 5$$

Therefore $z^* = 5$ and is taken on at $x = 2$, one end of the allowable interval.

It will be noted, in the flow chart and in the example, that although we are interested in the global maximum, we still must seek *all* relative optima using the classical optimization techniques under discussion. This is not true using more direct methods called "search" methods which will be discussed in Chapter 5.

4.6 NUMERICAL PROBLEMS OF ROOT FINDING

If one examines the computational scheme given in the flow chart in Figure 4-7, it will be seen that the only possibly difficult step may be in finding the roots of $f'(x) = 0$. In the simple example given in the previous section, this involved finding the roots of a quadratic equation. However, in general, this will involve finding the roots of nonlinear or transcendental equations, for which no computational procedure in closed form, analogous to that of solving a quadratic equation, may exist. Hence, numerical methods must be employed. There are a host of computational methods which exist

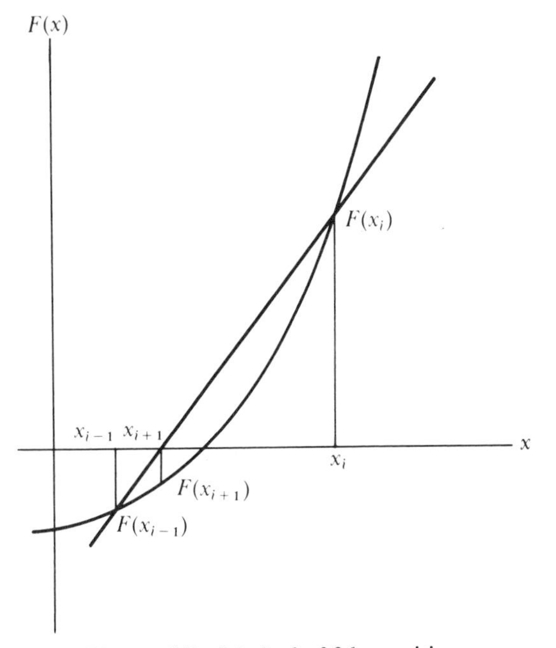

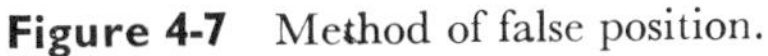

Figure 4-7 Method of false position.

for obtaining roots numerically. They are described in many texts on numerical analysis. See, for example, Hildebrand.[1] One such technique will be described here. It is called the Regula Falsi method, or method of false position.

The only assumption necessary for the use of the false position method is that the function $f'(x)$ be continuous and computable. A simple derivation of the method can be made by reference to Figure 4-7. We use the notation that $F(x) \equiv f'(x)$ and hence we are solving $F(x) = 0$. We assume that at any iteration we have available two points, x_{i-1}, x_i and the corresponding values $F(x_{i-1})$, $F(x_i)$. If a secant is passed through the points corresponding to x_{i-1} and x_i on the curve $F(x)$, the intersection of the secant with the x-axis yields the new approximation, x_{i+1}. It can be seen from Figure 4-7 that the slope, λ, of the secant is given by

$$\lambda = \frac{F(x_i) - F(x_{i-1})}{x_i - x_{i-1}} = \frac{F(x_i) - F(x_{i+1})}{x_i - x_{i+1}} = \frac{F(x_i)}{x_i - x_{i+1}} \qquad \textbf{(4-20)}$$

Solving equation (4-20) for x_{i+1} yields the false position method iteration formula

$$x_{i+1} = x_i - F(x_i)\frac{x_i - x_{i-1}}{F(x_i) - F(x_{i-1})}$$

Once the third point, x_{i+1} has been found, either x_{i-1} or x_i is dropped. The rule usually used is to drop that one which will still allow the root to be bracketed. In other words, we wish to have points such that $F(x_{i-1})$ and $F(x_i)$, at each stage of the calculation, will have opposite signs. This also applies to the initial pair of points. If this procedure is followed, convergence of the iteration scheme can be unconditionally guaranteed. An example of the method follows.

EXAMPLE

Find a root of $x^3 - 5x - 7 = 0$. We note that $F(2) = -9$, $F(3) = 5$. Hence, there is a root between $x = 2$ and $x = 3$. If $x_1 = 2$ and $x_2 = 3$ then

$$x_3 = 3 - 5\left(\frac{3 - 2}{5 + 9}\right) = 3 - 5\left(\frac{1}{14}\right) = 2.6$$

We then have $F(2.6) = -2.424$ and $F(3) = 5$, having discarded $x_1 = 2$.

$$x_4 = 3 - 5\left(\frac{3 - 2.6}{5 + 2.424}\right) = \frac{20.272}{7.424} = 2.73$$

We now have $F(2.73) = -0.303583$, $F(3) = 5$

$$x_5 = 3 - 5\left(\frac{3 - 2.73}{5 + 0.303583}\right) = \frac{14.56075}{5.303583} = 2.745$$

$$F(2.745) = -0.04136$$

This process can be continued to any desired accuracy.

4.7 OPTIMA OF CONVEX AND CONCAVE FUNCTIONS

Let us now examine a result which indicates why we singled out convex and concave functions as being of specific interest in optimization. An examination of a function such as the convex function in Figure 4-3a indicates that it has a single minimum over the interval in x that is shown. One might wonder whether this was, in some sense, generally true. We shall see that it is. In order to do this, we prove the following important Theorem.

Theorem 4.2. Let $f(x)$ be a convex function over a closed interval, $a \leq x \leq b$. Then any relative minimum of $f(x)$ in this interval is also the absolute minimum of $f(x)$ over the interval.

Proof: This proof is by contradiction. Let us assume that f has a local minimum at x^0 in the interval $[a, b]$. Let us assume further that the global minimum over this interval is taken on at a different point, x^* and that $f(x^0) > f(x^*)$. We defined a convex function as one for which

$$f[\lambda x_1 + (1 - \lambda)x_2] \leq \lambda f(x_1) + (1 - \lambda)f(x_2) \qquad \textbf{(4-21)}$$

for any x_1, x_2 in $[a, b]$ and all λ, $0 \leq \lambda \leq 1$. Therefore (4-21) must be satisfied by x_0 and x^*, since they are in $[a, b]$. Hence,

$$f[\lambda x^* + (1 - \lambda)x_0] \leq \lambda f(x^*) + (1 - \lambda)f(x_0) \qquad \textbf{(4-22)}$$

However, since by hypothesis, $f(x^*) < f(x^0)$, if we substitute $f(x^0)$ for $f(x^*)$ on the right hand side of (4-22), we can strengthen the inequality and therefore

$$f[\lambda x^* + (1 - \lambda)x^0] < \lambda f(x^0) + (1 - \lambda)f(x^0) = f(x^0) \qquad \textbf{(4-23)}$$

Now consider any neighborhood ε of x^0, such that $|x^* - x^0| > \varepsilon$. If λ is chosen so that $0 < \lambda < \dfrac{\varepsilon}{|x^* - x^0|}$ then we see that $x = \lambda x^* + (1 - \lambda)x^0$ is clearly in this neighborhood, ε of x^0. Therefore, $f(x) = f[\lambda x^* + (1 - \lambda)x^0] < f(x^0)$. This, however, contradicts the fact that x^0 is a relative minimum for $f(x)$. Hence, we have shown that any local minimum is also the global minimum.

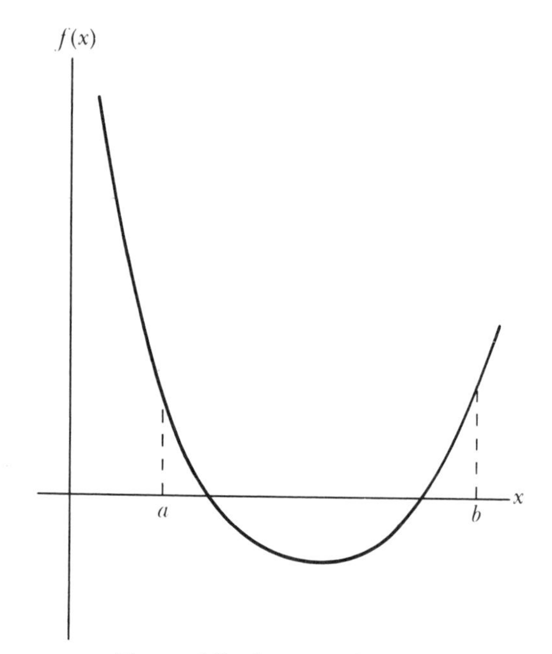

Figure 4-8 A convex function.

In a similar fashion, one can show that any local maximum of a concave function is also the global maximum.

Suppose now we wish to maximize a convex function. An examination of Figure 4-8 might lead one to speculate that if the function was not constrained to a finite interval, it might have no finite maximum. If a finite interval on x is considered, it appears as though the maximum would be at one or both of the end points of the interval. This is indeed the case as the following simple proof indicates.

Theorem 4.3. The global maximum of a convex function $f(x)$ over a closed interval $a \leq x \leq b$ will be taken on at either $x = a$ or $x = b$ or both.

Proof: First, we may note that by definition a convex function over a closed interval either has a minimum somewhere in the interval or it is a linear function over the interval. If it is a linear function, then clearly the theorem is true. Suppose it has a minimum in the interval. This minimum must be at one end point or in the interior of the interval. If the minimum is at one end point, for example $x = a$, as we increase x we cannot pass through a maximum at any point $x < b$, for then $f(x)$ would not be convex.

Hence, the maximum would have to be at $x = b$. The same argument holds for an interior minimum, except that we can either decrease or increase x to reach an end point. Therefore all maxima will be at end points. Hence, one or more of the end points will be the global maximum.

4.8 OPTIMIZATION OF FUNCTIONS OF SEVERAL VARIABLES: MATHEMATICAL BACKGROUND

We will now generalize the definitions and results given in Section 4.2 for functions of a single variable for functions of several variables. These basic results will be required in subsequent sections.

The notation $f(\bar{x})$ will be used to designate $f(x_1, x_2, \ldots, x_n)$. Hence $\bar{x} = (x_1, x_2, \ldots, x_n)$ will be a point or vector in an n-dimensional Euclidean space E^n. (See Chapter 3, Section 3.1.)

A function $f(\bar{x})$ takes on its *absolute maximum* (or global maximum) at a point $\bar{x}^*$ if $f(\bar{x}) \leq f(\bar{x}^*)$ for all values of $\bar{x}$ over which the function $f(\bar{x})$ is defined. We have implicitly assumed that the value of $\bar{x}$ at which $f(\bar{x})$ attains its maximum is actually in the set of values over which $\bar{x}$ is defined. This rules out certain anomalous cases that never occur in the solution of optimization problems which arise in practice. The definition of absolute (or global minimum) can be obtained from the preceding definition by reversing the sense of the inequality between $f(\bar{x})$ and $f(\bar{x}^*)$.

Let us now consider relative optima. Let us first recall that by a neighborhood δ of a point $\bar{y}$ in E^n, we mean all points $\bar{x}$ in E^n such that $|\bar{y} - \bar{x}| < \delta$, i.e.,

$$0 < (y_1 - x_1)^2 + (y_2 - x_2)^2 + \ldots + (y_n - x_n)^2 < \delta^2$$

With this in mind, let $f(\bar{x})$ be defined in some neighborhood δ about a point $\bar{x}^0$ in E^n. Then the function $f(\bar{x})$ is said to have a *strong relative maximum* at $\bar{x}^0$ if there exists an ε, $0 < \varepsilon < \delta$, such that for all $\bar{x}$ satisfying $0 < |\bar{x} - \bar{x}^0| < \varepsilon$, it is the case that $f(\bar{x}) < f(\bar{x}^0)$.

Intuitively, what the foregoing definition says is that if a function $f(\bar{x})$ has a strong relative maximum at some point $\bar{x}^0$ in an n-dimensional Euclidean space, E^n, then there is a hypersphere about $\bar{x}^0$ of radius ε, even though ε may be very small, such that for every point $\bar{x}$ in the interior of this hypersphere, $f(\bar{x})$ is *strictly less* than $f(\bar{x}^0)$.

A *weak relative maximum* may also be defined as follows: Let $f(\bar{x})$ be defined in some neighborhood δ about a point $\bar{x}^0$. Then the function $f(\bar{x})$ is said to have a weak relative maximum at $\bar{x}^0$ if there exists an ε, $0 < \varepsilon < \delta$ such that for all $\bar{x}$ satisfying

$0 < |\bar{x} - \bar{x}^0| < \varepsilon$, it is the case that $f(\bar{x}) \leq f(\bar{x}^0)$ and there is at least one point $\bar{x}$ in the interior of the hypersphere $|\bar{x} - \bar{x}^0| < \varepsilon$ such that $f(\bar{x}) = f(\bar{x}^0)$.

We will not usually distinguish between strong and weak relative maxima. They will both be called relative or local maxima.

The definitions of *strong and weak relative minima* are analogous to the definitions given for maxima except that the inequalities between $f(\bar{x})$ and $f(\bar{x}^0)$ are reversed. A related point should also be noted. If $f(\bar{x})$ has an absolute maximum at some point $\bar{x}^*$, then $-f(\bar{x})$ has an absolute minimum at $\bar{x}^*$. Similarly, if $f(\bar{x})$ has a relative maximum at some point $\bar{x}^0$, then $-f(\bar{x})$ has a relative minimum at $\bar{x}^0$.

In Figure 4-9 we have shown an example of a function $z = f(\bar{x}) = f(x_1, x_2)$. What is plotted are the contours of z in the two-dimensional plane. At the points designated A and B, two relative minima are shown. If we limit ourselves to the region of E^2 shown, then the point A is also the absolute or global minimum.

The definitions that we gave in Section 4.2 of convexity and concavity generalize in a straightforward fashion for functions of several variables. A function $f(\bar{x})$ is *convex* over some convex set X in E^n if for any two points $\bar{x}_1$ and $\bar{x}_2$ in X and for all λ, $0 \leq \lambda \leq 1$,

$$f[\lambda\bar{x}_1 + (1 - \lambda)\bar{x}_2] \leq \lambda f(\bar{x}_1) + (1 - \lambda)f(\bar{x}_2) \qquad \textbf{(4-24)}$$

In a similar fashion, a function $f(\bar{x})$ is *concave* over some convex set X in E^n if for any two points $\bar{x}_1$ and $\bar{x}_2$ in X and for all λ

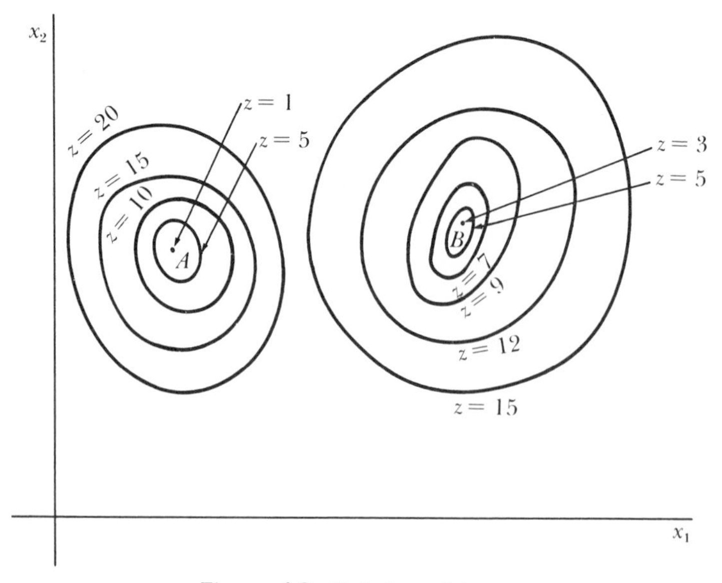

Figure 4-9 Relative minima.

$0 \leq \lambda \leq 1$

$$f[\lambda \bar{x}_1 + (1 - \lambda)\bar{x}_2] \geq \lambda f(\bar{x}_1) + (1 - \lambda)f(\bar{x}_2) \qquad \textbf{(4-25)}$$

In one dimension, a function was convex if the line segment drawn between any two points on its curve fell entirely on or above the curve. In n dimensions we have a similar situation. A function $z = f(\bar{x})$ is a hypersurface in n-dimensional space. It is convex if the line segment which connects any two points $[\bar{x}_1, z_1]$ and $[\bar{x}_2, z_2]$ on the surface of $f(\bar{x})$ lies entirely on or above the hypersurface. The reverse situation holds for concave functions.

Some results of interest concerning convex and concave functions are as follows.

Theorem 4.4. Let the functions $f_k(\bar{x})$, $k = 1, \ldots, p$ be convex functions over some convex set X in E^n. Then the function $f(\bar{x}) = \sum_{k=1}^{p} f_k(\bar{x})$ is also a convex function over X.

Proof: Choose any two points $\bar{x}_1$ and $\bar{x}_2$ in X and any λ; $0 \leq \lambda \leq 1$ and consider $f[\lambda \bar{x}_1 + (1 - \lambda)\bar{x}_2]$. Since

$$f(\bar{x}) = \sum_{k=1}^{p} f_k(\bar{x})$$

we can write

$$f[\lambda \bar{x}_1 + (1 - \lambda)\bar{x}_2] = \sum_{k=1}^{p} f_k[\lambda \bar{x}_1 + (1 - \lambda)\bar{x}_2] \qquad \textbf{(4-26)}$$

The $f_k(\bar{x})$ are assumed to be convex. Therefore,

$$\sum_{k=1}^{p} f_k[\lambda \bar{x}_1 + (1 - \lambda)\bar{x}_2] \leq \sum_{k=1}^{p} [\lambda f_k(\bar{x}_1) + (1 - \lambda)f_k(\bar{x}_2)]$$

$$= \lambda \sum_{k=1}^{p} f_k(\bar{x}_1) + (1 - \lambda) \sum_{k=1}^{p} f_k(\bar{x}_2) \qquad \textbf{(4-27)}$$

However, by definition $f(\bar{x}) = \sum_{k=1}^{p} f_k(\bar{x})$. Therefore we can write (4-27) as

$$\sum_{k=1}^{p} f_k[\lambda \bar{x}_1 + (1 - \lambda)\bar{x}_2] \leq \lambda f(\bar{x}_1) + (1 - \lambda)f(\bar{x}_2) \qquad \textbf{(4-28)}$$

Now, from (4-26) and (4-28) we have

$$f[\lambda \bar{x}_1 + (1 - \lambda)\bar{x}_2] \leq \lambda f(\bar{x}_1) + (1 - \lambda)f(\bar{x}_2)$$

which is what we wished to prove.

In short, the sum of convex functions is a convex function. Similarly, the sum of concave functions is also a concave function.

Theorem 4.5. If $f(\bar{x})$ is a convex function over the nonnegative orthant of E^n, then if $W = \{\bar{x} \mid f(\bar{x}) \leq b, \bar{x} \geq \bar{0}\}$ is not empty, W is a convex set.

Proof: In order to prove that W is a convex set, we need to show that if $\bar{x}_1$ and $\bar{x}_2$ are any points in W, their convex combination, $\hat{\bar{x}} = \lambda\bar{x}_1 + (1 - \lambda)\bar{x}_2$ for $0 \leq \lambda \leq 1$ is also in W. First, we note that $\hat{\bar{x}} = \lambda\bar{x}_1 + (1 - \lambda)\bar{x}_2 \geq 0$ if $\bar{x}_1$ and $\bar{x}_2$ are in W. Now we need to show that $\hat{\bar{x}}$ satisfies $f(\bar{x}) \leq b$, i.e., $f(\hat{\bar{x}}) \leq b$. However, we see that $f(\hat{\bar{x}}) = f[\lambda\bar{x}_1 + (1 - \lambda)\bar{x}_2]$. Since $f(\bar{x})$ is convex we have

$$f(\hat{\bar{x}}) = f[\lambda\bar{x}_1 + (1 - \lambda)\bar{x}_2] \leq \lambda f(\bar{x}_1) + (1 - \lambda)f(\bar{x}_2) \quad \textbf{(4-29)}$$

However, since $\bar{x}_1$ and $\bar{x}_2$ are in W, we know that $f(\bar{x}_1) \leq b$ and $f(\bar{x}_2) \leq b$ and that since $0 \leq \lambda \leq 1$, $\lambda f(x_1) \leq \lambda b$ and

$$(1 - \lambda)f(\bar{x}_2) \leq (1 - \lambda)b$$

Combining this with (4-29) we have

$$f(\hat{\bar{x}}) \leq \lambda f(\bar{x}_1) + (1 - \lambda)f(\bar{x}_2) \leq \lambda b + (1 - \lambda)b = b$$

Hence, $f(\hat{\bar{x}}) \leq b$ which is what we wished to prove.

Another characteristic of functions often referred to is known as *unimodality*. The definitions vary somewhat but the set given here are probably most generally used. A function is *unimodal* if there is *some* path from every point $\bar{x}$ to the optimal point along which the function continually increases or decreases. "Strong unimodality" requires that the straight line path from every point $\bar{x}$ to the optimal point be strictly increasing or decreasing. This is more restrictive than "unimodality." A still more restrictive form is "linear unimodality," in which *all* straight line paths between any two points in the space must yield objective function values which are unimodal. These differences are illustrated in Figure 4-10. The contours of $z = f(\bar{x})$ are shown for the various cases.

A strictly unimodal function will have just one local optimum which corresponds to the global optimum. However, it is also possible to consider a neighborhood of a local optimum in which a function will exhibit unimodal behavior locally. It is unfortunately the case that all present multivariable optimization methods find only local extrema or find global extrema for unimodal functions only.

We will now consider Taylor's theorem, previously discussed in Section 4.2, for functions of n variables. This theorem states that if $f(\bar{x})$ is continuous and has continuous first partial derivatives over an open convex set X in E^n, then for any two points $\bar{x}_1$ and

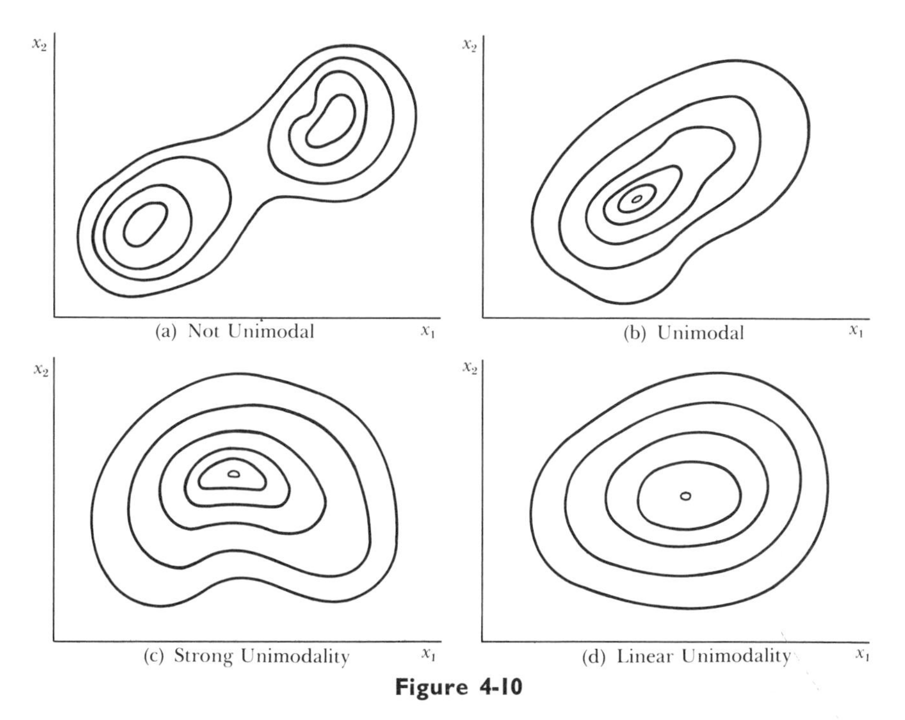

Figure 4-10

$\bar{x}_2 = \bar{x}_1 + \bar{h}$ in X, there exists a θ, $0 \leq \theta \leq 1$ such that

$$f(\bar{x}_2) = f(\bar{x}_1) + \bar{\nabla} f[\theta \bar{x}_1 + (1 - \theta)\bar{x}_2]\bar{h} \tag{4-30}$$

$\bar{\nabla} f$ is the *gradient vector* and is defined by

$$\bar{\nabla} f = \left(\frac{\partial f}{\partial x_1}, \frac{\partial f}{\partial x_2}, \ldots, \frac{\partial f}{\partial x_n}\right) \tag{4-31}$$

The notation $\bar{\nabla} f[\theta \bar{x}_1 + (1 - \theta)\bar{x}_2]$ indicates that the gradient is to be evaluated at the point $\bar{x} = \theta \bar{x}_1 + (1 - \theta)\bar{x}_2$ in E^n.

We can extend Taylor's theorem in the n-dimensional case just as we did in the one-dimensional case. If $f(\bar{x})$ is continuous and has continuous partial derivatives through second order, we can write

$$f(\bar{x}_2) = f(\bar{x}_1) + \bar{\nabla} f(\bar{x}_1)\bar{h} + \tfrac{1}{2}\bar{h}' H[\theta \bar{x} + (1 - \theta)\bar{x}_2]\bar{h} \tag{4-32}$$

H in (4-32) is the *Hessian matrix of* $f(\bar{x})$ and is defined as a matrix of the n^2 second partial derivatives of $f(\bar{x})$. In other words, it is the $n \times n$ matrix $H = \left\| \dfrac{\partial^2 f}{\partial x_i \, \partial x_j} \right\|$. The notation $H[\theta \bar{x}_1 + (1 - \theta)\bar{x}_2]$ indicates that the Hessian matrix is evaluated at the point $\bar{x} = \theta \bar{x}_1 + (1 - \theta)\bar{x}_2$ in E^n.

Taylor's theorem can be extended to any order if $f(\bar{x})$ has continuous partial derivatives of the requisite order. However, we shall not require this result.

One final result from analysis should be recalled. It is known as the *implicit function theorem.* Let us consider a set of m equations in n variables where $m < n$:

$$g_i(\bar{x}) = 0 \qquad i = 1, 2, \ldots, m \tag{4-33}$$

and

$$\bar{x} = (x_1, x_2, \ldots, x_n)$$

Under certain circumstances we may wish to use these m equations to eliminate some subset m of the n variables. It will facilitate the discussion if we consider performing this operation at some particular point, $\bar{x}^0$. It is not at all obvious that this elimination can always be carried out. For the sake of simplicity and without any loss in generality, suppose we wish to eliminate the first m variables. Therefore what we wish to know is whether or not there exist a set of m functions, ϕ_i, such that

$$x_i = \phi_i(x_{m+1}, x_{m+2}, \ldots, x_n) \qquad i = 1, 2, \ldots, m \tag{4-34}$$

In particular, we wish to know if the set of functions given by (4-34) exists at or in the neighborhood of $\bar{x}^0$. The theorem from analysis that supplies this result is known as the implicit function theorem. It may be stated as follows.

Theorem 4.6: *Implicit Function Theorem.* If the rank of the Jacobian matrix, J_m, evaluated at the point $\bar{x}^0$ is equal to m, then this is a necessary and sufficient condition for the existence of a set of m functions, ϕ_i, $i = 1, 2, \ldots, m$, which are unique, continuous, and differentiable in some neighborhood of $\bar{x}^0$.

The *Jacobian matrix,* J_m is defined as

$$J_m = \begin{bmatrix} \dfrac{\partial g_1}{\partial x_1} & \dfrac{\partial g_1}{\partial x_2} & \cdots & \dfrac{\partial g_1}{\partial x_m} \\ \dfrac{\partial g_2}{\partial x_1} & \dfrac{\partial g_2}{\partial x_2} & \cdots & \dfrac{\partial g_2}{\partial x_m} \\ \cdot & \cdot & & \cdot \\ \cdot & \cdot & & \cdot \\ \cdot & \cdot & & \cdot \\ \dfrac{\partial g_m}{\partial x_1} & \dfrac{\partial g_m}{\partial x_2} & & \dfrac{\partial g_m}{\partial x_m} \end{bmatrix} \tag{4-35}$$

The statement that the Jacobian matrix is evaluated at the point $\bar{x}^0$ signifies that each of the partial derivatives in (4-35) is evaluated at $\bar{x}^0$.

For simplicity, we assumed that the first m variables were to be eliminated, i.e., expressed in terms of the remaining $n - m$ variables. This, of course, is not necessarily the case. A more exact statement of the theorem would be that if one selected any m column vectors of the form

$$\begin{bmatrix} \dfrac{\partial g_1}{\partial x_j} \\ \dfrac{\partial g_2}{\partial x_j} \\ \cdot \\ \cdot \\ \cdot \\ \dfrac{\partial g_m}{\partial x_j} \end{bmatrix}$$

then, if the resulting Jacobian matrix has rank m, these particular m variables may be eliminated, i.e., the set of m functions ϕ_i exist. It is clear that it is not obvious which particular set of m vectors will give a matrix of rank m and hence, which variables may be eliminated. There are $\binom{m}{n} = \dfrac{n!}{m!\,(n-m)!}$ possible combinations of m columns out of a total set of n and hence that number of Jacobians. Every Jacobian that has rank m represents a set of variables that can be eliminated. An equivalent condition for the Jacobian matrix to have rank m is that the Jacobian be nonsingular, i.e., its determinant is not zero. (See Chapter 2, Section 2.8.)

4.9 OPTIMIZATION OF FUNCTIONS OF MORE THAN ONE VARIABLE

Given the background supplied by Section 4.3 we can proceed directly to considering the classical theory of relative maxima and minima.

If $f(\bar{x})$ has a relative minimum at a point $\bar{x}^0$, then we know from the definition of relative minimum, that there exists an $\varepsilon > 0$ such that for all points $\bar{x}$ in an ε-neighborhood of $\bar{x}^0$, $f(\bar{x}) \geq f(\bar{x}^0)$. In particular, let us consider points in this ε-neighborhood of $\bar{x}^0$ of the form $\bar{x} = \bar{x}^0 + h\bar{e}_j$, $0 < |h| < \varepsilon$. Then we may write

$$f(\bar{x}^0 + h\bar{e}_j) - f(\bar{x}^0) \geq 0 \qquad \textbf{(4-36)}$$

for all h, $0 < |h| < \varepsilon$. This may be divided by h to obtain

$$\frac{f(\bar{x}^0 + h\bar{e}_j) - f(\bar{x}^0)}{h} \geq 0 \qquad h > 0 \qquad \textbf{(4-37)}$$

$$\frac{f(\bar{x}^0 + h\bar{e}_j) - f(\bar{x}^0)}{h} \leq 0 \qquad h < 0 \qquad \textbf{(4-38)}$$

If we take the limit in (4-37) as $h \rightarrow 0$, we obtain the result that $\frac{\partial f(\bar{x}^0)}{\partial x_j} \geq 0$. Similarly, upon taking the limit as $h \rightarrow 0$ in (4-38) we obtain that $\frac{\partial f(\bar{x}^0)}{\partial x_j} \leq 0$. These two results taken together lead us to conclude that

$$\frac{\partial f(\bar{x}^0)}{\partial x_j} = 0 \qquad j = 1, 2, \ldots, n \tag{4-39}$$

An analogous argument, with inequalities reversed, can be made for maximization and it too yields equations (4-39). Hence, equations (4-39) tell us that if $\bar{x}^0$ is a point where $f(\bar{x})$ takes on a relative maximum or minimum, the first partial derivatives of $f(\bar{x})$ with respect to each of the n variables must vanish.

Just as in the case of a function of a single variable, what we have shown is that a *necessary* condition for $\bar{x}^0$ to be a relative maximum or minimum is that $\bar{x}^0$ be a stationary point. As in the case of the optimization of functions of a single variable, each stationary point must be evaluated to determine whether or not it is an optimum. To that end let us derive a sufficient condition for $f(\bar{x})$ to have a minimum at a point $\bar{x}^0$ where equation (4-39) is satisfied. We give this in the form of the following theorem.

Theorem 4.7. A sufficient condition for $f(\bar{x})$ to have a relative minimum at points $\bar{x}^0$ where $\frac{\partial f(\bar{x}^0)}{\partial x_j} = 0$, $i = 1,2 \ldots, n$ is that the Hessian matrix be positive definite.

Proof: From Taylor's theorem (Equation 3-30) we know that

$$f(\bar{x}^0 + \bar{h}) = f(\bar{x}^0) + \overline{\nabla} f(\bar{x}^0)\bar{h} + \tfrac{1}{2}\bar{h}'H[\theta\bar{x}^0 + (1 - \theta)(\bar{x}^0 + \bar{h})]\bar{h} \tag{4-40}$$

By hypothesis, the necessary conditions for the existence of a minimum are satisfied, i.e., $\frac{\partial f(\bar{x}^0)}{\partial x_j} = 0$, $i = 1,2 \ldots, n$ and therefore, $\overline{\nabla} f(\bar{x}^0) = 0$. Hence, we can rearrange (4-40) as

$$f(\bar{x}^0 + \bar{h}) - f(\bar{x}^0) = \tfrac{1}{2}\bar{h}'H[\theta\bar{x}^0 + (1 - \theta)(\bar{x}^0 + \bar{h})]\bar{h} \tag{4-41}$$

From (4-41) it is clear that $f(\bar{x}^0 + h) - f(\bar{x}^0)$ will have the same sign as $\bar{h}'H[\theta\bar{x}^0 + (1 - \theta)(\bar{x}^0 + \bar{h})]\bar{h}$. Since we have assumed the existence and continuity of the second partial derivatives of $f(\bar{x})$, it is clear that the second partial derivatives, $\frac{\partial^2 f(\bar{x}^0)}{\partial x_i \, \partial x_j}$ will have the same sign as the second partial derivatives,

$$\frac{\partial^2 f \, \partial x_j}{\partial x_i} [\theta\bar{x}^0 + (1 - \theta)(\bar{x}_0 + \bar{h})]$$

(which are the components of the Hessian matrix), providing we are in some suitable ε-neighborhood of $\bar{x}^0$, $0 < |\bar{h}| < \varepsilon$. This being the case, if $\bar{h}'H[\bar{x}^0]\bar{h}$ is positive, $f(\bar{x}^0 + \bar{h}) - f(\bar{x}^0)$ will also be positive for $0 < |\bar{h}| < \varepsilon$, and $f(\bar{x})$ will be a minimum at $\bar{x}^0$. We know from Chapter 2, Section 2.12 that a quadratic form $\bar{h}'H\bar{h}$ will be positive if and only if the Hessian is a positive definite matrix. Hence, we have proved the theorem.

As an illustration of the preceding theorem consider the following example.

EXAMPLE

Find the minimum of

$$f(\bar{x}) = 2x_1^2 + 3x_2^2 + 4x_3^2 - 8x_1 - 12x_2 - 24x_3 + 110$$

$$\frac{\partial f}{\partial x_1} = 4x_1 - 8 = 0$$

$$\frac{\partial f}{\partial x_2} = 6x_2 - 12 = 0$$

$$\frac{\partial f}{\partial x_3} = 8x_3 - 24 = 0$$

There is only one solution to the necessary conditions and that is obviously $\bar{x} = (x_1, x_2, x_3) = (2, 2, 3)$. Is this point a minimum? If we evaluate the second derivatives we have

$$\frac{\partial^2 f}{\partial x_1^2} = 4 \qquad \frac{\partial^2 f}{\partial x_1\, \partial x_2} = 0 \qquad \frac{\partial^2 f}{\partial x_1\, \partial x_3} = 0, \qquad \frac{\partial^2 f}{\partial x_2^2} = 6$$

$$\frac{\partial^2 f}{\partial x_2\, \partial x_3} = 0 \qquad \frac{\partial^2 f}{\partial x_3^2} = 8$$

The Hessian matrix at $\bar{x}^0 = (2, 2, 3)$ is therefore

$$\mathrm{H}[\bar{x}^0] = \begin{bmatrix} 4 & 0 & 0 \\ 0 & 6 & 0 \\ 0 & 0 & 8 \end{bmatrix}$$

which is clearly a positive definite matrix. Therefore, (2, 2, 3) is a minimum point.

Unfortunately, matters are considerably more complicated than the previous example might imply. First of all, the foregoing sufficient condition is of little practical importance because the determination of whether or not the Hessian matrix is positive definite requires considerable numerical computation in any problem arising from a question in science, engineering, or economics. This computation is generally not worth performing. More information about the character of a stationary point can usually be obtained from physical considerations. Secondly, the case of when the Hessian is not positive definite but rather semidefinite has not been considered. This turns out to be exceedingly complicated. For this reason and because of the computational complexity, in practice one rarely resorts to investigation of stationary points for sufficiency. For a very detailed theoretical exposition of the problems involved and a statement of the sufficiency conditions refer to Hancock.[2]

4.10 GLOBAL EXTREMA: MANY VARIABLES

Now that we have considered how to find stationary points, some of which may be local optima of $z = f(\bar{x})$, let us see how this information may be incorporated into a general computational procedure for finding global optima. This is similar to what was done in Section 4.5. As may be anticipated, the procedure for a function of several variables involves much more computation. The principles involved, however, remain the same. Let us first develop some useful notation.

Suppose we wish to solve

$$\text{Max } z = f(\bar{x}) \qquad \bar{a} \le \bar{x} \le \bar{b} \tag{4-42}$$

$\bar{x}$, $\bar{a}$, $\bar{b}$ are all n-component vectors, i.e., we write them in component form as: $a_j \le x_j \le b_j, j = 1,2 \ldots, n$. As we have already noted, the procedures of calculus will only find relative maxima and minima. However, we may well be interested in maxima or minima that occur when one or more of the variables are at their upper or lower bounds. It is readily seen that there are 2^n possible combinations of the a_j and b_j which the x_j may have. Let us designate these $p = 2^n$ combinations as the vectors $\bar{y}_1, \bar{y}_2, \ldots, \bar{y}_p$. For example, if $n = 3$,

$$\bar{y}_1 = [a_1, a_2, a_3], \bar{y}_2 = [b_1, a_2, a_3],$$

$$\bar{y}_3 = [a_1, b_2, a_3], \ldots, \bar{y}_p = [b_1, b_2, b_3].$$

With this notation, Figure 4-11 indicates how to find the global maximum of a continuous function $z = f(\bar{x})$ where the variables are constrained to lie in the intervals $\bar{a} \le \bar{x} \le \bar{b}$.

An example of the procedure shown in Figure 4-11 is as follows:

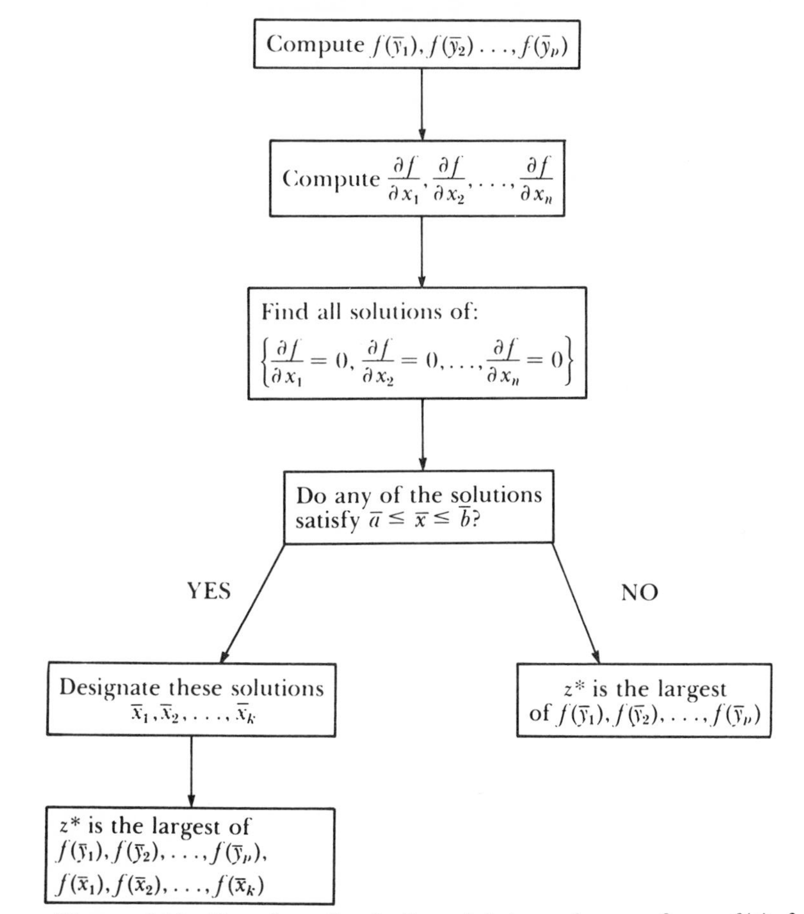

Figure 4-11 Procedure for finding global maximum of $z = f(\bar{x})$ for $\bar{a} \leq \bar{x} \leq \bar{b}$.

EXAMPLE

Find the global maximum of

$$f(\bar{x}) = 5(x_1 - 3)^2 - 12(x_2 + 5)^2 + 6x_1x_2. \quad \text{for} \quad \bar{a} \leq \bar{x} \leq \bar{b}$$

where $\bar{a} = [0, 0]$, $\bar{b} = [10, 5]$
First we calculate the combinations of boundary values of $\bar{x}$.

$$f(0, 0) = -225 \qquad f(0, 5) = -1155$$

$$f(10, 0) = -55 \qquad f(10, 5) = -655$$

Next we investigate possible interior stationary points.

$$\frac{\partial f}{\partial x_1} = 10(x_1 - 3) + 6x_2 = 0$$

$$\frac{\partial f}{\partial x_2} = -24(x_2 + 5) + 6x_1 = 0$$

Solving these equations yields only one solution

$$\bar{x}_1 = [x_1, x_2] = \left[\frac{120}{23}, \frac{-85}{23}\right]$$

It can be seen that $\bar{x}_1$ violates the restriction that $x_2 \geq 0$. Hence we ignore this solution. We see therefore that

$$z^* = \text{Max}\,[f(0, 0) = -255, f(0, 5) = -1155,$$
$$f(10, 0) = -55, f(10, 5) = -655] = -55$$

Therefore $z^* = -55$ and is taken on at $\bar{x} = [10, 0]$.

In problems of practical significance, a very common kind of simple constraint on the variables is that they be nonnegative, i.e., $\bar{x} \geq \bar{0}$. It can be seen that this is merely a special case of the method outlined. However, it is obvious that there are grave computational problems involved with the use of the method which is given in the flow chart in Figure 4-11. For one thing, the necessity for testing 2^n combinations of end conditions can be prohibitive. For a problem with 10 variables, the calculation of $2^{10} = 1024$ functional evaluations (on a computer, of course) is not unusually burdensome. However if there were 100 or more variables (not uncommonly the case in problems we shall consider later), $2^{100} \approx 10^{30}$ calculations is certainly prohibitive. Therefore, on this score alone, the procedure of Figure 4-11 is limited to problems with probably no more than 20 to 30 variables. The other major problem with the procedure is that in order to carry out what is described, one must find *all* stationary points of $f(\bar{x})$, i.e., solve the set of nonlinear simultaneous equations, $\frac{\partial f}{\partial x_j} = 0$, $j = 1, 2 \ldots, n$. This, in general, and for large numbers of variables, is not easy to do. For procedures used for doing this the reader may consult any standard text on numerical analysis, such as Hildebrand,[1] Ralston,[3] and Isaacson and Keller.[4] None of the methods are completely satisfactory.

In Chapter 5 we shall describe direct methods for maximizing or minimizing $z = f(\bar{x})$. While these involve less computational effort than the scheme we have just described, it should be borne in mind that with these methods one cannot be certain that the optimum found is a global optimum. In general, only a local optimum can be found. Hence, this criticism of classical methods is not completely fair. However, it is true that direct maximization or minimization, even if it leads to a local optimum, is easier to do than using the necessary conditions of classical optimization to derive a set of simultaneous nonlinear equations which are difficult to solve.

4.11 OPTIMA OF CONVEX AND CONCAVE FUNCTIONS

If we are maximizing $z = f(\bar{x})$ and we know no other characteristics of $f(\bar{x})$ (even if we know that it is continuous and differentiable), we can only hope, by methods to be described in future chapters, to find a local maximum or minimum. However, as in the case of functions of a single variable, considerably more can be said if the functions being maximized or minimized are convex or concave. To this end we shall prove the following theorems.

Theorem 4.8. Let $f(\bar{x})$ be a convex function over a closed convex set X in E^n. Then any local minimum of $f(\bar{x})$ in X is also the global minimum of $f(\bar{x})$ over X.

Proof: This proof is by contradiction. Let us assume that f has a local minimum at $\bar{x}^0$, where $\bar{x}^0 \in X$. Let us assume further that the global minimum over X occurs at a different point, $\bar{x}^* \neq \bar{x}^0$, and that $f(\bar{x}^0) > f(\bar{x}^*)$. Previously, we defined a convex function as one for which

$$f[\lambda\bar{x}_1 + (1 - \lambda)\bar{x}_2] \leq \lambda f(\bar{x}_1) + (1 - \lambda)f(\bar{x}_2) \qquad \textbf{(4-43)}$$

for any $\bar{x}_1$, $\bar{x}_2$ in X and all λ, $0 \leq \lambda \leq 1$. Therefore (4-43) must also be satisfied by $\bar{x}^0$ and $\bar{x}^*$ since they are in X. Hence,

$$f[\lambda\bar{x}^* + (1 - \lambda)\bar{x}^0] \leq \lambda f(\bar{x}^*) + (1 - \lambda)f(\bar{x}^0) \qquad \textbf{(4-44)}$$

However, by hypothesis, $f(\bar{x}^0) > f(\bar{x}^*)$. Therefore, if we substitute $f(\bar{x}^0)$ for $f(\bar{x}^*)$ on the right-hand side of equation (4-44), we strengthen the inequality and therefore

$$f[\lambda\bar{x}^* + (1 - \lambda)\bar{x}^0] < \lambda f(\bar{x}^0) + (1 - \lambda)f(\bar{x}^0) = f(\bar{x}^0) \qquad \textbf{(4-45)}$$

Now let us consider any ε-neighborhood, of $\bar{x}^0$ such that $|\bar{x}^* - \bar{x}^0| > \varepsilon$. If λ is chosen so that $0 < \lambda < \dfrac{\varepsilon}{|\bar{x}^* - \bar{x}^0|}$ then we see that $\bar{x} = \lambda\bar{x}^* + (1 - \lambda)\bar{x}^0$ is in this ε-neighborhood, of $\bar{x}^0$ and that from (4-45)

$$f(\bar{x}) = f[\lambda\bar{x}^* + (1 - \lambda)\bar{x}^0] < f(\bar{x}^0)$$

This, however, contradicts the hypothesis that $\bar{x}^0$ is a local minimum for $f(\bar{x})$. Hence, we have shown that any local minimum is also the global minimum.

In a similar fashion to the preceding theorem we can show that any local maximum of a concave function $f(\bar{x})$ over a closed convex set X in E^n is also the global maximum over X.

In the previous theorem we considered minimizing a convex function over a convex set. Let us now consider the maximization problem, i.e., finding the global maximum of a convex function over a convex set. We present the result in the following theorem.

Theorem 4.9. Let X be a closed convex set bounded from below and let $f(\bar{x})$ be a convex function over X in E^n. If $f(\bar{x})$ has global maxima at points $\bar{x}^0$, with $|\bar{x}^0|$ finite, then one or more $\bar{x}^0$ are extreme points of X.

Proof: Let $\bar{x}^0$ be the point in X at which $f(\bar{x})$ has its global maximum.

Case 1: Suppose $\bar{x}^0$ is in the interior of X. For this case we claim that $f(\bar{x})$ is constant over all of X. We prove this by contradiction. Suppose it is not true. Then let $\hat{\bar{x}}$ in X be any point such that $f(\hat{\bar{x}}) \neq f(\bar{x}^0)$. Since $\bar{x}^0$ is the global maximum, then $f(\hat{\bar{x}}) < f(\bar{x}^0)$. Now since X is a convex set and $\bar{x}^0$ is in the interior of X, there must obviously exist a point $\bar{x}_1$ in X and $0 < \lambda < 1$ such that

$$\bar{x}^0 = \lambda\bar{x}_1 + (1 - \lambda)\hat{\bar{x}} \tag{4-46}$$

However, $f(\bar{x})$ is a convex function. Therefore

$$f(\bar{x}^0) = f[\lambda\bar{x}_1 + (1 - \lambda)\hat{\bar{x}}] \leq \lambda f(\bar{x}_1) + (1 - \lambda)f(\hat{\bar{x}}) \tag{4-47}$$

Since $f(\bar{x}^0) > f(\hat{\bar{x}})$, if we substitute $f(\bar{x}^0)$ for $f(\hat{\bar{x}})$ in (4-47) we may write

$$\lambda f(\bar{x}_1) + (1 - \lambda)f(\bar{x}^0) > f(\bar{x}^0) \tag{4-48}$$

From (4-48) we have

$$\lambda f(\bar{x}_1) > \lambda f(\bar{x}^0) \quad \text{and since } \lambda > 0$$

we have $f(\bar{x}_1) > f(\bar{x}^0)$ which is a contradiction. Therefore, we have shown that if $\bar{x}^0$ is in the interior of X, $f(\bar{x})$ is constant, and hence the theorem is trivially true for this case.

Case 2: Suppose $\bar{x}^0$ is on the boundary of X. We must show that $\bar{x}^0$ is an extreme point. If $\bar{x}^0$ is on the boundary of X, a convex set, and is not an extreme point then there exists a supporting hyperplane to X at $\bar{x}^0$. (See Chapter 3, Section 3.5.) Let S_1 be that supporting hyperplane and consider the intersection of X and S_1, i.e., $T_1 = S_1 \cap X$. Since the intersection of closed convex sets is a closed convex set, T_1 is a closed convex set. Furthermore, T_1 is not empty and contains at least one extreme point of X since X is bounded from below. Now let us examine T_1. T_1 lies in a space of one lower dimension than that of the space in which S_1 lies. Let us then consider T_1 as a set in a space of dimension $n - 1$. If $\bar{x}^0$ is in the interior of T_1, then we apply Case 1 to prove that $f(\bar{x})$ is constant over T_1 and hence it assumes a maximum at an extreme point of X. If $\bar{x}^0$ is on the boundary of T_1, we reapply the preceding argument in this case and consider the intersection of the supporting hyperplane S_2 and T_1, i.e., $T_2 = S_2 \cap T_1$ at the point $\bar{x}^0$. This is a closed convex set and is not empty and the foregoing argument can be repeated. At most we can apply the argument $n-1$ times and if so, we obtain $T_n = \{\bar{x}^0\}$ and so obviously, $\bar{x}^0$ is an extreme point of X. This concludes the proof.

4.12 CONSTRAINED OPTIMIZATION

In the previous sections we considered the problem of maximizing or minimizing a function of several variables where the variables were unconstrained. (We did however, consider "mild" constraints of the upper and lower bound variety, $\bar{a} \leq \bar{x} \leq \bar{b}$ in Section 4.10.) What we would like to consider now is the classic method of dealing with the optimization of functions when certain additional constraints and restrictions, usually in the form of equations, are placed on the variables.

It is obvious that this is an important class of problems. A little observation and reflection will reveal that all optimization problems of the real world are, in fact, constrained problems. Suppose one has an expression for the output of a chemical reaction vessel in which some set of chemical reactions are taking place and one wishes to maximize this output. It is also necessary to take into account material balance restrictions on the reactants and products, the laws governing flow of materials into and out of the reaction vessel, and other conditions. All of these are additional constraints on the variables of the function to be optimized.

Consider another example. Suppose we wish to maximize the profits from the operation of some equipment that can manufacture several different products. Suppose we know the demand for each of these products over the next month. The problem is to determine which products and how much of each to manufacture. There are *restrictions*, however, that must be taken into account relating to the total amount of time available on each unit of equipment, how long it takes to change over from one product to the next, and so forth. Properly phrased in mathematical language, this becomes a constrained optimization problem.

In one sense, constrained problems should be no more difficult than unconstrained problems. One could argue, for example, that if we wish to maximize $f(\bar{x})$ where $\bar{x} = [x_1, x_2, \ldots, x_n]$ and we have constraints $g_i(\bar{x}) = b_i$, $i = 1, 2, \ldots, m$ and $m < n$, we could use the equations $g_i(\bar{x}) = b_i$, $i = 1, 2, \ldots, m$ to eliminate the first m variables, $x_1, x_2, \ldots, x_m$. We could then proceed with solving the resulting unconstrained optimization problem,

$$\operatorname{Max} f(x_1, x_2, \ldots, x_n) = \operatorname{Max} h(x_{m+1}, x_{m+2}, \ldots, x_n) \quad \textbf{(4-49)}$$

where $h(x_{m+1}, \ldots, x_n)$ was obtained from $f(\bar{x})$ by using the $g_i(\bar{x}) = b_i$, $i = 1, 2, \ldots, m$ to eliminate $x_1, x_2, \ldots, x_m$.

As long as the conditions of the implicit function theorem are satisfied, we know this can be done, in principle. Unfortunately, as is the case for most existence theorems, we are given no guidance as to how to carry out this calculation. This may be impossible to do by any known means. Let us consider some examples.

EXAMPLE 1

$$\operatorname{Min} f(\bar{x}) = 2x_1^2 + 3x_2^2$$

Subject to

$$x_1 + 3x_2 = 10$$

We could easily solve $x_1 + 3x_2 = 10$ for $x_1 = 10 - 3x_2$ and then obtain

$$\operatorname{Min} f(\bar{x}) = \operatorname{Min} [2(10 - 3x_2)^2 + 3x_2^2] = \operatorname{Min} h(x_2)$$

This has a minimum and it could be readily obtained.

EXAMPLE 2

$$\operatorname{Min} f(\bar{x}) = 2x_1^2 + 3x_2^2$$

Subject to

$$x_1 \sin x_1 + 5x_2 \sin x_2 = 10$$

We now see that it is not possible to use the constraint equation to eliminate either x_1 or x_2 and if we are to solve the problem, the constraint must be considered in some sort of explicit form.

While the technique we are about to describe can deal with such situations, it has very considerable computational problems associated with it. The theory we shall develop is important because of its underlying connection with other computational methods for nonlinear optimization problems, although the theory itself does not lead to an efficient computational method.

The technique we shall describe relates to what is known as the Lagrange multiplier method. It is most easily illustrated first for the case of optimization of a function of two variables subject to a single equality constraint. This will then be generalized to any number of variables.

Suppose we wish to maximize $f(x_1, x_2)$ subject to the constraint $g(x_1, x_2) = b$. Suppose we can obtain an explicit representation of x_2 from $g(x_1, x_2) = b$, i.e., $x_2 = h(x_1)$. If a function such as $h(x_1)$ exists, we know from the implicit function theorem that it is differentiable and has a continuous first derivative. Then, as was indicated in the first example we can obtain an unconstrained single variable problem which is as follows:

$$\operatorname{Max} f[x_1, h(x_1)] \tag{4-50}$$

We recall that a necessary condition for f to have a maximum at some point $\bar{x}^0 = (x_1^0, x_2^0)$ is that the first derivative of f with respect to x_1 be zero at $\bar{x}^0$. First we note that

$$\frac{d}{dx_1} f(x_1, x_2) = \frac{\partial f(x_1, x_2)}{\partial x_1} + \frac{\partial f(x_1, x_2)}{\partial x_2}\frac{dx_2}{dx_1} \tag{4-51}$$

Let us recall, however, that $x_2 = h(x_1)$. If we substitute this into (4-51) and evaluate the derivatives at $\bar{x}^0$, we obtain

$$\frac{d}{dx_1} f(x_1^0, x_2^0) = \frac{\partial f(x_1^0, x_2^0)}{\partial x_1} + \frac{\partial f(x_1^0, x_2^0)}{\partial x_2}\frac{dh(x_1)}{dx_1} \tag{4-52}$$

and we know that the expression in (4-52) must vanish if $\bar{x}^0$ is to be a stationary point. Since the entire purpose of the Lagrange multiplier method is to avoid finding $h(x_1)$, let us consider how to eliminate $\frac{dh}{dx_1}$ from (4-52). We can recall that $g(x_1, x_2) = b$. Therefore we can write:

$$\frac{dg(x_1, x_2)}{dx_1} = \frac{\partial g(x_1, x_2)}{\partial x_1} + \frac{\partial g(x_1, x_2)}{\partial x_2}\frac{dh(x_1)}{dx_1} = 0 \tag{4-53}$$

since $x_2 = h(x_1)$. We can rearrange (4-53) to obtain

$$\frac{dh(x_1)}{dx_1} = -\frac{\partial g(x_1, x_2)}{\partial x_1} \Big/ \frac{\partial g(x_1, x_2)}{\partial x_2} \tag{4-54}$$

If we now substitute from (4-54) evaluated at $\bar{x}^0$ into (4-52) and set (4-52) equal to zero, we have

$$\frac{\partial f(x_1^0, x_2^0)}{\partial x_1} - \left(\frac{\partial f(x_1^0, x_2^0)}{\partial x_2}\right)\left(\frac{\partial g(x_1^0, x_2^0)}{\partial x_1} \Big/ \frac{\partial g(x_1^0, x_2^0)}{\partial x_2}\right) = 0 \tag{4-55}$$

If we now define a quantity λ as

$$\lambda = \frac{\partial f(x_1^0, x_2^0)}{\partial x_2} \Big/ \frac{\partial g(x_1^0, x_2^0)}{\partial x_2} \tag{4-56}$$

and substitute from (4-56) into (4-55) we have

$$\frac{\partial f(x_1^0, x_2^0)}{\partial x_1} - \lambda \frac{\partial g(x_1^0, x_2^0)}{\partial x_1} = 0 \qquad \textbf{(4-57)}$$

From the definition of λ we have

$$\frac{\partial f(x_1^0, x_2^0)}{\partial x_2} - \lambda \frac{\partial g(x_1^0, x_2^0)}{\partial x_2} = 0 \qquad \textbf{(4-58)}$$

and the original constraint must also be satisfied, i.e.,

$$g(x_1^0, x_2^0) = b \qquad \textbf{(4-59)}$$

The foregoing equations (4-57) to (4-59) represent a set of necessary conditions for the existence of a solution to our original problem. We have three equations in three variables x_1, x_2, λ which are to be solved for a stationary point $\bar{x}^0$ and λ. It will be noted that the value of this approach is that we do not have to determine the function $h(x_1)$, even though we assumed that it existed when we made use of the implicit function theorem in the derivation.

The simplest way of arriving at the necessary preceding conditions for any specific problem is to construct what is known as the *Lagrangian function.* The Lagrangian function is defined as

$$F(x_1, x_2, \lambda) = f(x_1, x_2) + \lambda[b - g(x_1, x_2)] \qquad \textbf{(4-60)}$$

If we differentiate F with respect to x_1, x_2, λ and equate the resulting expressions to zero, we obtain (4-57) to (4-59). We see this as follows:

$$\frac{\partial F(x_1, x_2, \lambda)}{\partial x_1} = \frac{\partial f(x_1, x_2)}{\partial x_1} - \lambda \frac{\partial g(x_1, x_2)}{\partial x_1} = 0$$

$$\frac{\partial F(x_1, x_2, \lambda)}{\partial x_2} = \frac{\partial f(x_1, x_2)}{\partial x_2} - \lambda \frac{\partial g(x_1, x_2)}{\partial x_2} = 0 \qquad \textbf{(4-61)}$$

$$\frac{\partial F(x_1, x_2, \lambda)}{\partial \lambda} = b - g(x_1, x_2) = 0$$

Let us consider some examples to illustrate some of the points in our discussion of constrained optimization problems.

EXAMPLE 3

$$\text{Min } z = 3x_1^2 + 4x_2^2$$

Subject to $2x_1 - 3x_2 = 10$. If we form the Lagrangian function we obtain

$$F(x_1, x_2, \lambda) = 3x_1^2 + 4x_2^2 + \lambda[10 - 2x_1 + 3x_2]$$

$$\frac{\partial F(\bar{x}, \lambda)}{\partial x_1} = 6x_1 - 2\lambda = 0$$

$$\frac{\partial F(\bar{x}, \lambda)}{\partial x_2} = 8x_2 + 3\lambda = 0 \tag{4-62}$$

$$\frac{\partial F(\bar{x}, \lambda)}{\partial \lambda} = 10 - 2x_1 + 3x_2 = 0$$

Solving equations (4-62) we have

$$x_1 = \frac{80}{43} \qquad x_2 = \frac{-90}{43} \qquad \lambda = \frac{240}{43}$$

It will be noted that one could not have applied this method directly if the restriction that $\bar{x} \geq \bar{0}$ had also been imposed, as $x_2 < 0$ at the minimum point.

EXAMPLE 4

This is a problem in chemical equilibrium. a and b are known constants.

$$\text{Min} f(\bar{x}) = a - x_1 - x_2$$

Subject to

$$(a - x_1 - x_2)^2 - bx_1x_2^3 = 0$$

$$F(x_1, x_2, \lambda) = a - x_1 - x_2 + \lambda[-(a - x_1 - x_2)^2 + bx_1x_2^3]$$

$$\frac{\partial F}{\partial x_1} = -1 + 2\lambda(a - x_1 - x_2) + b\lambda x_2^3 = 0$$

$$\frac{\partial F}{\partial x_2} = -1 + 2\lambda(a - x_1 - x_2) + 3b\lambda x_1 x_2^2 = 0 \tag{4-63}$$

$$\frac{\partial F}{\partial \lambda} = -(a - x_1 - x_2)^2 + bx_1x_2^3 = 0$$

Solving the first two equations (by subtraction) yields $x_2 = 3x_1$. Substituting this value into the third equation yields

$$27bx_1^4 - 16x_1^2 + (6a + 2)x_1 - a^2 = 0$$

If one solves this for x_1 and then computes $x_2 = 3x_1$ a solution will be obtained. Roots of the preceding equation can be obtained by a number of methods.

It is a relatively straightforward matter to generalize the derivation given here for two variables and one constraint to the case of many variables and constraints. Suppose we wish to maximize a function $f(\bar{x})$ where $\bar{x} = (x_1, x_2, \ldots, x_n)$ which is subject to m equality constraints of the form

$$g_i(\bar{x}) = b_i \qquad i = 1, 2, \ldots, m$$

We can start out in a similar fashion to the way we did previously in Equation (4-51). We know that at a relative maximum, $\bar{x}^0$, a necessary condition is that a stationary point of $f(\bar{x})$ exists, i.e., the first partial derivatives of $f(\bar{x})$ be zero. Let us write this in the form of a total differential:

$$df(\bar{x}) = \sum_{j=1}^{n} \frac{\partial f(\bar{x}^0)}{\partial x_j} dx_j = 0 \tag{4-64}$$

This will not suffice in this case because a relative maximum could occur on the boundary of one or more of the constraints. At such a point it is not necessarily true that all the first partial derivatives of $f(\bar{x})$ are zero. Let us then consider the total differential of the constraints:

$$g_i(\bar{x}) = b_i \qquad i = 1, 2, \ldots, m \tag{4-65}$$

From (4-65) we have

$$dg_i(\bar{x}) = \sum_{j=1}^{n} \frac{\partial g_i(\bar{x})}{\partial x_j} dx_j = 0 \qquad i = 1, 2, \ldots, m \tag{4-66}$$

It is more convenient not to follow the procedure we used with only one constraint. Let us instead consider the expression we obtain if we multiply each of the functions in (4-66) by an associated *Lagrange multiplier*, λ_i, and subtract this from the total differential $df(\bar{x})$. We then obtain

$$df(\bar{x}) - \sum_{i=1}^{m} \lambda_i \, dg_i(\bar{x}) = \sum_{j=1}^{n} \left[\frac{\partial f(\bar{x})}{\partial x_j} - \sum_{i=1}^{m} \lambda_i \frac{\partial g_i(\bar{x})}{\partial x_j} \right] dx_j = 0 \tag{4-67}$$

Without loss of generality, let us assume that we can eliminate the first m variables, $x_1, x_2, \ldots, x_m$. Therefore, we can express these variables in terms of the remaining $n - m$ variables, which can then be regarded as independent variables. Therefore the dx_j,

$$j = m + 1, \ldots, n$$

can be regarded as independent variables and if as equation (4-67) indicates, the expression on the right is to vanish, then it must be true that

$$\frac{\partial f(\bar{x})}{\partial x_j} - \sum_{i=1}^{m} \lambda_i \frac{\partial g_i(\bar{x})}{\partial x_j} = 0 \qquad j = m + 1, \ldots, n \tag{4-68}$$

Since the terms in (4-68) are zero, if we subtract these from (4-67) we then obtain

$$\sum_{j=1}^{m}\left[\frac{\partial f(\bar{x})}{\partial x_j} - \sum_{i=1}^{m}\lambda_i\frac{\partial g_i(\bar{x})}{\partial x_j}\right] dx_j = 0 \qquad \textbf{(4-69)}$$

Since the $dx_j, j = 1, \ldots, m$ are not independent but are determined uniquely by the values of the dx_j, $j = m + 1, \ldots, n$, in order to satisfy equation (4-69), the coefficients of the dx_j, $j = 1, 2, \ldots, m$ must also be zero. Therefore,

$$\frac{\partial f(\bar{x})}{\partial x_j} - \sum_{i=1}^{m}\lambda_i\frac{\partial g_i(\bar{x})}{\partial x_j} = 0 \qquad j = 1, 2, \ldots, m \qquad \textbf{(4-70)}$$

If we now combine our results, we see that at a relative maximum (or minimum) $\bar{x}^0$, the following must be satisfied:

$$\begin{aligned} &\frac{\partial f(\bar{x}^0)}{\partial x_j} - \sum_{i=1}^{m}\lambda_i\frac{\partial g_i(\bar{x}^0)}{\partial x_j} = 0 \qquad j = 1, 2, \ldots, n \\ &g_i(\bar{x}) - b_i = 0 \qquad i = 1, 2, \ldots, m \end{aligned} \qquad \textbf{(4-71)}$$

Equations (4-71) are $m + n$ equations in $m + n$ variables. They are necessary conditions for the existence of a solution to the problem

$$\text{Max } z = f(\bar{x})$$

Subject to

$$g_i(x) = b_i \qquad i = 1, 2, \ldots, m \qquad \textbf{(4-72)}$$

The necessary conditions given in equations (4-71) can be simply generated by defining, as before, a Lagrangian function:

$$F(\bar{x}, \bar{\lambda}) = f(\bar{x}) + \sum_{i=1}^{m}\lambda_i[b_i - g_i(\bar{x})] \qquad \textbf{(4-73)}$$

If we now take partial derivatives we obtain (4-71):

$$\begin{aligned} &\frac{\partial F(\bar{x}, \bar{\lambda})}{\partial x_j} = \frac{\partial f(\bar{x})}{\partial x_j} - \sum_{i=1}^{m}\lambda_i\frac{\partial g_i(\bar{x})}{\partial x_j} = 0 \qquad j = 1, 2, \ldots, n \\ &\frac{\partial F(\bar{x}, \bar{\lambda})}{\partial \lambda_i} = b_i - g_i(\bar{x}) = 0 \qquad i = 1, 2, \ldots, m \end{aligned} \qquad \textbf{(4-74)}$$

In order to derive the foregoing results we had to assume that m of the variables or differentials could be expressed in terms of the remaining $n - m$. This is equivalent to the assumption that the rank of the $m \times n$ matrix

$$\begin{bmatrix} \frac{\partial g_1}{\partial x_1} & \cdots & \frac{\partial g_1}{\partial x_n} \\ \cdot & & \cdot \\ \cdot & & \cdot \\ \cdot & & \cdot \\ \frac{\partial g_m}{\partial x_1} & \cdots & \frac{\partial g_m}{\partial x_n} \end{bmatrix}$$

is equal to m at $\bar{x}^0$. If this fails to be true, a more complex situation results. This is discussed in Carathéodary[5] and in Hadley.[6]

It is also possible to derive sufficient conditions for a relative maximum or minimum of the problem in (4-72). However, they are quite complex and are of no computational or practical significance. They are discussed in Hancock.[2]

4.13 EXAMPLES OF CONSTRAINED OPTIMIZATION

EXAMPLE 1

In what direction is the rate of change of $z = f(\bar{x})$ a maximum?

In order to solve this problem, we recall from multidimensional calculus the definition of a *directional derivative* (Apostol[7]). The directional derivative of $f(\bar{x})$ at $\bar{x}^0$ in the direction $\bar{u}$ is

$$D_{\bar{u}}f(\bar{x}^0) = \lim_{\lambda \to 0} \frac{f(\bar{x}^0 + \lambda\bar{u}) - f(\bar{x}^0)}{\lambda}$$

when the limit exists. Furthermore, it is well known that the directional derivative, i.e., the derivative of $f(\bar{x})$ with respect to a direction $\bar{u}$, can be expressed in terms of ordinary partial derivatives (Apostol[7]) as follows:

$$D_{\bar{u}}f(\bar{x}^0) = \bar{\nabla}f(\bar{x}^0)\bar{u} \qquad |\bar{u}| = 1 \tag{4-75}$$

In component form, (4-75) can be written

$$D_{\bar{u}}f(\bar{x}^0) = \sum_{j=1}^{n} \frac{\partial f(\bar{x}^0)}{\partial x_j} u_j \qquad |\bar{u}| = 1 \tag{4-76}$$

We can now address ourselves to the problem of determining the direction $\bar{u}$ such that the rate of change of $z = f(\bar{x})$ at a point $\bar{x}^0$ is a maximum. We wish to maximize

$$\sum_{j=1}^{n} \frac{\partial f(\bar{x}^0)}{\partial x_j} u_j$$

subject to the constraint,

$$g(\bar{u}) = \sum_{j=1}^{n} u_j^2 = 1$$

We first form the Lagrangian,

$$F(\bar{u}, \lambda) = \sum_{j=1}^{n} \frac{\partial f(\bar{x}^0)}{\partial x_j} u_j + \lambda\left(1 - \sum_{j=1}^{n} u_j^2\right) \tag{4-77}$$

Differentiating (4-77) we obtain

$$\frac{\partial F}{\partial u_j} = \frac{\partial f(\bar{x}^0)}{\partial x_j} - 2\lambda u_j = 0 \qquad j = 1, 2, \ldots, n \tag{4-78}$$

$$\frac{\partial F}{\partial \lambda} = 1 - \sum_{j=1}^{n} u_j^2 = 0 \tag{4-79}$$

From (4-78) we obtain

$$u_j = \frac{1}{2\lambda}\frac{\partial f(\bar{x}^0)}{\partial x_j} \qquad i = 1, 2, \ldots, n \tag{4-80}$$

Substituting into (4-79) we have

$$1 - \sum_{j=1}^{n} \frac{1}{4\lambda^2}\left(\frac{\partial f(\bar{x}^0)}{\partial x_j}\right)^2 = 0 \tag{4-81}$$

Equation (4-81) can be simplified to

$$\lambda^2 = \tfrac{1}{4}\,|\overline{\nabla}f(\bar{x}^0)|^2 \tag{4-82}$$

Equation (4-82) can be solved to yield two solutions:

$$\lambda = \pm\tfrac{1}{2}\,|\overline{\nabla}f(\bar{x}^0)| \tag{4-83}$$

Substituting into (4-80) from (4-83) yields

$$\bar{u} = \frac{\pm\overline{\nabla}f(\bar{x}^0)}{|\overline{\nabla}f(\bar{x}^0)|} \tag{4-84}$$

It is a simple matter to verify that the plus sign gives the rate of maximum increase of $z = f(\bar{x})$ and the minus sign the rate of maximum decrease of $z = f(\bar{x})$. Hence, what we have shown in (4-84) is that the gradient vector gives the direction of maximum increase of the function $z = f(\bar{x})$.

EXAMPLE 2

In most elementary texts on inventory theory or operations research an expression for the "economic order quantity" of an item which is held in inventory is derived. This usually takes into account:

1. Cost of placing an order
2. Cost of holding item in inventory
3. Expected demand for an item

Let us now assume that, as is usually the case, there are many items to be held in inventory and that economic order quantities will be calculated for each item. Let us further assume that a constraint has been placed on the average value of the inventory to be held. We will assume that this is an equality constraint.

Let x_j, $j = 1, 2, \ldots, n$ be the order sizes to be determined. Our basic data consists of the following:

P_I = inventory cost as percentage of item cost
V_j = cost of the jth item
S_j = demand for the jth item.
C = fixed cost of placing an order
M = allowed average value of inventory

We wish to minimize the total costs of ordering and holding inventory. If total cost is C_T we have

$$\text{Min } C_T = \sum_{j=1}^{n} \left[\frac{CS_j}{x_j} + \frac{P_I V_j x_j}{2}\right] \tag{4-85}$$

Subject to

$$\sum_{j=1}^{n} \frac{V_j x_j}{2} = M$$

We form the Lagrangian,

$$F(\bar{x}, \lambda) = \sum_{j=1}^{n} \left[\frac{CS_j}{x_j} + \frac{P_I V_j x_j}{2}\right] + \lambda\left[M - \sum_{j=1}^{n} \frac{V_j x_j}{2}\right]$$

$$\frac{\partial F}{\partial x_j} = \frac{-CS_j}{x_j^2} + \frac{P_I V_j}{2} - \frac{\lambda V_j}{2} = 0 \qquad j = 1, 2, \ldots, n \tag{4-86}$$

$$\frac{\partial F}{\partial \lambda} = M - \sum_{j=1}^{n} \frac{V_j x_j}{2} = 0 \tag{4-87}$$

From (4-86) we have

$$x_j^2 = \frac{2CS_j}{V_j(P_I - \lambda)} \qquad j = 1, 2, \ldots, n$$

or

$$x_j = \left[\frac{2CS_j}{V_j(P_I - \lambda)}\right]^{\frac{1}{2}} \qquad j = 1, 2, \ldots, n \tag{4-88}$$

We now substitute (4-88) into (4-87) and we have

$$\sum_{j=1}^{n} \frac{v_j}{2} \left[\frac{2CS_j}{V_j(P_I - \lambda)}\right]^{\frac{1}{2}} = M$$

which upon rearrangement becomes

$$\frac{1}{(P_I - \lambda)^{\frac{1}{2}}} \sum_{j=1}^{n} \left(\frac{V_j CS_j}{2}\right)^{\frac{1}{2}} = M \tag{4-89}$$

However, we can rewrite (4-88) as

$$x_j = \frac{1}{(P_I - \lambda)} \left(\frac{2CS_j}{V_j}\right)^{\frac{1}{2}} \qquad j = 1, 2, \ldots, n$$

or

$$\frac{1}{(P_I - \lambda)^{\frac{1}{2}}} = x_j \left(\frac{V_j}{2CS_j}\right)^{\frac{1}{2}} \qquad j = 1, 2, \ldots, n \qquad \textbf{(4-90)}$$

Substituting (4-90) into (4-89) we then have

$$x_j \left(\frac{V_j}{2CS_j}\right)^{\frac{1}{2}} \sum_{j=1}^{n} \left(\frac{V_j CS_j}{2}\right)^{\frac{1}{2}} = M \qquad j = 1, 2, \ldots, n$$

from which we obtain

$$x_j = \frac{M\left(\dfrac{2CS_j}{V_j}\right)^{\frac{1}{2}}}{\displaystyle\sum_{j=1}^{n} \left(\frac{V_j CS_j}{2}\right)^{\frac{1}{2}}} \qquad j = 1, 2, \ldots, n \qquad \textbf{(4-91)}$$

The reader may wish to consider how it might be shown that the values of x_j given by (4-91) do indeed give a minimum total cost.

Can this problem be solved without the use of a Lagrange multiplier?

4.14 INEQUALITY CONSTRAINTS

Classical optimization deals with equality constraints. Until comparatively recently, very little consideration was given to how to deal with problems of the kind

$$\begin{aligned} &\text{Max } z = f(\bar{x}) \\ &g_i(\bar{x}) \leq b_i \qquad i = 1, 2, \ldots, m \end{aligned} \qquad \textbf{(4-92)}$$

We did consider constraints of the kind

$$x_j \geq 0 \quad \text{or} \quad x_j \leq a_j$$

However, these are not dealt with easily, as we noted in previous sections. In Chapter 8 we shall develop, in the "Kuhn-Tucker" theory, a set of necessary and sufficient conditions for problems of the kind given by (4-92). For the present, we can consider some simple and fairly obvious extensions of how to use Lagrange multipliers and certain kinds of inequality constraints. The methods to be described are inefficient computationally. Hence, their use is restricted to problems with a very small number of constraints and variables.

The simplest extension involves solving

$$\begin{aligned} \text{Max } z &= f(\bar{x}) \\ g_i(\bar{x}) &= b_i \qquad i = 1, 2, \ldots, m \\ \bar{x} &\geq \bar{0} \end{aligned} \tag{4-93}$$

In (4-93) we have added the constraints, $x_j \geq 0, j = 1, 2, \ldots, n$. We continue to seek the global maximum and we assume that at the maximal point, each component of $\bar{x}$ is finite.

An algorithm for solving the problem of (4-93) is given next. It will be noted that it is a slightly more complicated variant of the method given in Figure 4-11, where no equality constraints were included.

Algorithm for Equality Constraints and Nonnegative Variables

1. Determine all interior† solutions by finding all solutions to

$$\begin{aligned} \frac{\partial f(x)}{\partial x_j} - \sum_{i=1}^{m} \frac{\partial g_i(\bar{x})}{\partial x_j} &= 0 \qquad j = 1, 2, \ldots, n \\ g_i(\bar{x}) - b_i &= 0 \qquad i = 1, 2, \ldots, m \end{aligned} \tag{4-94}$$

z is evaluated for each of these to find the largest.

2. Set $x_1 = 0$ and repeat solving (4-94) as in Step 1. Set $x_2 = 0$, $x_3 = 0, \ldots, x_n = 0$, each in turn until we have solved (4-94) n times, one time for each variable set equal to zero.

3. Repeat Step 2, setting two variables, x_k, x_l at a time equal to zero, over all possible combinations of k, l. For each combination we resolve equations (4-94). There will be $n!/2(n-2)!$ possible combinations.

4. Repeat Step 3, setting three variables at a time equal to zero; then we set four variables at a time equal to zero, and so forth, until $n - m$ variables have been set equal to zero.

5. The largest value of z found from all the solutions obtained in the previous Steps is the global maximum.

It is clear that in the preceding algorithm, equations (4-94) would have to be solved an impractically large number of times if there were many variables. For example, if there were ten variables and four constraints, we would have to solve the 14 simultaneous nonlinear equations (4-94) 638 times, and also evaluate z at all possible combinations of six variables set equal to zero at the last stage. This is a very large amount of computation. As the number

---

† Solutions for which all $x_j > 0, j = 1, 2, \ldots, n$.

of variables and constraints increases even slightly, it becomes completely out of the realm of possibility.

Another extension of Lagrangian methods is to the solution of the following problem:

$$\begin{aligned} &\text{Max } z = f(\bar{x}) \\ &g_i(\bar{x}) \leq b_i \qquad i = 1, 2, \ldots, m \end{aligned} \tag{4-95}$$

We may make the following observation about this problem which will prove to be useful. Suppose we convert the inequalities in (4-95) to equalities, i.e.,

$$g_i(\bar{x}) + y_i = b_i \qquad i = 1, 2, \ldots, m \tag{4-96}$$

where

$$y_i \geq 0, \qquad i = 1, 2, \ldots, m$$

By using (4-96) we can now form the Lagrangian

$$F = f(\bar{x}) + \sum_{i=1}^{m} \lambda_i[b_i - y_i - g_i(\bar{x})] \tag{4-97}$$

Let us consider

$$\frac{\partial F}{\partial y_i} \qquad i = 1, 2, \ldots, m$$

$$\frac{\partial F}{\partial y_i} = \lambda_i = 0 \qquad i = 1, 2, \ldots, m \tag{4-98}$$

at a stationary point. What this says is that if $y_i > 0$ (i.e., the constraint $g_i(\bar{x}) < b_i$ or is not active), then $\lambda_i = 0$. In other words, we ignore the inequality constraint. However, if $y_i = 0$, the constraint is active, i.e., we are on the boundary and λ_i may not be zero.

Another useful observation to make about (4-95) is the following. If we solve Max $z = f(\bar{x})$ and find that the optimal $\bar{x}$, say $\bar{x}^*$, also satisfies $g_i(\bar{x}) \leq b_i$, $i = 1, 2, \ldots, m$, then we also have the solution to (4-95). This follows from the fact that if $z = f(\bar{x}^*)$, the addition of one or more constraints can only cause z to be less than or equal to z^*. It cannot increase z. If it did this, it would contradict the fact that z^* was the optimal z value for the original problem.

We are now in a position to use the foregoing observations to construct the following method for solving (4-95).

Algorithm for Inequality Constraints

1. Solve the problem:

$$\text{Max } z = f(\bar{x})$$

We ignore the inequality constraints at this stage. Call the optimal solution $\bar{x}_0$.

2. If $\bar{x}_0$ satisfies

$$g_i(\bar{x}) \leq b_i \qquad i = 1, 2, \ldots, m$$

then we are done. $\bar{x}^0$ is the global maximum.

3. If one or more of the constraints are not satisfied, select any constraint, say the first, and allow it to become an active constraint, i.e., set $y_1 = 0$. Now solve the problem

$$\text{Max } z = f(\bar{x})$$

Subject to

$$g_1(\bar{x}) = b$$

Call the optimal solution $\bar{x}_1$. If $\bar{x}_1$ satisfies all the constraints, we are done. If not, we must repeat this procedure, allowing one constraint at a time to be active.

4. If Step 3 did not yield an optimum, we now allow all combinations of two constraints at a time to be active and resolve all these problems. As soon as a solution is found that satisfies *all* the constraints, we are done.

5. If Step 4 did not yield an optimal we allow three constraints at a time to be active, and so forth.

It is obvious that this procedure involves a great deal of computation. However, we have minimized this as much as possible by attempting to ignore as many constraints as possible.

EXERCISES

Convex Functions—One Variable

1. Prove that $|x|$ is a convex function.

Optimization of Functions of a Single Variable

2. Find all stationary points and relative optima of

$$f(x) = x^3 + 6x^2 + 3x - 10$$

3. The potential energy, $U(r)$, of a diatomic molecule as a function of distance, r, is given by

$$U(r) = \frac{B}{r^2} - \frac{ze^2}{r}$$

For our purposes it is sufficient to note that $B > 0$, $e > 0$, and $z \geq 1$. Find the distance, r, at which the potential energy is a minimum.

4. Find a root of $f(x) = e^x - 5x^2 + 3$.

5. We wish to minimize $f(x) = -x^2 + 4x + 2$, $0 \leq x \leq 10$. What is the minimum number of evaluations of $f(x)$ that we need to determine in order to find the minimum value of $f(x)$? Find this minimum value of $f(x)$.

6. Find the dimension of the cube of largest volume that can be placed inside a sphere of radius, R.

Convex Functions of Several Variables

7. Suppose that $f(\bar{x})$ and $g(\bar{x})$ are both convex functions. Is it generally true that the product $f(\bar{x})g(\bar{x})$ is a convex function? Give a proof or provide a counter example.

8. Prove that the global maximum of a continuous function $f(\bar{x})$ must be a local maximum.

9. Is $f(\bar{x}) = x_1^2 + 2x_1x_2 + x_2^2$ a convex function? Prove your statement.

Constrained Optimization

10. Find the maximum of $f(\bar{x}) = 2x_1x_2 + 3x_1x_3$ given that $x_1^2 + x_2^2 = 3$ and $x_1x_3 = 2$.

11. Find the maximum of $f(\bar{x}) = x_1^2x_2^2x_3^2$ given that

$$x_1^2 + x_2^2 + x_3^2 = c^2$$

12. The method of least squares is based on the principle that the "best" value of a quantity that can be deduced from a set of measurements or observations is that for which the sum of the squares of the deviations of the observed values from this best value is a minimum. Suppose we have made a set of p observations of k variables, $[x_{i1}, x_{i2}, \ldots, x_{ik}]$, $i = 1, 2, \ldots, p$ and that y depends on these k variables. Suppose we wish to find the hypersurface

$$\hat{y} = f(\bar{a}; \bar{x})$$

which satisfies the least squares criterion, i.e., we wish to find the vector of parameters $\bar{a} = [a_1, a_2, \ldots, a_s]$ which will minimize

$$z = \sum_{i=1}^{p} (y_i - \hat{y}_i)^2$$

(1) Find the necessary conditions for $\bar{a}$ to be the set of parameters which minimize z.

(2) Suppose $\hat{y} = \sum_{j=1}^{k} a_j x_j + b$. Give the expression for the parameters $\bar{a}$ in terms of the observed values of x_j.

Inequality Constraints

13. Minimize

$$f(\bar{x}) = x_1^2 + 3x_1x_2 + x_3^2$$

Subject to:

$$x_1 + x_2 = 10$$
$$x_1 \geq 0, x_2 \geq 0, x_3 \geq 0$$

14. Minimize

$$f(\bar{x}) = x_1^2 + 2x_2^2$$

Subject to:

$$x_1x_2 \geq 5, x_1 \geq 0, x_2 \geq 0$$

15. Minimize

$$f(\bar{x}) = x_1^2 + 3x_2^2 - 4x_1 - 6x_2$$

Subject to:

$$x_1 + 2x_2 \leq 4$$
$$x_1 \geq 0, x_2 \geq 0$$

16. A company manufactures three grades of a certain product. They all use the same two raw materials. Each unit of grade 1 requires 6 lb. of raw material A and 4 lb. of raw material B. Each unit of grade 2 requires 2 lb. of raw material A and 4 lb. of raw material B. Each unit of grade 3 requires 3 lb. of raw material A and 3 lb. of raw material B. There is a limit upon the amounts of each of these raw materials that is available each week. The amount of A that is available is 1600 lb. and the amount of B that is available is 2400 lb. We shall assume that our company is in the unusual position of selling all it can manufacture. However, the selling price varies with the relative amounts of each of the three grades sold. It has been found that the unit profits for each of the three grades are as follows:

$$1:\quad 1000 - x_1 - 2x_3$$
$$2:\quad 1800 - 2x_1 - x_2$$
$$3:\quad 1500 - x_2 - 3x_3$$

where x_1, x_2, x_3 are the amounts of each of the three grades produced each week. What are the values of x_1, x_2, x_3 which will maximize the total weekly profit of the company?

17. Minimize

$$z = 6x_1x_2x_3$$

Subject to:

$$2x_1^2 + 6x_2^2 + 3x_3^2 \geq 20$$

$$x_1, x_2, x_3 \geq 0$$

REFERENCES

1. Hildebrand, F. B.: *Introduction to Numerical Analysis*. McGraw-Hill, New York (1956).
2. Hancock, H.: *Theory of Maxima and Minima*. Dover Publications, New York (1960). (Reprint of original 1917 edition.)
3. Ralston, A.: *A First Course in Numerical Analysis*. McGraw-Hill, New York (1965).
4. Isaacson, E., and Keller, H. B.: *Analysis of Numerical Methods*. John Wiley and Sons, New York (1966).
5. Carathéodary, C.: *Calculus of Variations and Partial Differential Equations of the First Order, Vol. II*. Holden-Day, San Francisco (1967).
6. Hadley, G.: *Nonlinear and Dynamic Programming*. Addison-Wesley, Reading, Massachusetts (1964).
7. Apostol, T. M.: *Mathematical Analysis*. Addison-Wesley, Reading, Massachusetts (1957).

Chapter 5

SEARCH TECHNIQUES: UNCONSTRAINED PROBLEMS

5.1 INTRODUCTION

In this chapter we shall concern ourselves with the general problem of maximizing or minimizing a function of a single variable or a function of more than one variable by procedures known as search methods. We previously treated these same problems in Chapter 4 by classical optimization techniques, which make use of the calculus to locate the optimal value of x in maximizing or minimizing $f(x)$ or the optimal vector of values, $\bar{x}$ in maximizing or minimizing $f(\bar{x})$.

There are a number of drawbacks to the use of these classical optimization techniques which seriously limit their use in practice. Generally speaking, the classical methods require either the continuity or the differentiability of $f(\bar{x})$ or both. These conditions are not always met in practical problems that must be solved. It may even be possible that whether the function is continuous or differentiable is not known. It is frequently the case that even less than this is known. In fact, almost nothing may be known about how the objective function depends on the variables in question, except the mathematical form of the objective function. While an elegant theory is available for treating convex or concave functions, most functions are neither, and many are sufficiently complex as to defy simple characterization. A few examples will illustrate this. Consider the following actual problem from the field of chemistry:

$$\text{Min } g(x)$$

$$g(x) = \left[\frac{x^2(x^2 + \frac{3}{2} + \sigma)}{(x^2 + \frac{1}{2})^2} - 1.5625\right]^2$$

$$\sigma = \frac{1}{\theta_1}\left\{\frac{8\theta_2^2}{\pi\theta_1} - (x^2+1)\frac{e^{-x^2}}{\sqrt{\pi}} - x[x^2+\tfrac{3}{2}][1+P(x)]\right\}$$

$$\theta_1 = \frac{e^{-x^2}}{\sqrt{\pi}} + x[1+P(x)] \qquad \textbf{(5-1)}$$

$$\theta_2 = \frac{xe^{-x^2}}{\sqrt{\pi}} + [x^2+\tfrac{1}{2}][1+P(x)]$$

$$P(x) = \frac{2}{\sqrt{\pi}}\int_0^x e^{-u^2}\,du$$

We wish to find the value of x that minimizes $g(x)$ in equation (5-1). One would be hard pressed to have any intuitive feeling about the neighborhood of x in which the solution may lie. Hence we see that, even for functions of a single variable, characterization may be difficult.

In the case of functions of more than one variable, the interaction of variable effects makes it difficult for the unaided mind to grasp all the possibilities inherent in a multivariable situation. An example of such a multivariable problem would be the following:

$$\text{Min } S = \sum_{i=1}^{n} \frac{[y_i - A\cos(Bx_i+E) - C\cosh(Dx_i+E)]^2}{a_i[A\cos(Bx_i+E) + C\cosh(Dx_i+E)]} \qquad \textbf{(5-2)}$$

The problem in equation (5-2) is to find A, B, C, D, and E, given the $2n$ values of x_i and y_i, so as to minimize S. This is a typical statistical curve fitting problem that could arise in any of the natural or social sciences.

It is important to recognize the rather frustrating nature of almost all complex situations of this sort. What we are saying is that until we choose a particular value of x in equation (5-1) or a particular set of values of A, B, C, D, and E in equation (5-2) we have no idea what the value of $g(x)$ or S may be, and worse yet, we do not see very clearly how we should change these values to make g or S smaller. It is in situations of this sort that one resorts to search methods. In contrast with the classical methods, the only requirement that the objective function must satisfy is that it be computable.

We shall classify the search methods to be discussed in the following sections as follows.

Optimization Search Methods (Unconstrained)

I. One dimensional search methods
   A. Simultaneous methods
      1. Exhaustive search
      2. Random search

B. Sequential methods
   1. Dichotomous search
   2. Equal interval search
   3. Fibonacci search

II. Multidimensional search methods

A. Simultaneous methods
   1. Exhaustive search
   2. Random search

B. Sequential methods
   1. Multivariate grid search
   2. Univariate search
   3. Powell's method
   4. Method of steepest descent
   5. Fletcher-Powell method
   6. Direct search

This classification scheme, and the enumeration of methods it contains, by no means exhausts all the possibilities that exist. Many other methods and combinations of methods have been discussed in papers in the optimization literature.

5.2 THE ONE-DIMENSIONAL SEARCH PROBLEM

In contrast with the methods discussed in previous chapters, the one-dimensional search methods place no restrictions such as continuity or differentiability on the function $f(x)$. The only requirement is that $f(x)$ can be computed either from a formula or a series of formulas; even a table of values will suffice.

The search problem, by any of the methods to be discussed, is simply to find the value of x in an interval $a \leq x \leq b$ at which the function $f(x)$ takes on its global extreme value (maximum or minimum). This statement, however, requires considerable modification. Since we shall be using numerical schemes, we shall not know the exact value of x for which, say, $f(x)$ is a maximum. Instead, what we shall determine is a value of x, for example x^*, and that we have determined two values x_L and x_U such that $x_L \leq x^* \leq x_U$. Another common way of phrasing this is that the value determined, x^*, and the true value, $\hat{x}$, both lie within the same "interval of uncertainty." In order to simplify the presentation and comparison of the methods, we shall assume that the interval $[a, b]$ has been mapped onto the unit interval $[0, 1]$ by a suitable linear transformation. What this means, in terms of the value of x^* obtained in the search, is that if the original interval was $[a, b]$ and a value of $0 \leq x^* \leq 1$ is

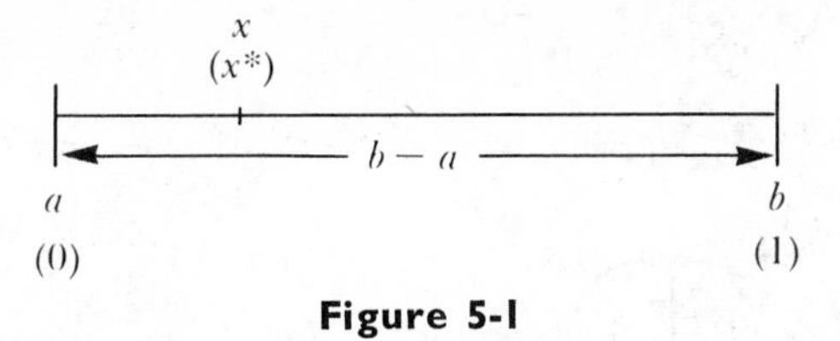

Figure 5-1

obtained, the value of x on the interval $[a, b]$ is

$$x = a + x^*(b - a) \tag{5-3}$$

This can be easily seen by referring to Figure 5-1. If we adopt this simplification, then the initial interval of uncertainty for each of the methods will be of length 1. In other words, when we begin, all we know is that the optimal value of x lies somewhere in the interval $[0, 1]$.

One further general point relates to the distinction made between simultaneous and sequential searches. In a simultaneous search method all points at which the function is to be evaluated are selected in advance, or what is more important, the selection of the $(n + 1)^{st}$ point in no way depends upon the selection of the nth point at which the function is evaluated. By contrast, in sequential search, the points at which the function is to be evaluated cannot be selected *a priori*, since the particular sequence of values of x to be chosen depends upon the values of $f(x)$ that are observed at prior points.

5.3 SIMULTANEOUS METHODS: ONE-DIMENSIONAL SEARCH

The simplest (in concept) of the simultaneous search methods is what may be called "exhaustive search." This is clearly a misnomer, since it may, in some cases, be far from exhaustive. The algorithm can be described as follows.

Exhaustive Search Algorithm

1. Subdivide the interval $[0, 1]$ into $\frac{\Delta x}{2}$ equally spaced intervals.
2. Evaluate the function $f(x)$ at each of the points $\frac{k\,\Delta x}{2}$, $k = 1, \ldots, \frac{2}{\Delta x} - 1$.
3. Select the maximum (minimum) value of $f(x)$.

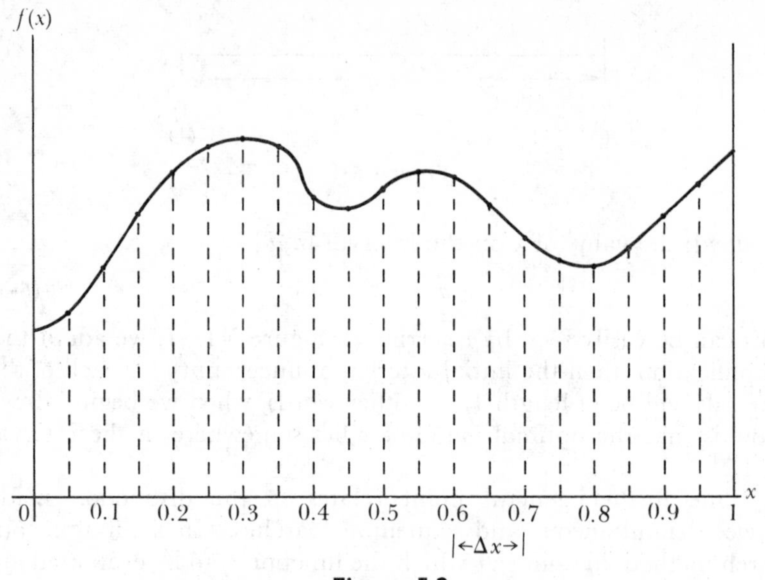

Figure 5-2

By selecting an interval size of $\frac{\Delta x}{2}$, we insure that in $\frac{2}{\Delta x} - 1$ functional evaluations we have reduced the length of the interval of uncertainty to Δx. It is assumed, of course, that Δx has been suitably chosen so that no "narrow" maxima or minima will be missed. Figure 5-2 indicates graphically the salient features of this method. There are two chief difficulties with this approach. It is very costly in terms of computation, especially if the function is a complicated one and requires lengthy computation for evaluation. In addition, one must be very conservative with respect to the choice of the interval size Δx, in order to avoid inadvertently missing the global extreme value.

The other simultaneous search method that we shall consider is that of random search. The method to be described here is to simply generate a random number, r, in the unit interval and to evaluate the function at that point. This procedure is continued for a specified number of times. The largest value of $f(x)$, if we are maximizing, is retained at each stage of the calculation.

This procedure was actually used by a student of one of the authors in a computer program to solve the problem of equation (5-1). However, the problem was written in the root-finding form, i.e., find a root of

$$f(x) = \frac{x^2(x^2 + \frac{3}{2} + \sigma)}{(x^2 + \frac{1}{2})^2} = 1.5625 \tag{5-4}$$

Initially, numbers were generated on the interval (0, 10) in order to gain insight as to a starting value. The following was then obtained:

Table 5.1 *Initial Estimates*

Number of random numbers generated	x	$f(x)$
100	0.47655	1.4565
1000	0.51559	1.5868
5000	0.50496	1.5524

These results suggested that x was in the neighborhood of 0.5. 30,000 numbers were then generated in the interval (0, 1) with the result that the best approximation yielded a value of 0.50803 for x and 1.56245 for $f(x)$. The true value of x known correct to four places is 0.5080.

In summary the random search algorithm for maximization is as follows.

Random Search Algorithm

1. Given the problem $\max f(x)$, set a limit N on the number of random trials to be used. Initially set $f(x) = -10^{50}$.
2. Generate a random number $\hat{x}$; calculate $f(\hat{x})$.
3. Is $f(\hat{x}) > f(x)$? If yes, go to 4, otherwise to 5.
4. Set $f(x)$ equal to $f(\hat{x})$.
5. Is the number of trials $< N$? If yes, go to 2, otherwise stop.

Simultaneous search methods are generally very inefficient. One may well ask why they are used. If the function being maximized is a purely mathematical one that requires only a calculation, then there is probably no justification for ever using them. However, consider a situation in real life, where the time required to do a calculation does not allow sequential calculations, i.e., not performing a calculation until previous calculations are known. Instead, the available time must be utilized to perform several calculations parallelly, or simultaneously, and the best result chosen. This justifies the use of simultaneous search. In other situations one would definitely choose a sequential search method.

5.4 ONE-DIMENSIONAL SEQUENTIAL METHODS

As was mentioned earlier, a characteristic of sequential methods is that the arguments for which $f(x)$ will be evaluated cannot be known in advance. Instead, the sequence of argument values depends upon the previously observed values of $f(x)$. However, all

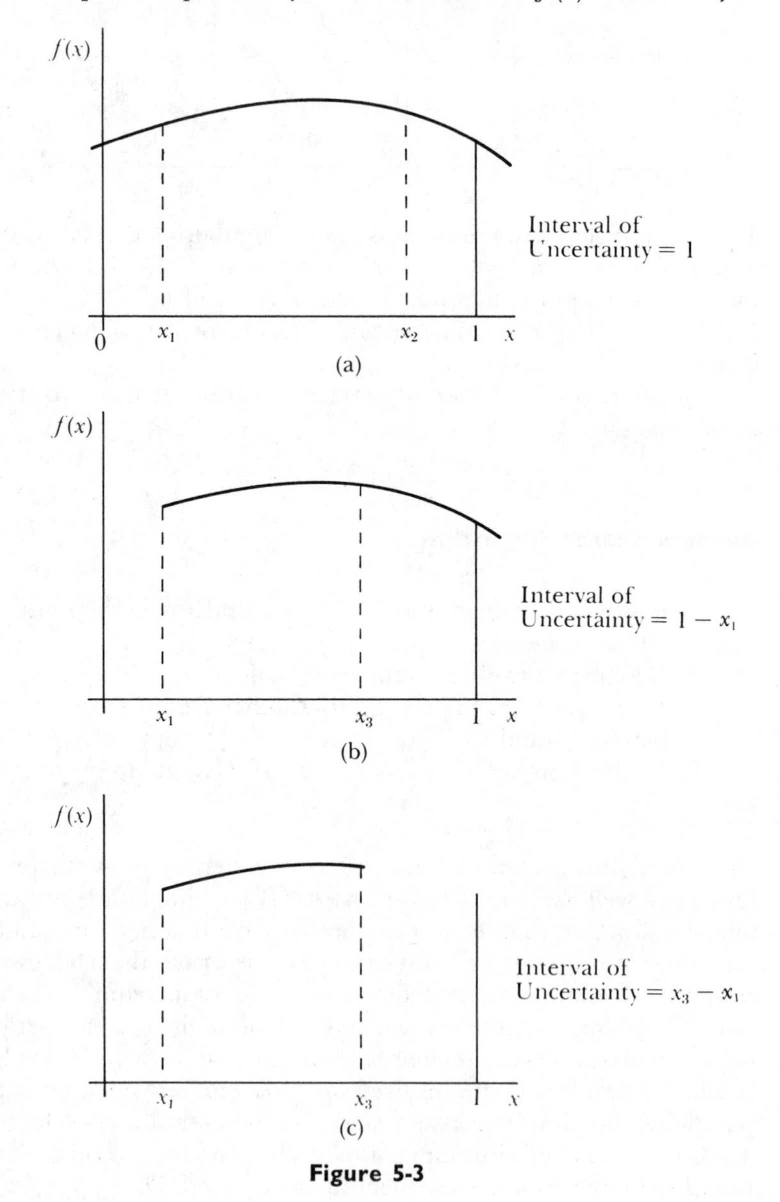

Figure 5-3

sequential methods do have the following interesting property: that although the arguments cannot be known in advance, the number of function evaluations required to reach a specified interval of uncertainty can be known in advance. We will perform this calculation for each of the sequential methods we describe.

All sequential methods require that $f(x)$ be a *unimodal function* (see Section 4.8), i.e., exhibit only one maximum or minimum over the search interval. The function need not be either differentiable or continuous. The assumption of unimodality is required because all these sequential methods are involved with, in one form or another, the computation of the function at one or more arguments. With the unimodality assumption we are then in a position to *reduce* the interval of uncertainty by eliminating a portion of the original interval. If we refer to Figure 5-3, this can be grasped more easily.

In Figure 5-3a, we have shown a unimodal function. Without reference to any particular search scheme at this point, suppose we have arbitrarily chosen points x_1, x_2 in the interval $[0, 1]$ and suppose $f(x_2) > f(x_1)$. Initially, the interval of uncertainty $= 1$. Since the function is known to be unimodal, we can drop the interval $(0, x_1)$ as is shown in Figure 5-3b. The interval of uncertainty is now $1 - x_1$. Now suppose we select the point x_3 and $f(x_3) > f(x_2)$. We are then justified in dropping the interval $(x_3, 1)$ resulting in Figure 5-3c where the interval of uncertainty is $x_3 - x_1$. It is the *relative efficiency* of the placing of these points that gives rise to the various search methods to be described.

Of course, one could use any of the search methods to be described on a function that was not unimodal. However, in such a situation, one could never be sure that one had not discarded a subinterval that contained an extreme value.

5.5 DICHOTOMOUS SEARCH

It is fairly clear that in any kind of scheme of the general character that was described in the previous section, we shall require at least two function evaluations, at different values in the interval $[0, 1]$, in order to reduce the interval of uncertainty. Clearly, one function evaluation does not give us sufficient information to reduce the interval of uncertainty.

In dichotomous search we first evaluate the function at two points separated by some small distance, ε, about the midpoint of the interval. In other words, we compute $f\left(\frac{1}{2} + \frac{\varepsilon}{2}\right)$ and $f\left(\frac{1}{2} - \frac{\varepsilon}{2}\right)$. Having done this we can reject the smaller subinterval of length

$(1 - \varepsilon)/2$ by examining the values of $f\left(\frac{1}{2} + \frac{\varepsilon}{2}\right)$ and $f\left(\frac{1}{2} - \frac{\varepsilon}{2}\right)$. If $f\left(\frac{1}{2} + \frac{\varepsilon}{2}\right) > f\left(\frac{1}{2} - \frac{\varepsilon}{2}\right)$ then we reject the interval $\left(0, \frac{1 - \varepsilon}{2}\right)$. Otherwise, we reject the interval $\left(\frac{1 - \varepsilon}{2}, 1\right)$. Having done this, we have reduced the interval of uncertainty from 1 to $(1 + \varepsilon)/2$. If we now repeat this procedure, by placing two points ε apart at the midpoint of the remaining interval, we will then have reduced the interval of uncertainty to

$$\frac{1}{2}\left(\frac{1 + \varepsilon}{2}\right) + \frac{\varepsilon}{2} = \frac{1 + 3\varepsilon}{4}$$

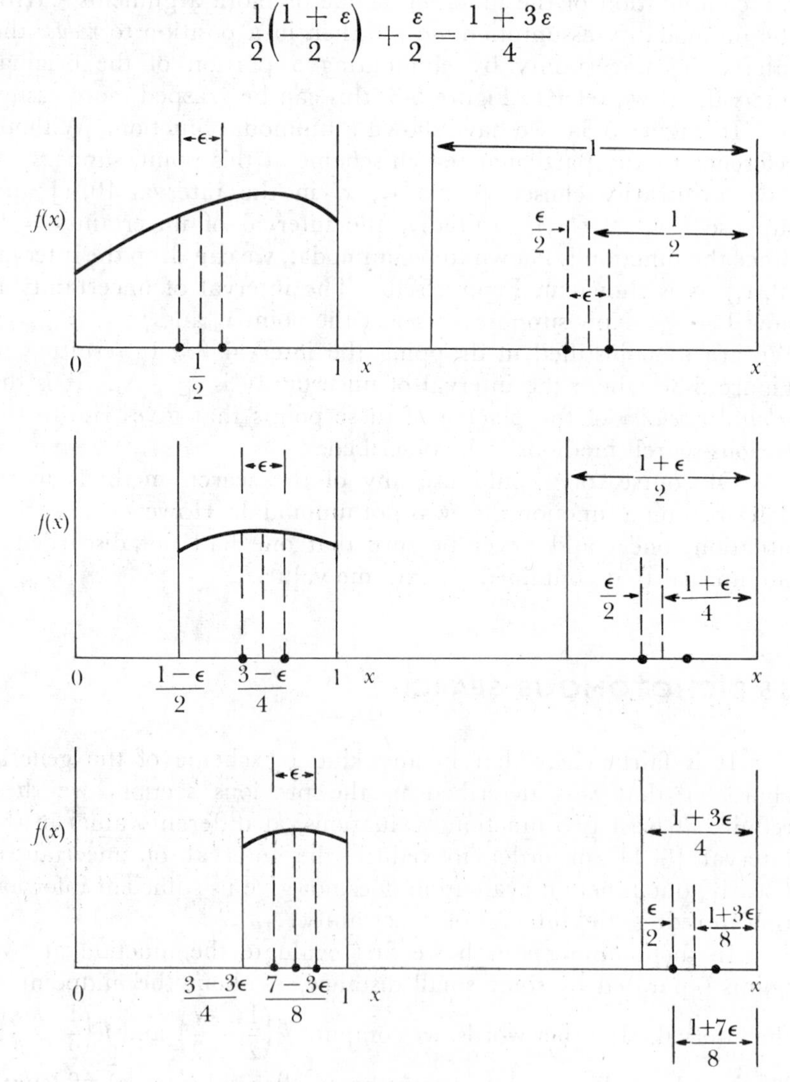

Figure 5-4

If we repeat this again, the interval of uncertainty will be reduced to $(1 + 7\varepsilon)/8$, and so forth. It is easily shown that after n repetitions of this process we can locate the optimum within an interval of uncertainty of $2^{-n} + (1 - 2^{-n})\varepsilon$. Figure 5-4 illustrates this process.

EXAMPLE 5.1

Find the minimum of $f(x) = x^2 - 6x + 2$ in the interval $0 \leq x \leq 10$ by means of dichotomous search. (Although in our discussions of the algorithms we assumed that x was in the interval [0, 1] for convenience, there is no need to make this transformation in practice).

The midpoint is $x = 5$. We choose $\varepsilon = 0.5$. Therefore, we compute:

$$f\left(5 + \frac{0.5}{2}\right) = f(5.25) = (5.25)^2 - 6(5.25) + 2 = -1.9375$$

$$f\left(5 - \frac{0.5}{2}\right) = f(4.75) = (4.75)^2 - 6(4.75) + 2 = -3.9375$$

We, therefore, discard the interval (5.25, 10). We now determine the midpoint of [0, 5.25] which is 2.625 and evaluate:

$$f(2.875) = -6.9844$$
$$f(2.375) = -6.6094$$

We, therefore, discard the interval (0, 2.375). We now determine the midpoint of [2.375, 5.25] which is 3.8125 and evaluate:

$$f(4.0625) = -5.8711$$
$$f(3.5625) = -6.6836$$

We, therefore, discard the interval (4.0625, 5.2500) and determine the midpoint of [2.3750, 4.0625] which is 3.2188 and evaluate:

$$f(3.4688) = -6.7800$$
$$f(2.9688) = -6.9990$$

We, therefore, discard the interval (3.4688, 4.0625) and determine the midpoint of [2.3750, 3.4688] which is 2.9219, and so forth.

The exact answer is $x = 3$ and $f_{\min} = -7.0$. Schematically, what we have done is shown in Figure 5-5.

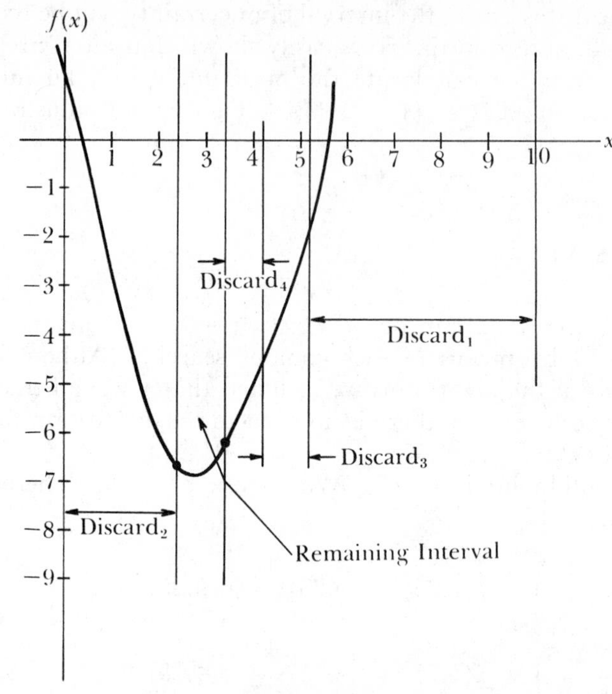

Figure 5-5

5.6 EQUAL INTERVAL SEARCH

The method we consider here is related to the dichotomous search method of the previous section except that we now (arbitrarily) impose the constraint that the points at which the function $f(x)$ is to be evaluated shall be *equally* spaced over the interval of uncertainty.

There is no particular advantage of an equal interval search as compared with an unequal interval search other than a "natural" tendency to subdivide into equal intervals. In fact, as we shall see later, there are distinct disadvantages to equal interval search methods in terms of economy. Nevertheless, they are often used.

The simplest equal interval method would use two points equally spaced in the interval of uncertainty. Initially the function would be evaluated at $x = \frac{1}{3}$ and $x = \frac{2}{3}$ on the interval $[0, 1]$. Two cases could arise for a unimodal function and these are shown in Figure 5-6. In either case, we retain that $\frac{2}{3}$ of the interval $[0, 1]$ which must obviously contain the extreme value. The key assumption of unimodality of $f(x)$ is seen most clearly here.

With the remaining interval, which will be $[0, \frac{2}{3}]$ or $[\frac{1}{3}, 1]$ this process is again repeated. We continue the process until a desired

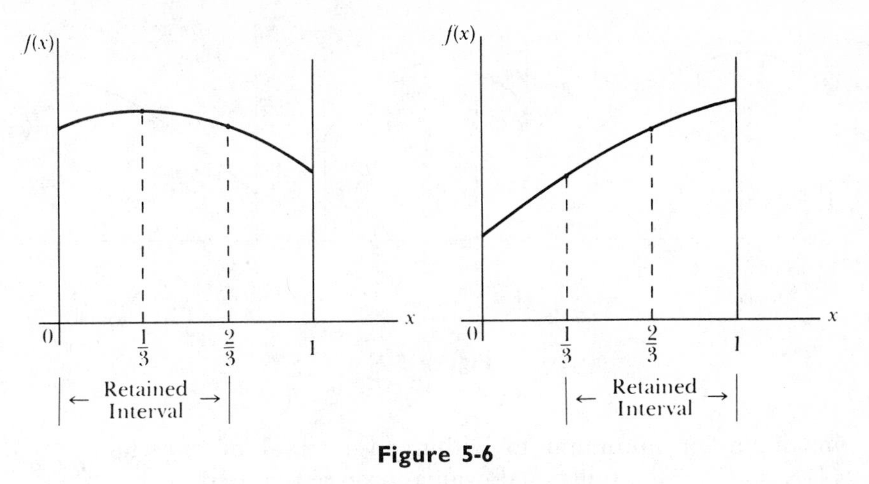

Figure 5-6

interval of uncertainty is achieved. Since at each stage we retain $\frac{2}{3}$ of the interval, it is clear that after m repetitions of this process, we have located the extreme value to within $(\frac{2}{3})^m$. In other words, after m steps, the interval of uncertainty is $(\frac{2}{3})^m$. In order to do this, $2m$ functional evaluations have been necessary.

It is obvious that we could have other equal interval searches. For example, we could use a three-point, a four-point, and so on, equal interval search. An interesting question then arises. Are any of these more efficient than any others? This question has been posed and answered in Kiefer.[1] In this paper it is shown that the three-point equal interval search is the most "economical" scheme. What this means precisely is that it requires the fewest number of functional evaluations, of any of the *equal interval schemes*, in order to achieve a given interval of uncertainty. We will now describe this scheme.

In the three-point equal interval search method, the function is evaluated initially at $x = \frac{1}{4}$, $x = \frac{1}{2}$, and $x = \frac{3}{4}$. Now we retain $\frac{1}{2}$ of the original interval, namely that half which contains as its center the value of x at which the largest (in the case of maximization) value of the function has been obtained. There are three possibilities that might arise. They are shown in Figure 5-7.

What is retained, as Figure 5-7 shows, is $\frac{1}{2}$ of the original interval. For that remaining interval we now repeat the process of evaluating the function at $\frac{1}{4}$, $\frac{1}{2}$, and $\frac{3}{4}$ of the total interval length and so forth. Part of the economy of this method is that there will always be one function value from the preceding iteration and hence that value need not be recalculated. For example, if we retained the interval $[0, \frac{1}{2}]$ we would require function evaluations at $x = \frac{1}{8}$, $x = \frac{1}{4}$, and $x = \frac{3}{8}$ and we would already have $f(x)$ evaluated at $x = \frac{1}{4}$. It is clear that after m stages we will have located the

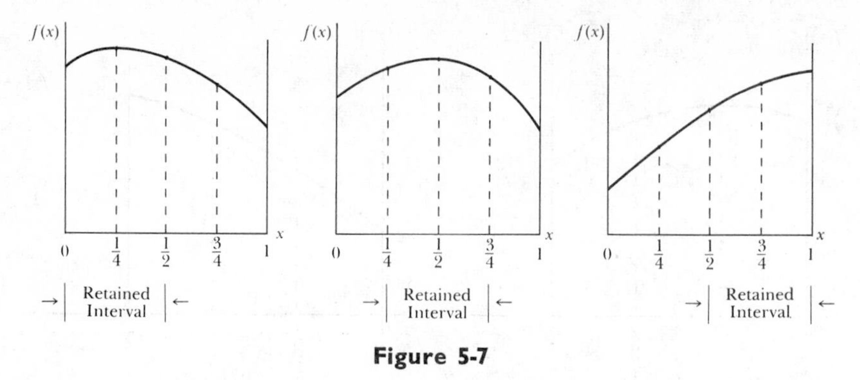

Figure 5-7

maximum or minimum to within an interval of uncertainty of $(\frac{1}{2})^m$, and $2m + 3$ functional evaluations are required.

One might wish to know how many functional evaluations are required to obtain a given interval of uncertainty. This is easily calculated. If ΔL = desired interval of uncertainty then

$$(\tfrac{1}{2})^m = \Delta L$$

or

$$m = -\frac{1}{\ln 2} \ln \Delta L$$

We know we require $2m + 3$ functional evaluations. If we designate n_f the number of functional evaluations, then

$$n_f = 2m + 3 = 2 - \left(\frac{\ln \Delta L}{\ln 2}\right) + 3 = -2.8854 \ln \Delta L + 3$$

$$n_f = 3 - 2.8854 \ln \Delta L$$

The following is an example of the three-point equal interval search:

EXAMPLE 5.2

Find the minimum of $f(x) = x^2 - 6x + 2$ in the interval $0 \leq x \leq 10$ by means of the three-point equal interval search method. This is the same problem solved in Example 5.1.

We evaluate $f(x)$ at $\frac{1}{4}$, $\frac{1}{2}$, and $\frac{3}{4}$ of the interval.

$$f(2.5) = -6.750$$
$$f(5.0) = -3.000$$
$$f(7.5) = 13.250$$

We, therefore, retain the interval [0, 5]. We now evaluate the function at $x = 1.25$, $x = 2.50$, and $x = 3.75$. We already have $f(2.50)$. Therefore, we calculate:

$$f(1.25) = -3.9375$$
$$f(2.50) = -6.750$$
$$f(3.75) = -6.4375$$

We, therefore, retain the interval [2.5, 5]. We now evaluate the function at $x = 3.125$, $x = 3.750$, and $x = 4.375$.

$$f(3.125) = -6.9844$$
$$f(3.750) = -6.4375$$
$$f(4.375) = -5.1094$$

We, therefore, retain the interval [2.50, 3.75] and repeat, and so on.

5.7 FIBONACCI SEARCH

There is a certain similarity that one might note between the dichotomous search method and the three-point equal interval search. In the case of the former, one rejects slightly less than half of the remaining interval at any stage and in the case of the latter, one rejects exactly half of the interval at any stage. One might guess that this is the best one could hope for, assuming no prior knowledge of the nature of the function, except that it is unimodal. It turns out that this is not the case. As we have seen, if we restrict ourselves to *equal interval searches*, then the three-point search is best. However, if one asks, what is the best strategy, in order to minimize functional evaluations (an optimization problem), no matter how the points are selected, the answer is a curious and interesting one that is based on a set of numbers, which were known for a long time as Fibonacci numbers. This was first shown in two papers (Kiefer[2]) and (Johnson[3]). It is also discussed in Bellman.[4]

Fibonacci numbers are named after Leonardo of Pisa, also called Fibonacci, who lived from 1180 to 1225 and who did some very early work in the field of infinite series.

Fibonacci numbers are defined as follows:

$$F_0 = F_1 = 1$$
$$F_n = F_{n-1} + F_{n-2}, \qquad n > 1 \tag{5-5}$$

The first few Fibonacci numbers are given in the following table:

n	0	1	2	3	4	5	6	7	8	9	10	11	12	13
F_n	1	1	2	3	5	8	13	21	34	55	89	144	233	377

Let us now see what role these Fibonacci numbers play in the development of this optimal search method.

Consider our initial interval [0, 1] which we shall designate L_1. Suppose we wish to maximize a unimodal function over this interval. L_1, then, is our initial interval of uncertainty.

Let

$$\Delta_2 = L_1 \frac{F_{n-2}}{F_n} \tag{5-6}$$

We now choose the two points at which to evaluate the function as

$$x_1 = 0 + \Delta_2 = \Delta_2$$
$$x_2 = 1 - \Delta_2$$

Because of the way the Fibonacci numbers are defined, i.e.,

$$F_n = F_{n-1} + F_{n-2}$$

Δ_2 is never more than $\frac{1}{2}$ the length of the original interval. Therefore, either $(0, \Delta_2)$ or $(1 - \Delta_2, 1)$ will be eliminated by examining $f(x_1)$ and $f(x_2)$. Hence, after one stage and two functional evaluations, we have a new interval $L_2 = L_1 - \Delta_2$. This yields

$$L_2 = L_1 - \Delta_2 = L_1 - L_1\left(\frac{F_{n-2}}{F_n}\right) = L_1\left(1 - \frac{F_{n-2}}{F_n}\right)$$

$$= L_1\left(\frac{F_n - F_{n-2}}{F_n}\right) = L_1\left(\frac{F_{n-1}}{F_n}\right) \tag{5-7}$$

Let the new end points of L_2 be designated a_2 and b_2 (a_1 and b_1 were 0 and 1 respectively). We now define the new distance Δ_3 similarly to Δ_2 in (5-6) as

$$\Delta_3 = L_2 \frac{F_{n-3}}{F_{n-1}} \tag{5-8}$$

The reason for the effectiveness of the Fibonacci search can now be seen, for it is easily demonstrated that the distance Δ_3 is actually the distance between x_1 and x_2 and therefore one of the preceding points at which the function was evaluated will be at one end or the other of L_2. We see this as follows:

$$x_2 - x_1 = 1 - \Delta_2 - \Delta_2 = L_1 - 2\Delta_2 = L_1 - 2L_1 \frac{F_{n-2}}{F_n}$$

$$= L_1\left[1 - 2\frac{F_{n-2}}{F_n}\right] = L_1\left[\frac{F_n - 2F_{n-2}}{F_n}\right]$$

$$= L_1\left[\frac{(F_n - F_{n-2}) - F_{n-2}}{F_n}\right] = \frac{L_1(F_{n-1} - F_{n-2})}{F_n} = \frac{L_1F_{n-3}}{F_n} \tag{5-9}$$

From (5-7) we know that

$$L_2 = L_1\left(\frac{F_{n-1}}{F_n}\right) \tag{5-7}$$

Combining (5-7) and (5-9) we have

$$x_2 - x_1 = \frac{L_2F_{n-3}}{F_{n-1}} = \Delta_3$$

which is what we wished to show. Therefore, one of the preceding functional evaluations, either $f(x_1)$ or $f(x_2)$, will be at one end of the new interval L_2 and thus we have to make only *one* new functional evaluation for the next stage. We place this new point x_3 at a distance Δ_3 from the end of the interval L_2 opposite the point we already have from the preceding trial. This is determined as follows:

If $f(x_1) \geq f(x_2)$ then $a_2 = a_1 = 0,\ b_2 = x_2$

$$x_3 = a_2 + \Delta_3$$

If $f(x_1) \leq f(x_2)$ then $a_2 = x_1,\ b_2 = b_1 = 1$

$$x_3 = b_2 - \Delta_3$$

We calculate the interval of uncertainty as we did previously. After this third functional evaluation it is now

$$L_3 = L_2 - \Delta_3 = L_2 - L_2\frac{F_{n-3}}{F_{n-1}} = L_2\left(1 - \frac{F_{n-3}}{F_{n-1}}\right)$$

$$= L_2\left(\frac{F_{n-1} - F_{n-3}}{F_{n-1}}\right) = L_2\left(\frac{F_{n-2}}{F_{n-1}}\right) = L_1\left(\frac{F_{n-1}}{F_n}\right)\left(\frac{F_{n-2}}{F_{n-1}}\right)$$

$$= L_1\left(\frac{F_{n-2}}{F_n}\right) \tag{5-10}$$

It is clear that we can continue this process and after n functional evaluations we will have

$$L_n = L_1\left(\frac{F_0}{F_n}\right) \tag{5-11}$$

From (5-11) it is a simple rearrangement to arrive at a measure of effectiveness of the Fibonacci search, i.e.,

$$\frac{L_n}{L_1} = \frac{F_0}{F_n} \tag{5-12}$$

Since $L_1 = 1$, we see that we can achieve an interval of uncertainty of L_n which is given by $F_0/F_n = \dfrac{1}{F_n}$. For example in order to reduce the interval of uncertainty to <1% we require 11 Fibonacci search steps, since $\dfrac{1}{F_{11}} = \dfrac{1}{144} < 1\%$.

EXAMPLE 5.3

Minimize $f(x) = x^2 - 6x + 2$ in the range $0 \le x \le 10$ to 3% of x.

Since $\dfrac{1}{F_8} = \dfrac{1}{34} < 3\%$, we start with F_8. and $L_1 = 10$, $a_1 = 0$, $b_1 = 10$

$$\Delta_2 = L_1 \frac{F_{n-2}}{F_n} = L_1 \frac{F_6}{F_8} = 10\left(\frac{13}{34}\right) = 3.82$$

$$x_1 = a_1 + \Delta_2 = 0 + 3.82 = 3.82$$

$$x_2 = b_1 - \Delta_2 = 10 - 3.82 = 6.18$$

$$f(x_1) = f(3.82) = -6.328$$

$$f(x_2) = f(6.18) = 3.112$$

We, therefore, discard the interval from 6.18 to 10. We now have

$$L_2 = 6.18,\ a_2 = 0,\ b_2 = 6.18$$

$$\Delta_3 = L_2 \frac{F_5}{F_7} = (6.18)\left(\frac{8}{21}\right) = 2.35$$

$$x_3 = 0 + 2.35 = 2.35$$

$$f(x_3) = -6.5775$$

Now we discard the interval (3.82, 6.18)

$$L_3 = 3.82,\ a_3 = 0,\ b_3 = 3.82$$

$$\Delta_4 = L_3 \frac{F_4}{F_6} = (3.82)\left(\frac{5}{13}\right) = 1.47$$

$$x_4 = 0 + 1.47 = 1.47$$

$$f(x_4) = -4.659$$

Now we discard the interval (0, 1.47)

$$L_4 = 2.35, a_4 = 1.47, b_4 = 3.82$$

$$\Delta_5 = L_4 \frac{F_3}{F_5} = (2.35)\left(\frac{3}{8}\right) = 0.881$$

$$x_5 = 3.82 - 0.881 = 2.94$$

$$f(x_5) = -6.996$$

Now we discard (1.47, 2.35)

$$L_5 = 1.47, a_5 = 2.35, b_5 = 3.82$$

$$\Delta_6 = L_5 \frac{F_2}{F_4} = (1.47)\left(\frac{2}{5}\right) = 0.588$$

$$x_6 = 3.82 - 0.59 = 3.23$$

$$f(x_6) = -6.947$$

Now we discard (3.23, 3.82)

$$L_6 = 0.88, a_6 = 2.35, b_6 = 3.23$$

$$\Delta_7 = L_6 \frac{F_1}{F_3} = (0.88)\left(\frac{1}{3}\right) = 0.293$$

$$x_7 = 2.35 + 0.29 = 2.64$$

$$f(x_7) = -6.870$$

Now we discard (2.35, 2.64)

$$L_7 = 0.59, a_7 = 2.64, b_7 = 3.23$$

$$\Delta_7 = L_7 \frac{F_0}{F_2} = (0.59)\left(\frac{1}{2}\right) = 0.295$$

$$x_8 = 3.23 - 0.295 = 2.94$$

$$f(x_8) = -6.9664$$

The exact solution is $x = 3, f(x) = -7$. We have determined that x lies in [2.94, 3.23] and $f(x)$ is ~ -6.996. We have indeed determined the location of x to less than 3% of x.

5.8 THE MULTIDIMENSIONAL SEARCH PROBLEM

In the case of one-dimensional optimization, we could be reasonably certain that by using either classical methods or search methods, an extreme value of $f(x)$ could be found. However, the problem of finding maxima or minima of a function of more than one

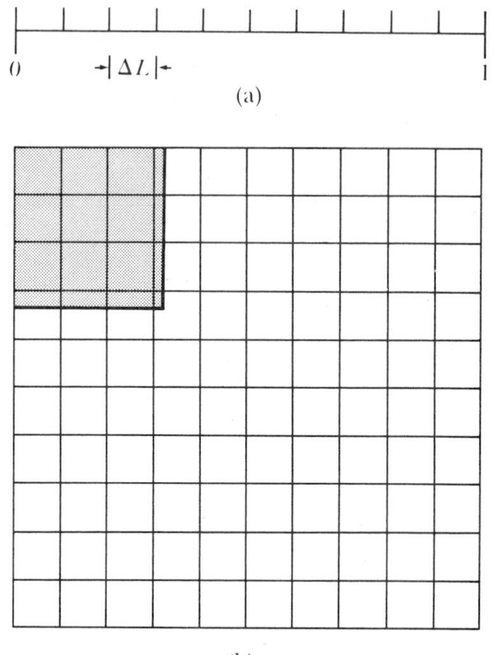

Figure 5-8

variable is a problem of extraordinarily greater difficulty than might appear at first glance. The interaction of variable effects on an objective function

$$f(\bar{x}) = f(x_1, x_2, \ldots, x_n)$$

is extremely difficult to anticipate. One author, Richard Bellman, has referred to the "curse of dimensionality," i.e., the difficulty of an n-dimensional optimization problem seems to increase in some hyperexponential fashion with the number of dimensions. These difficulties for search methods are, in some respects, more acute than for other methods. This can be intuitively seen if we consider how much "information" is gained by a functional evaluation as the number of dimensions of the problem (and therefore the space) is increased.

Consider first the unit interval [0, 1], and suppose as a result of some unidimensional search method, we are left with an interval of uncertainty, $\Delta L = 0.1$. In other words, of all possible 10 intervals, we have found one that contains our optimum. Therefore, we have narrowed the optimum to within $\frac{1}{10}$ of the original interval.

Now consider a unit square subdivided into ten intervals on a side, as was the unit interval. It is shown in Figure 5-8. The unit interval alone is also shown in Figure 5-8.

In Figure 5-8a we see an interval ΔL which is $\frac{1}{10}$ of the unit interval. In Figure 5-8b we see an *area* which is $\frac{1}{10}$ of the original area. Something rather strange and paradoxical seems to have happened. First, a point located within that area is not nearly as "precisely" located as was a point within ΔL in the one-dimensional case. Second, each side of the "square of uncertainty" is equal to $\sqrt{0.1} \simeq 0.316$. This means that in each dimension or variable there is an interval of uncertainty of 0.316, whereas in the one-dimensional case there was only an interval of 0.1. In order to achieve the equivalent interval to the one-dimensional case, we should require here an "area of uncertainty" of 0.01.

As the dimensionality increases, the situation becomes worse. In three-space, a volume of $\frac{1}{10}$ of the original volume comprises a significantly larger portion of the original volume, and is a cube equal to $\sqrt[3]{0.1}$ on a side which is 0.465 units in each variable. We see that the interval of uncertainty *in each variable* has increased more than fourfold over the one-dimensional case. To see how catastrophic this increase is, if we had a function of 100 variables, the n-dimensional cube containing 10 per cent of the volume would measure $(0.1)^{\frac{1}{100}}$ units on a side. $(0.1)^{\frac{1}{100}} \simeq 0.977$! The interval of uncertainty in each variable is now close to 1, yet this hypercube contains only 10 per cent of the total volume. The point is now very imprecisely located.

Given the characteristic discussed previously, or what might be termed the extreme "density" of hyperspace, or correspondingly, the extreme lack of precision for each independent variable, it is quite difficult to find economical multidimensional search methods. The difficulty, of course, resides in the fact that we must seek minute fractions, in terms of volume, of an n-dimensional space, if we hope to find an optimal value of a function $f(\bar{x})$. This gives rise to a great increase in the number of functional evaluations that must be employed or iterative stages that must be undertaken.

In what follows, we shall describe some of the more commonly used methods or other methods which the authors feel are useful.

5.9 SIMULTANEOUS METHODS—MULTIDIMENSIONAL SEARCH

As in the one dimensional case, the simplest (in concept) of the simultaneous search methods is exhaustive search. It is also the most inefficient of the multidimensional search methods. It is the n-dimensional analogue of the method presented in Section 5.3. What one does is simply to evaluate the objective function at all the nodes of the lattice, or grid of points, generated over the region of interest, when one suitably subdivides each variable. A more detailed description is as follows.

1. If each variable, x_j, lies in the interval, $[a_j, b_j]$, then we subdivide each of these intervals into $\frac{\Delta x_j}{2}$ equally spaced subintervals.
2. Evaluate the function $f(\bar{x})$ at each of the $\prod_{j=1}^{n} \frac{\Delta x_j}{2}$ grid points.
3. Select the optimal value of $f(\bar{x})$.

It would be hard to overestimate the inefficiency of this procedure. For example, in a problem with 10 variables, with each variable subdivided into 20 subintervals, we would have to carry out 10^{20} functional evaluations! Despite this characteristic, it has been often used in practice, where the number of variables is not large, say 3 or 4, and the function to be optimized is exceedingly intractable mathematically. However, for large numbers of variables and significant and numerous departures from unimodality, exhaustive search has serious, if not impossible, drawbacks.

The only other simultaneous search method that we shall describe is that of random search. This is the n-dimensional analogue of what is discussed in Section 5.3. Suppose we seek to maximize $f(\bar{x})$ over the unit interval [0, 1]. We would generate n numbers, r_n, uniform over the interval [0, 1] and evaluate the function at that point in E^n. This procedure is continued for some specified number of times. If we were maximizing, the largest value of $f(\bar{x})$ would be retained at each stage of the calculation.

As an example of this random search technique, consider the following problem:

EXAMPLE 5.4

Min $[f^2 + g^2]$ where

$$f(x_1, x_2) = x_1^2 x_2^2 - 2x_1^3 - 5x_2^3 + 10$$

$$g(x_1, x_2) = x_1^4 - 8x_2 + 1$$

or equivalently, find a root of

$$f(x_1, x_2) = 0$$

$$g(x_1, x_2) = 0$$

In Table 5.2, we show the results of this calculation.

Referring to Table 5.2, we see that 5000 pairs of random numbers were first generated between [0, 1] and [−1, 0]. The results show that x_1 is close to 1, suggesting that the upper bound of the interval could be limiting. In this way, if no notion of the

Table 5.2 *First Root*

Interval	Number of Sets of Random Numbers	Best Approximation
$0 \leq x_1, x_2 \leq 1$ $-1 \leq x_1, x_2 \leq 0$	5000	$x_1 = 0.98965$ $x_2 = 0.24510$
$0 \leq x_1 \leq 2$ $0 \leq x_2 \leq 1$	5000	$x_1 = 1.5954$ $x_2 = 0.93697$
$1.58 \leq x_1 \leq 1.6$ $0.92 \leq x_2 \leq 0.94$	50,000	$x_1 = 1.59593$ $x_2 = 0.93592$

location of the extreme value (or root) is known, this technique may lead one to the proper range. The next 5000 trials were generated with $0 \leq x_1 \leq 2$ and $0 \leq x_2 \leq 1$ and now we note that with the widened bounds on x_1, the value of x_1 seems not to be constrained by the upper bound on x_1. The subsequent calculation was performed to refine the values obtained. The approximate location of the root by other means was known to be $(x_1, x_2) = (1.59602, 0.93607)$.

It was known that this problem had a second root (or minimum). Table 5.3 shows the result of a similar calculation for this root. The approximate location of this root was known to be at $(-1.97352, 2.02117)$.

The particular random search technique described here has actually been used in practice for up to five-dimensional problems and probably could be used with somewhat larger dimensions. It is particularly useful as a technique for the approximate location of roots or optima, which can then be refined (if the derivatives exist) by one of the standard numerical techniques such as the Newton-Raphson method. See Isaacson and Keller.[14] More details concerning this random search method, and additional examples, can be found in Mettrey.[5] For a somewhat different approach to a random search method see Brooks.[6]

Table 5.3 *Second Root*

Interval	Number of Sets of Random Numbers	Best Approximation
$2 \leq x_1, x_2 \leq 10$ $-10 \leq x_1, x_2 \leq -2$	5000	$x_1 = -2.24603$ $x_2 = 2.28206$
$-2.3 \leq x_1 \leq -2$ $2 \leq x_2 \leq 2.3$	50,000	$x_1 = -2.0022$ $x_2 = 2.1338$
$-2 \leq x_1 \leq -1.9$ $2 \leq x_2 \leq 2.3$	50,000	$x_1 = -1.9739$ $x_2 = 2.02253$

5.10 MULTIDIMENSIONAL SEQUENTIAL METHODS

As is true in one-dimensional sequential methods, a characteristic of sequential methods is that the arguments for which $f(\bar{x})$ will be evaluated cannot be known in advance. Instead, as we have seen in Section 5.4, the sequence of argument values depends upon the previously observed values of $f(\bar{x})$. However, in one-dimensional search methods, although the arguments could not be known in advance, the number of function evaluations required to reach a specified interval of uncertainty could be calculated in advance. We made such calculations for each of the one-dimensional search methods. Multidimensional search methods do not allow any such simple calculation. It is not, in general, possible to know how many function evaluations will be required nor even whether an optimum point can be found. Certainly these methods will be effective for strictly unimodal functions. They may also be effective if they are confined to a region of interest in which the function is locally unimodal. Most of the practical methods to be described in subsequent sections of this chapter require that the function exhibit unimodality, at least locally. All of the methods are more efficient when the objective function is either strongly or linearly unimodal. The methods can be used with functions that are not unimodal but there are severe problems in dealing with local extrema and stationary points that are not either local maxima or minima. (See Chapter 4.)

5.11 MULTIVARIATE GRID SEARCH

This is a relatively inefficient but possibly useful method for functions of several variables as long as the dimensionality of $\bar{x}$ is not too great; for example, $n < 10$. The notion behind the search method is extremely simple. What is desired is to have some idea of how the function is responding simultaneously to changes in the variables in all dimensions about whatever point in the n-dimensional space we are. The algorithm is as follows:

Algorithm For Multivariate Grid Search

1. Place a grid of some preselected size, Δx_j, $j = 1, \ldots, n$ in each variable x_j, in the region of n-space, $a_j \leq x_j \leq b_j$, $j = 1, \ldots, n$ over which we seek to optimize $f(\bar{x})$.

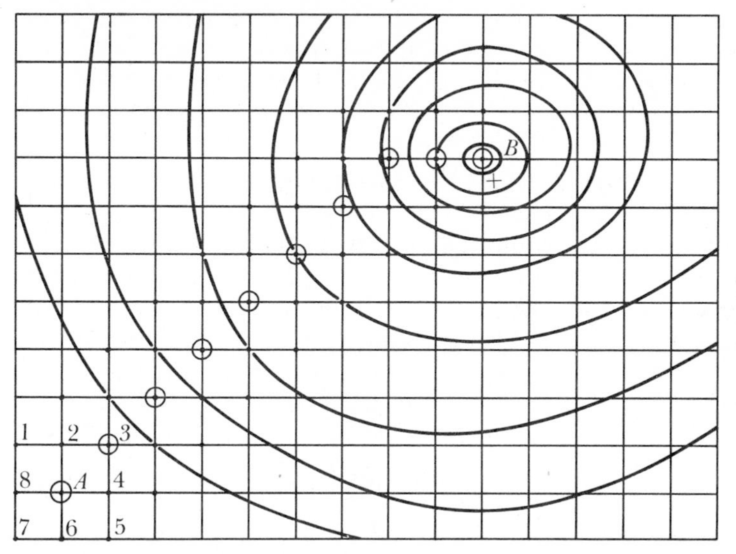

Figure 5-9 Multivariate grid search.

2. By some means select some node of this grid as the starting point.

3. Evaluate the function, $f(\bar{x})$ at the $3^n - 1$ surrounding points, where n is the dimensionality of $\bar{x}$.

4. Select the point with the greatest (least) functional value, $f(\bar{x})$. This point, $\hat{\bar{x}}$, becomes the new starting point.

5. Repeat Steps 3 and 4 until the central point yields the greatest (least) value of $f(\bar{x})$.

6. Reduce the grid size by halving or some other process and return to Step 3. This entire process of grid size reduction is continued until some preselected tolerance is reached.

It is a simple matter to illustrate this method in two dimensions. Figure 5-9 shows an example of how this search might proceed. The initial point for the search is labeled A. The $3^2 - 1$ points surrounding A are numbered 1–8. If the contours shown are in the direction of $f(\bar{x})$ increasing towards B, then if we are maximizing, we would choose point 3 as the new point to surround with grid points. Successive choices are shown in Figure 5-9 as circled points. The surrounding points are also shown. In the neighborhood of the point B, which is very close to the maximum, designated with the cross, we would have to resort to grid size reduction in order to get as close as desired to the maximal point.

The chief difficulty with the multivariate grid search method is its inefficiency. At each iteration of the calculation, as we have noted, one must compute $3^n - 1$ function evaluations. In 2 dimensions, $3^2 - 1 = 8$ function evaluations. In 3- space this amounts to

$3^3 - 1 = 26$. However, if we had a 10-dimensional problem, this would amount to $3^{10} - 1 = 59{,}048$ function evaluations at each iteration! One can readily see the inefficiency in terms of computational requirements.

5.12 UNIVARIATE SEARCH METHOD

Because of the computational load associated with changing all variables simultaneously, at least in the particular way indicated in the previous section, one might logically consider ways to reduce the total amount of computation. One simple way is merely to change one variable at a time, hence the name "univariate" search.

Usually, one will change the variables, one by one, in some sequence. For simplicity, let us assume that the variables are changed in their natural order, i.e., $x_1, x_2, \ldots, x_n$. (If this is not desired, they can always be renumbered). The guiding idea behind univariate search is to change one variable at a time so that the function is maximized (or minimized) in each of the co-ordinate directions. The algorithm can be described as follows:

Univariate Search Method

1. Start at some arbitrary starting point, $\bar{x}_0$, within the feasible space of solutions.

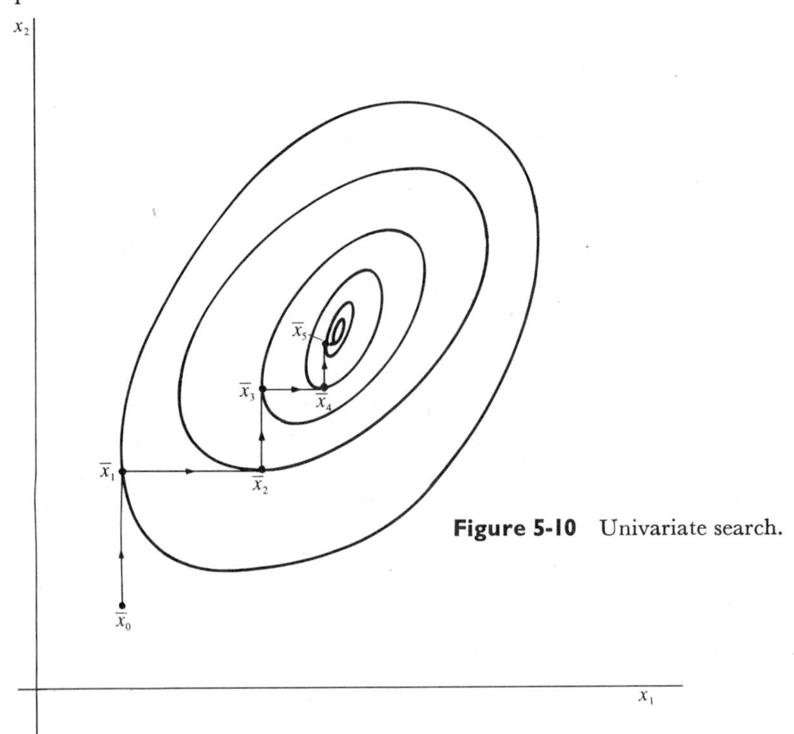

Figure 5-10 Univariate search.

2. We find $\bar{x}_1$, the next point, by performing a maximization with respect to the variable, x_1, i.e.,

$$\bar{x}_1 = \bar{x}_0 + \lambda_1 \bar{e}_1$$

where $\bar{e}_1 = [1, 0, \ldots, 0]$. λ_1 is a scalar such that $f(\bar{x}_0 + \lambda_1 \bar{e}_1)$ is maximized, i.e., we solve the problem $\max\limits_{\lambda_1} f(\bar{x}_0 + \lambda_1 \bar{e}_1)$.

3. The general step corresponding to 2 is as follows:

Find the point $\bar{x}_{k+1}$ by performing a maximization with respect to the variable, x_k, i.e.,

$$\bar{x}_{k+1} = \bar{x}_k + \lambda_{k+1} \bar{e}_{k+1}, \qquad k = 0, 1, \ldots, n-1$$

such that $f(\bar{x}_k + \lambda_{k+1} \bar{e}_{k+1})$ is maximized.

4. Steps 2 and 3 are completed in turn through consecutive iterations until the quantities, $|\lambda_k|$ are less than some tolerance value.

A simple illustration of this procedure in two dimensions is shown in Figure 5-10.

With reference to Figure 5-10, we start at the point $\bar{x}_0$. Next we increase the variable x_1 alone so that $f(\bar{x}_0 + \lambda_1 \bar{e}_1)$ is maximized. This takes us to the point $\bar{x}_1$. If $\bar{x}_0 = [x_{01}, x_{02}]$ and $\bar{x}_1 = [x_{11}, x_{12}]$ then $\lambda_1 = x_{11} - x_{01}$, i.e., the point $\bar{x}_1$ was chosen, or alternatively λ_1 was chosen so as to make $f(\bar{x}_1) = f(\bar{x}_0 + \lambda_1 \bar{e}_1)$ as large as possible. Next, we held x_1 constant and increased x_2 so as to arrive at $\bar{x}_2$, and so forth.

The problem whose contours are shown in Figure 5-10 is characterized by having little interaction among the variables. Hence, univariate search appears to be quite successful. However, the univariate search method can be very ineffective and in fact may fail for problems where there is significant interaction among the variables, e.g., where, geometrically speaking, there may be deep narrow valleys or narrow ridges in the contour representation of the function. An example of such a surface is shown in Figure 5-11. Therefore, the univariate search method can only be recommended for problems in which it is known that there are not strong interactions of the variables or for which a transformation of variables could be made to reduce the extent of the interaction. However, this knowledge of the extent of variable interaction may not often be available in advance of the solution of the problem.

Two numerical examples follow to illustrate the use of univariate search and also the problems associated with it.

EXAMPLE 5.5

Minimize

$$P = 3x_1^2 + 4x_2^2 - 5x_1 x_2 - 2x_1$$

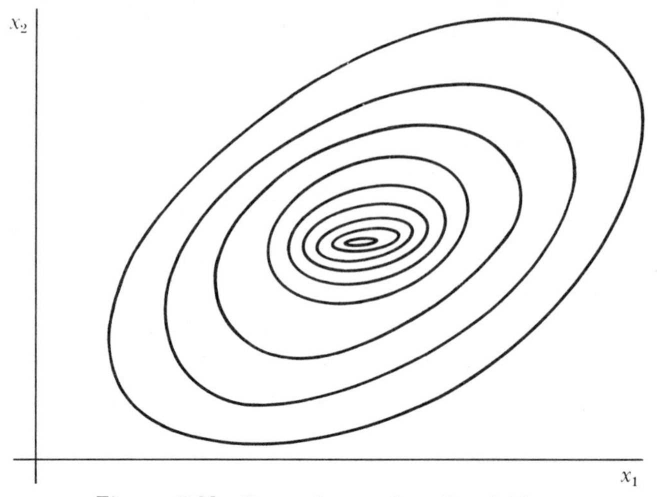

Figure 5-11 Strong interaction of variables.

Table 5.4 *Univariate Search*

POINT	FUNCTION MINIMIZED	VARIABLE FOUND	BEST CURRENT POINT
$\bar{x}_0 = (5,5)$	$P(x_1, 5)$	$x_1 = 4.5$	(4.5, 5)
$\bar{x}_1 = (4.5, 5)$	$P(4.5, x_2)$	$x_2 = 2.8125$	(4.5, 2.8125)
$\bar{x}_2 = (4.5, 2.8125)$	$P(x_1, 2.8125)$	$x_1 = 2.6771$	(2.6771, 2.8125)
$\bar{x}_3 = (2.6771, 2.8125)$	$P(2.6771, x_2)$	$x_2 = 1.6732$	(2.6771, 1.6732)
$\bar{x}_4 = (2.6771, 1.6732)$	$P(x_1, 1.6732)$	$x_1 = 1.7277$	(1.7277, 1.6732)
$\bar{x}_5 = (1.7277, 1.6732)$	$P(1.7277, x_2)$	$x_2 = 1.0798$	(1.7277, 1.0798)
$\bar{x}_6 = (1.7277, 1.0798)$	$P(x_1, 1.0798)$	$x_1 = 1.2332$	(1.2332, 1.0798)
$\bar{x}_7 = (1.2332, 1.0798)$	$P(1.2332, x_2)$	$x_2 = 0.7708$	(1.2332, 0.7708)
$\bar{x}_8 = (1.2332, 0.7708)$	$P(x_1, 0.7708)$	$x_1 = 0.9757$	(0.9757, 0.7708)
$\bar{x}_9 = (0.9757, 0.7708)$	$P(0.9757, x_2)$	$x_2 = 0.6098$	(0.9757, 0.6098)
$\bar{x}_{10} = (0.9757, 0.6098)$	$P(x_1, 0.6098)$	$x_1 = 0.8415$	(0.8415, 0.6098)
$\bar{x}_{11} = (0.8415, 0.6098)$	$P(0.8415, x_2)$	$x_2 = 0.5259$	(0.8415, 0.5259)
$\bar{x}_{12} = (0.8415, 0.5259)$	$P(x_1, 0.5259)$	$x_1 = 0.7716$	(0.7716, 0.5259)
$\bar{x}_{13} = (0.7716, 0.5259)$	$P(0.7716, x_2)$	$x_2 = 0.4822$	(0.7716, 0.4822)
$\bar{x}_{14} = (0.7716, 0.4822)$	$P(x_1, 0.4822)$	$x_1 = 0.7352$	(0.7352, 0.4822)
$\bar{x}_{15} = (0.7352, 0.4822)$	$P(0.7352, x_2)$	$x_2 = 0.4595$	(0.7352, 0.4595)
$\bar{x}_{16} = (0.7352, 0.4595)$	$P(x_1, 0.4595)$	$x_1 = 0.7162$	(0.7162, 0.4595)
$\bar{x}_{17} = (0.7162, 0.4595)$	$P(0.7162, x_2)$	$x_2 = 0.4476$	(0.7162, 0.4476)
$\bar{x}_{18} = (0.7162, 0.4476)$	$P(x_1, 0.4476)$	$x_1 = 0.7063$	(0.7063, 0.4476)
$\bar{x}_{19} = (0.7063, 0.4476)$	$P(0.7063, x_2)$	$x_2 = 0.4414$	(0.7063, 0.4414)
$\bar{x}_{20} = (0.7063, 0.4414)$	$P(x_1, 0.4414)$	$x_1 = 0.7012$	(0.7012, 0.4414)
$\bar{x}_{21} = (0.7012, 0.4414)$	$P(0.7012, x_2)$	$x_2 = 0.4382$	(0.7012, 0.4382)
$\bar{x}_{22} = (0.7012, 0.4382)$	$P(x_1, 0.4382)$	$x_1 = 0.6985$	(0.6985, 0.4382)
$\bar{x}_{23} = (0.6985, 0.4382)$	$P(0.6985, x_2)$	$x_2 = 0.4366$	(0.6985, 0.4366)

Suppose we start at $(x_1, x_2) = (5, 5)$. If we fix $x_2 = 5$ and solve the problem Min $P(x_1, 5)$, either by calculus or a one-dimensional search method, we find that Min $P(x_1, 5)$ occurs at $x_1 = 4.5$. Next, we minimize $P(4.5, x_2)$ with respect to x_2 and we find that the minimum occurs at $x_2 = 2.8125$. If we continue this in sequence, the following results are obtained. Note that at each stage only a one-dimensional optimization problem need be solved. Table 5.4 illustrates how slowly the method may converge in the neighborhood of the optimum. The correct co-ordinates of the minimum point to 4 decimal places are (0.6956, 0.4348).

An example of what may happen in a case where no maximum exists is given in Example 5.6.

EXAMPLE 5.6

Suppose we wish to maximize

$$P = 10x_1^2 - 5x_2^2 + 3x_1x_2 - 4x_1 + 2x_2 - 500$$

Any properly formulated physical problem should have a maximum, if we expect to find one. For example, x_1 might represent the selling price of a product and x_2 the advertising expenditures and P represents profit. However, errors of problem formulation might have led us to the foregoing expression. This function has neither a maximum nor a minimum as is easily shown by calculus.

$$\frac{\partial P}{\partial x_1} = 20x_1 + 3x_2 - 4$$

$$\frac{\partial P}{\partial x_2} = -10x_2 + 3x_1 + 2$$

$$\frac{\partial^2 P}{\partial x_1^2} = 20 \qquad \frac{\partial^2 P}{\partial x_2^2} = -10 \qquad \frac{\partial^2 P}{\partial x_1\,\partial x_2} = 3$$

$$\left(\frac{\partial^2 P}{\partial x_1^2}\right)\left(\frac{\partial^2 P}{\partial x_2^2}\right) - \left(\frac{\partial^2 P}{\partial x_1\,\partial x_2}\right)^2 = (20)(-10) - 9 < 0$$

Therefore, there is neither a maximum nor a minimum. A stationary point does exist, which is of no interest here. Let us now see what may happen if we attempt to use the univariate search method.

Let $\bar{x}_0 = (5, 5)$ and maximize $P(5, x_2)$. A search will show that $P(5, x_2)$ is maximized when $x_2 = 1.7$. Let us now attempt to maximize $P(x_1, 1.7)$. This has an unbounded maximum since

$$P(x_1, 1.7) = 10x_1^2 + 1.1x_1 - 505.1$$

It is clear that as x_1 increases $P(x_1, 1.7)$ becomes unbounded. Hence, we have learned that we have a function with no finite maximum.

Other difficulties may arise when trying to use the univariate search method. A function with a ridge with strong interaction of the variables may completely defeat this search technique. Hence, as previously mentioned, it can only be used where the amount of variable interaction is not great.

5.13 POWELL'S METHOD

The method described in the previous section had a simplicity which was appealing. It was a relatively simple matter to change one variable at a time and minimize (or maximize) the objective function with respect to one variable, i.e., in a single direction parallel to the coordinate planes. It also had the appealing characteristic of not requiring the calculation of derivatives. However, we noted that the method would not always converge to a solution in the case of strongly interacting variables. M. J. D. Powell has proposed a simple variation of the univariate search method which is vastly superior to it (Powell[7]). Powell's method has the characteristic that when it is applied to a quadratic form, it chooses *conjugate directions* (defined in the following paragraph) in which to move. Hence, the rate of convergence is fast when the method is used to minimize a general function. In addition, Powell has added a modification to the basic procedure to ensure a reasonable rate of convergence when the initial approximation is quite poor.

Let us first examine the notion of conjugate directions. Most of the newer unconstrained minimization methods use this notion in one form or another.

If a quadratic function to be minimized is

$$h(\bar{x}) = \bar{x}'A\bar{x} + \bar{b}'\bar{x} + c$$

then the directions $\bar{p}$ and $\bar{q}$ are defined to be conjugate directions if

$$\bar{p}'A\bar{q} = 0 \tag{5-13}$$

For a unique minimum to be defined it is necessary for the matrix A to be positive definite. There is no loss in generality if we also assume that A is symmetric.

The way in which Powell chooses his conjugate directions is as follows. Each iterative step begins with a search (one-dimensional) in n linearly independent directions, assuming $\bar{x} = (x_1, x_2, \ldots, x_n)$. Let us call these directions $\bar{r}_1, \bar{r}_2, \ldots, \bar{r}_n$ and let us assume we start at a point $\bar{x}_0$. Initially, these directions are chosen to be the co-ordinate directions, i.e.,

$$\bar{r}_1 = (x_1, 0, \ldots, 0), \quad \bar{r}_2 = (0, x_2, \ldots, 0), \ldots, \quad \bar{r}_n = (0, 0, \ldots, x_n)$$

Hence the start of the first iteration corresponds to an iteration of the univariate search method of the previous section, in which we changed one variable at a time. In Powell's method, we generate conjugate directions by making each iteration develop a new direction, $\bar{r}$ and using for the next iteration the linearly independent directions $\bar{r}_2, \bar{r}_3, \ldots, \bar{r}_n, \bar{r}$. The way in which $\bar{r}$ is defined guarantees that, if a positive definite quadratic function is being minimized, after n iterations all the directions are mutually conjugate. From this it can be proved that the minimum of the quadratic function has been found.

Even though we shall modify it later on, the following gives the details of the basic iteration procedure we have discussed somewhat vaguely. Assume we are minimizing $f(\bar{x})$.

1. For $s = 1, 2, \ldots, n$ determine λ_s so that $f(\bar{x}_{s-1} + \lambda_s\bar{r}_s)$ is minimized and define $\bar{x}_s = \bar{x}_{s-1} + \lambda_s\bar{r}_s$.
2. For $s = 1, 2, \ldots, n - 1$, replace $\bar{r}_s$ by $\bar{r}_{s+1}$.
3. Replace $\bar{r}_n$ by $(\bar{x}_n - \bar{x}_0)$.
4. Choose λ so that $f(\bar{x}_n + \lambda\{\bar{x}_n - \bar{x}_0\})$ is minimized and replace $\bar{x}_0$ by $\bar{x}_n + \lambda(\bar{x}_n - \bar{x}_0)$.

We shall now prove, following Powell, that this procedure will converge to the minimum of a quadratic function. We shall require two lemmas for the proof.

Lemma 5.1. If $\bar{q}_1, \bar{q}_2, \ldots, \bar{q}_m, m \leq n$, are mutually conjugate directions, then the minimum of the positive definite quadratic function,

$$h(\bar{x}) = \bar{x}'A\bar{x} + \bar{b}'\bar{x} + c$$

where $\bar{x}$ is any point in the m-dimensional space which also contains $\bar{x}_0$ and the directions $\bar{q}_1, \bar{q}_2, \ldots, \bar{q}_m$, may be found by searching along each of the directions once only.

Proof: The required minimum is the point

$$\bar{x}_0 + \sum_{i=1}^{m} \alpha_i\bar{q}_i$$

where the parameters α_i, $i = 1, 2, \ldots, m$ have been determined so as to minimize

$$h\left(\bar{x}_0 + \sum_{i=1}^{m} \alpha_i\bar{q}_i\right) = \sum_{i=1}^{m} [\alpha_i^2\bar{q}_i'A\bar{q}_i + \alpha_i\bar{q}_i'(2A\bar{x}_0 + \bar{b})] + h(\bar{x}_0) \qquad \textbf{(5-14)}$$

There are no terms in $\alpha_i\alpha_j$, $i \neq j$ because the $\bar{q}_i$ are conjugate directions. Hence we see that the result of searching along a direction $\bar{q}_i$ is to find α_i which minimizes

$$\alpha_i^2 \bar{q}_i' A \bar{q}_i + \alpha_i \bar{q}_i'(2A\bar{x}_0 + \bar{b}) \tag{5-15}$$

and therefore, the value of α_i obtained is independent of the terms in equation (5-14). Therefore, we have shown that searching in each of the directions $\bar{q}_i$ just once will find the absolute minimum of $h(\bar{x})$ in E^m.

Lemma 5.2. If $\bar{x}_0$ is the minimum in a space containing the direction $\bar{q}$, and $\bar{x}_1$ is also the minimum in such a space, then the direction $(\bar{x}_1 - \bar{x}_0)$ is conjugate to $\bar{q}$.

Proof: By definition, $\dfrac{\partial}{\partial \lambda} h(\bar{x}_0 + \lambda \bar{q}) = 0$ at $\lambda = 0$. Therefore, since

$$h(\bar{x}_0 + \lambda \bar{q}) = \lambda^2 \bar{q}' A \bar{q} + \lambda \bar{q}'(2A\bar{x}_0 + \bar{b}) + h(\bar{x}_0)$$

then

$$\frac{\partial}{\partial \lambda} h(\bar{x}_0 + \lambda \bar{q}) = 2\lambda \bar{q}' A \bar{q} + \bar{q}'(2A\bar{x}_0 + \bar{b}) = 0, \quad \lambda = 0 \tag{5-16}$$

Similarly

$$2\lambda \bar{q}' A \bar{q} + \bar{q}'(2A\bar{x}_1 + \bar{b}) = 0, \quad \lambda = 0 \tag{5-17}$$

By combining (5-16) and (5-17) we obtain

$$\bar{q}' A(\bar{x}_1 - \bar{x}_0) = 0$$

which proves the lemma.

We can now prove the convergence of a quadratic function in n iterations to a minimum. The proof is by induction. Therefore, it will be assumed that k iterations have been completed and that the directions $\bar{r}_{n-k+1}, \bar{r}_{n-k+2}, \ldots, \bar{r}_n$, defined for the $(k+1)^{\text{st}}$ iteration, are mutually conjugate. These were the last k directions of search. Therefore, by Lemma 5.1, the starting point for the $(k+1)^{\text{st}}$ iteration, $\bar{x}_0$, is the minimum in a space containing these directions. Applying Lemma 5.1 again, the point $\bar{x}_n$ is also the minimum in such a space. If we now apply Lemma 5.2, the new direction given by our procedure is conjugate to $\bar{r}_{n-k+1}, \bar{r}_{n-k+2}, \ldots, \bar{r}_n$. This proves the general step of the induction.

The point $\bar{x}_0$, which is defined for the second iteration, and also $\bar{x}_n$, are both minima in the direction $\bar{r}_n$. Therefore, the second iteration yields a pair of conjugate directions, thus allowing us to start the induction. After n iterations all the directions of search are mutually conjugate. Hence, by Lemma 5.1, the required minimum will have been found.

In actual practice, Powell has modified the foregoing basic procedure to be certain that the rate of convergence is reasonable, even if the initial approximation is a very poor one. A detailed justification of the modification is given in his paper, (Powell[7]). The modified algorithm is as follows. We assume that we are minimizing $f(\bar{x})$.

Powell's Method

1. For $s = 1, 2, \ldots, n$ calculate λ_s so that $f(\bar{x}_{s-1} + \lambda_s \bar{r}_s)$ is a minimum and define $\bar{x}_s = \bar{x}_{s-1} + \lambda_s \bar{r}_s$.

2. Find the integer m, $1 \leq m \leq n$ so that $f(\bar{x}_{m-1}) - f(\bar{x}_m)$ is a maximum and define $\Delta = f(\bar{x}_{m-1}) - f(\bar{x}_m)$.

3. Calculate $f_3 = f(2\bar{x}_n - \bar{x}_0)$ and define $f_1 = f(\bar{x}_0)$ and $f_2 = f(\bar{x}_n)$.

4. If either $f_3 \geq f_1$ or

$$(f_1 - 2f_2 + f_3)(f_1 - f_2 - \Delta)^2 \geq \tfrac{1}{2}\Delta(f_1 - f_3)^2$$

use the old directions $\bar{r}_1, \bar{r}_2, \ldots, \bar{r}_n$ for the next iteration and use $\bar{x}_n$ for the next $\bar{x}_0$.

5. If neither condition in Step 4 holds, define $\bar{r} = (\bar{x}_n - \bar{x}_0)$ and calculate λ so that $f(\bar{x}_n + \lambda\bar{r})$ is a minimum. Use $\bar{r}_1, \bar{r}_2, \ldots, \bar{r}_{m-1}, \bar{r}_{m+1}, \bar{r}_{m+2}, \ldots, \bar{r}_n, \bar{r}$ as the new directions and $\bar{x}_n + \lambda\bar{r}$ as the starting point for the next iteration.

At various stages of Powell's method, one must find the minimum of $f(\bar{x} + \lambda\bar{r})$ with respect to λ. This will be designated λ^*. The points $\bar{x}$ and $\bar{r}$ will be known and we must therefore solve a one-dimensional minimization problem. Powell has suggested the use of a quadratic defined by three function values. A simple quadratic interpolation procedure is as follows. We shall assume that we seek the minimum of $F(\lambda) = f(\bar{x} + \lambda\bar{r})$ for known $\bar{x}$ and $\bar{r}$. We shall designate the components of $\bar{r}$ by r_i.

Quadratic Interpolation for a Minimum

1. Calculate $M = \max_i |r_i|$. Normalize each component of r_i by dividing each component of $\bar{r}$ by M.

2. If $F(1) > F(0)$, calculate $F(\lambda)$ for $\lambda = \frac{1}{2}, \frac{1}{4}, \ldots$ until $F(\lambda) < F(0)$. Let $a = 0$, $b = \lambda$, $c = 2\lambda$ and go to Step 4.

3. Calculate $F(\lambda)$ for $\lambda = 0, 1, 2, 4, 8, \ldots, a, b, c$. Terminate the calculation when $\lambda = c$ when the current value of $F(\lambda)$ is greater than the previously computed value of $F(\lambda)$.

4. Calculate λ_m from

$$\lambda_m = \frac{1}{2}\frac{F(a)(c^2 - b^2) + F(b)(a^2 - c^2) + F(c)(b^2 - a^2)}{F(a)(c - b) + F(b)(a - c) + F(c)(b - a)}$$

5. If $F(\lambda_m) < F(b)$ we accept λ_m as the desired minimum value, λ^*. If $F(\lambda_m) \geq F(b)$ we accept b as the minimum value, λ^*.

For a detailed justification of this interpolation procedure, which is quite simple, the reader may refer to Powell.[7] It is also possible to use a cubic interpolation formula, if it is warranted. One is presented in an important paper by Davidon[8] which influenced Powell's method as well as several others.

5.14 METHOD OF STEEPEST DESCENT

The methods for maximization or minimization presented in the previous sections have not required that the functions to be optimized be differentiable. That was part of their appeal. If we now assume that $f(\bar{x})$ does possess first derivatives, at least, then a method is available to us, which is quite old and quite popular. Indeed, it is one of the oldest of the unconstrained optimization procedures. It dates back to Cauchy.[9] It has been "rediscovered" and represented in many papers since that time.

If $f(\bar{x})$ is assumed to be continuous and differentiable, then its first partial derivatives are known to exist. If we now recall the definition of the *gradient vector*, $\overline{\nabla} f(\bar{x})$ (see Section 4.8), we shall see that the gradient plays an important role in the seeking of local maxima and minima.

Given a function $f(\bar{x})$, whose first partial derivatives exist, it is known that the gradient $\overline{\nabla} f(\bar{x})$ is a vector pointing in the direction of the greatest rate of increase of $f(\bar{x})$. (See Chapter 4, Section 4.13.) At any given point, $\bar{x}_0$, the vector $\overline{\nabla} f(\bar{x}_0)$ is normal to the contour (whose value of $f(\bar{x})$ is constant) that passes through the point $\bar{x}_0$. The negative gradient points in a direction opposite to that of the gradient and hence in a direction of the greatest rate of decrease of $f(\bar{x})$.

Let us now see how information provided by the gradient can be utilized in the method of steepest descent.

Method of Steepest Descent

1. We start at some initial point, which we shall designate $\bar{x}_0$. This should be the best estimate of the minimum being sought that is available.

2. The general iteration step begins here. We designate this the sth iteration ($s = 0, 1, 2, 3, \ldots$). Calculate $\overline{\nabla} f(\bar{x}_s)$.

3. We shall now move in the direction, $-\overline{\nabla} f(\bar{x}_s)$. In order to do so, we need to calculate a step size, λ_s. In other words we wish

to choose λ so as to minimize $f(\bar{x}_s - \lambda \overline{\nabla} f(\bar{x}_s))$. (We can use the same quadratic interpolation procedure given in Section 5.13 or any other method.) Formally,

$$f(\bar{x}_s - \lambda_s \overline{\nabla} f(\bar{x}_s)) = \operatorname*{Min}_{\lambda} f(\bar{x}_s - \lambda \overline{\nabla} f(\bar{x}_s))$$

4. We now calculate the next point $\bar{x}_{s+1}$ as

$$\bar{x}_{s+1} = \bar{x}_s - \lambda_s \overline{\nabla} f(\bar{x}_s)$$

5. Terminate the calculation if $f(\bar{x}_s) - f(\bar{x}_{s+1}) \leq \varepsilon$, where ε is some preassigned tolerance. If $f(\bar{x}_s) - f(\bar{x}_{s+1}) > \varepsilon$, return to Step 2.

It is possible to use other criteria to terminate the calculation in Step 5 of this procedure given. A variety of such "stopping rules" have been suggested by various authors. Some others that have been proposed are

$$\operatorname*{Max}_{j} \left| \frac{\partial f(\bar{x}_s)}{\partial x_j} \right| < \varepsilon$$

or

$$\sum_{j=1}^{n} \left(\frac{\partial f(\bar{x}_s)}{\partial x_j} \right)^2 < \varepsilon$$

The authors prefer the rule that is given under Step 5. No matter what rule is used to terminate the calculation, one needs to select the tolerance ε, which is some preselected "small number." It is obvious that the smaller ε is, the more precisely will the location of the minimum be found. However, a correspondingly greater number of computational steps will be required to reach this tolerance. Only experience and some experimentation can guide the user who is faced with solving some particular problem.

Figure 5-12 shows graphically the possible succession of points that one would obtain in the application of the method of steepest descent (or ascent in the case of maximization). It is obvious that because of the amount of calculation required in multivariable optimization, this is most conveniently done on a digital computer. However, the course of a possible problem solution can be shown conveniently as in Figure 5-12. In the problem illustrated in Figure 5-12 we have plotted the contours of a function, $f(x_1, x_2)$ that we wish to minimize. If we start at the point designated $\bar{x}_0$ and move in the direction of the negative gradient, we will move in a direction perpendicular to a tangent to the contour of $f(x_1, x_2)$ at $\bar{x}_0$ and in a direction of decreasing f. We move in this direction until $f(x_1, x_2)$ can no longer be decreased. This takes us to $\bar{x}_1$. Any further movement along the line determined by $\bar{x}_0$ and $\bar{x}_1$ will *increase* f, which we do not wish to do. Now we redetermine the

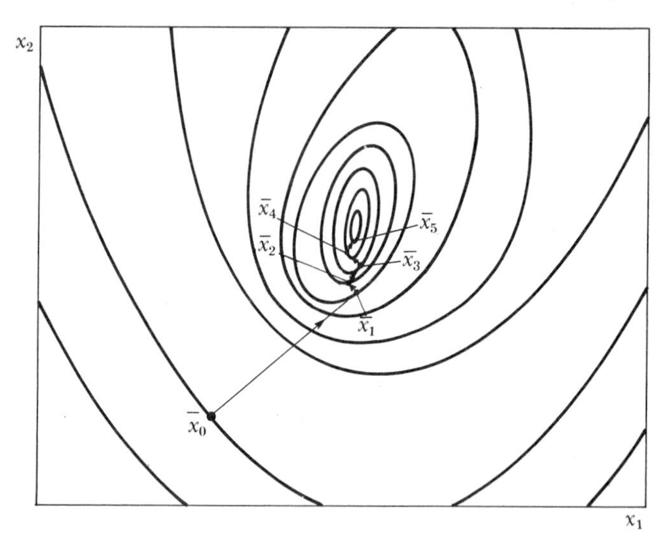

Figure 5-12 Method of steepest descent.

negative gradient or gradient at $\bar{x}_1$ and repeat the previous procedure. This takes us to $\bar{x}_2$. Continuing, we proceed to $\bar{x}_3$, $\bar{x}_4$, and so forth until we reach the minimum to whatever tolerance is deemed desirable.

There are a number of useful observations which can be made after a consideration of the details of the calculations involved in the method of steepest descent. They are perhaps even more readily perceived by examining Figure 5-12. First, it is obvious that successive steps or directions in which to move, which are generated by this algorithm, are orthogonal or perpendicular to each other. One of the implications of this fact is that, if the contours of the function being optimized are hyperspheres in n-dimensions, the optimum would be found in one step. As Figure 5-12 indicates, significant departures from such a configuration produces a kind of "zig-zag" approach to the optimum. A perhaps not so obvious characteristic of the method of steepest descent which this conclusion leads to is the following. If we have a problem, for example, Min $z_1 = f(\bar{x})$ and a transformation $\bar{x} = h(\bar{y})$ can be found such that the problem is now converted to Min $z_1 = f[h(\bar{y})] = g(\bar{y})$, where $g(\bar{y})$ now has hyperspherical contours, we can reduce the computation involved. Of course, such transformations cannot always be easily produced. In any case, it is clear that the more the contours of the function to be optimized depart from hypersphericity, the greater will be the number of computational steps required to approach the optimum.

5.15 THE FLETCHER-POWELL METHOD

The attentive and thoughtful reader may have noted that some of the difficulties or shortcomings of the method of steepest descent, while not identical to, are similar to those of the univariate search method. We recall that Powell's method was a response to and a tremendous improvement over that of the simple univariate search method. It will be recalled that Powell's method made use of conjugate directions. We will now do something very similar in an effort to improve the method of steepest descent. We introduced Powell's method earlier because it made no use of gradients and hence did not need to assume that the function to be optimized could be differentiated. Almost all of the newer and more efficient unconstrained optimization methods of the past ten years are dependent upon the use of conjugate directions, which we have already described in Section 5.13. The method of *conjugate gradients* for solving a set of simultaneous linear equations having a symmetric positive definite matrix of coefficients was first proposed by Hestenes and Stiefel in 1952.[10] Fletcher and Powell, in a paper in the British *Computer Journal*,[11] used these ideas along with some of the ideas of Davidon's very fruitful approach to develop an extremely successful method which is an enormous improvement over simple steepest descent procedures.

The Fletcher-Powell method is constructed in such a way that, if the function to be optimized is quadratic in n variables, then the Fletcher-Powell iteration scheme converges to the optimal solution in exactly n iterations. It also gives reasonably good results for functions which are not quadratic forms.

The Fletcher-Powell method is a gradient method, or more properly, a modified gradient method. Essentially what is involved is the use of certain information which is generated at each iteration to construct the Hessian matrix of the function. This is information which is not used by the ordinary steepest descent method. If one is actually dealing with a function which is quadratic and unimodal and for which the second partial derivatives exist, then the Hessian is constructed after n iterations (if there are n variables) of the Fletcher-Powell algorithm. When more complicated or different functions are being optimized, then the Fletcher-Powell method generates an approximation to the Hessian in the neighborhood of the optimum. The use of the Hessian, or an approximation to the Hessian, indicates that we are attempting to use second-order information or an approximation to second-order information. The method of steepest descent, on the other hand, uses only gradient or first-order information. It is not surprising, therefore, that the use of second-order approximations to the Hessian yields more accurate results.

In order to illustrate the basic idea of the computational procedure, let us consider a quadratic function of n variables, i.e.,

$$f(\bar{x}) = \bar{c}'\bar{x} + \bar{x}'D\bar{x} \quad \textbf{(5-18)}$$

where $\bar{c}$ and $\bar{x}$ are n-vectors and D is an n-square matrix. We shall assume further that f is a unimodal function. Therefore, for this case, $\bar{\nabla} f = \bar{0}$ at the optimal point, i.e.,

$$\bar{\nabla} f(\bar{x}) = \bar{c} + 2D\bar{x} = 0 \quad \textbf{(5-19)}$$

If we solve (5-19) for the optimal point, $\bar{x}^*$ we obtain

$$\bar{x}^* = -\tfrac{1}{2}D^{-1}\bar{c} \quad \textbf{(5-20)}$$

Unfortunately, in the general case we cannot perform this simple procedure because we usually do not have a simple quadratic form. Hence, in effect, we don't have enough information about $\bar{c}$ and D to make use of this technique. However, if we knew $\bar{\nabla} f$ at several points, we might be able to piece together "approximations" that would enable us to do, in effect, what we do in equation (5-20). If, in the case of maximization, we could also do this in such a way that the function increases at each iteration, we could then, hopefully, have a useful computational algorithm. This, in simplified language, is what the Fletcher-Powell algorithm attempts to do.

The basic procedure involves three general steps. They are as follows:

Step 1: Compute the gradient $\bar{\nabla} f(\bar{x}_s)$ at some point $\bar{x}_s$.

Step 2: Determine a direction, $\bar{r}_s$, along which to make a desired move.

Step 3: Make the move along this direction to some new point $\bar{x}_{s+1}$.

This general procedure is repeated until the gradient at some particular point becomes sufficiently small. Let us now examine this procedure in somewhat more detail. Step 1 requires no clarification. Let us then examine Steps 2 and 3. We express $\bar{r}_s$ as follows:

$$\bar{r}_s = -H_s \bar{\nabla} f(\bar{x}_s) \quad \textbf{(5-21)}$$

where H_s is some positive definite matrix if we are minimizing and a negasive definite matrix if we are maximizing. We shall assume that we are considering a *minimization* algorithm. The point $\bar{x}_{s+1}$ is determined from

$$\bar{x}_{s+1} = \bar{x}_s + \lambda_s \bar{r}_s \quad \textbf{(5-22)}$$

and λ_s is chosen from

$$f(\bar{x}_s + \lambda_s \bar{r}_s) = \underset{\lambda}{\mathrm{Min}}\, f[\bar{x}_s + \lambda \bar{r}_s] \quad \textbf{(5-23)}$$

It is obvious that this procedure bears a strong resemblance in many respects to the method of steepest descent. It can be readily seen that if $H_s = I$, the direction to move will be along the negative gradient as in the method of steepest descent. If we use an H_s which differs from the identity matrix, we are moving along a path different from the gradient (or negative gradient) path. It is for this reason that this general procedure is sometimes called a "deflected gradient" technique.

We now consider how Fletcher and Powell construct the sequence of matrices, H_s, $s = 1, 2, \ldots$. The following equation is used to generate the sequence:

$$H_{s+1} = H_s + A_s + B_s \tag{5-24}$$

If the iteration process is to converge in n steps then

$$H_n = \tfrac{1}{2}D^{-1} \tag{5-25}$$

Using equation (5-25) we can determine the role that the matrices A_s and B_s play in the Fletcher-Powell method. If we sum from $s = 0$ to $n - 1$ over equation (5-24) we obtain

$$\sum_{s=0}^{n-1} H_{s+1} = \sum_{s=0}^{n-1} H_s + \sum_{s=0}^{n-1} A_s + \sum_{s=0}^{n-1} B_s \tag{5-26}$$

Simplifying (5-26) yields

$$H_n - H_0 = \sum_{s=0}^{n-1} A_s + \sum_{s=0}^{n-1} B_s \tag{5-27}$$

Since we require from (5-25) that $H_n = \tfrac{1}{2}D^{-1}$, then from (5-27) we obtain

$$D^{-1} = 2H_0 + 2\sum_{s=0}^{n-1} A_s + 2\sum_{s=0}^{n-1} B_s \tag{5-28}$$

What we would like to be able to do is to construct the sequence B_s, so that as the calculation proceeds the effect of H_0, the initial matrix, is cancelled out, i.e.,

$$\sum_{s=0}^{n-1} B_s = -H_0 \tag{5-29}$$

If we do this we have as a result

$$\sum_{s=0}^{n-1} A_s = D^{-1} \tag{5-30}$$

Our initial estimate of D^{-1} is H_0. As can be seen from equations (5-29) and (5-30), as the number of iterative steps in the Fletcher-Powell process increases, we are attempting to improve our initial estimate of D^{-1} and approach a more accurate representation. This is the function of equation (5-24).

The way in which we approximate a second-order representation with only first-order information, which was mentioned earlier in this section, is by using a calculation of the gradient at two different points. It will be seen in the statement of the algorithm that we compute

$$\bar{y}_s = \bar{\nabla} f(\bar{x}_{s+1}) - \bar{\nabla} f(\bar{x}_s) \tag{5-31}$$

This is the basic equation for using gradient or first-order information at two points to approximate a second-order result. The vector $\bar{y}_s$ is then used to generate the matrices A_s and B_s by means of

$$\bar{\beta}_s = \lambda_s \bar{r}_s \tag{5-32}$$

$$A_s = \frac{\bar{\beta}_s \bar{\beta}_s'}{\bar{\beta}_s' \bar{y}_s} \tag{5-33}$$

$$B_s = \frac{-H_s \bar{y}_s \bar{y}_s' H_s}{\bar{y}_s' H_s \bar{y}_s} \tag{5-34}$$

The following is a step-by-step description of the Fletcher-Powell algorithm.

The Fletcher-Powell Method

1. Assuming a minimization algorithm, we start with a positive definite matrix, H_0 and some initial point, $\bar{x}_0$. For convenience, H_0 can be chosen to be the identity matrix, I.

2. The general iteration step begins here. We designate this the sth iteration ($s = 0, 1, 2, 3, \ldots$). Calculate the gradient vector, $\bar{\nabla} f(\bar{x}_s)$.

3. Calculate a direction in which to move. This is given by

$$\bar{r}_s = -H_s \bar{\nabla} f(\bar{x}_s)$$

4. In order to move in the direction $\bar{r}_s$, we need to calculate a step size, λ_s. Therefore, we wish to choose λ so as to minimize $f(\bar{x}_s + \lambda \bar{r}_s)$. This can make use of the quadratic interpolation method given in section 5.13 or any other method. Formally,

$$f(\bar{x}_s + \lambda_s \bar{r}_s) = \min_{\lambda} f(\bar{x}_s + \lambda \bar{r}_s)$$

5. Calculate: $\bar{\beta}_s = \lambda_s \bar{r}_s$

$$\bar{x}_{s+1} = \bar{x}_s + \bar{\beta}_s$$

6. Calculate:

$$\bar{y}_s = \bar{\nabla} f(\bar{x}_{s+1}) - \bar{\nabla} f(\bar{x}_s)$$

7. Calculate the two matrices A_s and B_s:

$$A_s = \frac{\bar{\beta}_s \bar{\beta}'_s}{\bar{\beta}'_s \bar{y}_s}$$

$$B_s = \frac{-H_s \bar{y}_s \bar{y}'_s H_s}{\bar{y}'_s H_s \bar{y}_s}$$

8. Calculate the next approximation in the sequence of H matrices:

$$H_{s+1} = H_s + A_s + B_s$$

9. Terminate the calculation if

$$f(\bar{x}_s) - f(\bar{x}_{s+1}) \leq \varepsilon$$

If $f(\bar{x}_s) - f(\bar{x}_{s+1}) > \varepsilon$, return to Step 2 using H_{s+1} as the new H_s.

The algorithm given can be shown to converge with no particular difficulty. One needs to assume that the function has a minimum (or maximum) on the interior of the region under consideration. Since the function decreases (or increases) monotonically at each iteration, this minimum will be reached. The reader interested in the details of this argument may consult Fletcher and Powell.[11]

5.16 PATTERN SEARCH

One of the most significant aspects of our present day ability to deal with optimization problems is the existence of digital computers. The use of computers certainly enables an analyst to solve very much larger and vastly more complex problems than by using time-honored classical techniques of optimization. However, the existence of our ability to perform enormous numbers of calculations in a very brief period of time has also led some people to suggest that methods of approach to optimization might be devised which are inherently "simpler" in concept than some of the classical methods and which are peculiarly suited to the simple-minded ability of a digital computer to carry out vast numbers of arithmetical operations fairly rapidly. Furthermore, the very simplicity of such an approach, if one can be devised, may have some significant advantages over some of the classical methods.

We will present one such method of a class called "direct search" methods. This particular method is called pattern search and is due to Hooke and Jeeves.[13] It has been widely used and has achieved some popularity. While we may emphasize some of the supposed advantages of direct search methods in this section, it should be clearly recognized that no one is in a position to state categorically that any one optimization method is clearly superior to all others for all conceivable classes of problems. It is probably the case that the problem solver will always desire to have a variety of methods at his disposal. Which one will be best suited to his particular problem will be determined by experience.

The description of direct search philosophy which follows and the description of the particular form of direct search which will be described in detail, viz., pattern search, is drawn from the paper of Hooke and Jeeves.

The general form of direct search is as follows. A point $\bar{x}_0$ is selected arbitrarily to be the first "base point." A second point $\bar{x}_1$ is selected by some method or strategy (see subsequent discussion) and is compared with $\bar{x}_0$. If $\bar{x}_1$ is "better" than $\bar{x}_0$, it becomes the new base point. If we are minimizing, for example, "better" means that $f(\bar{x}_1) < f(\bar{x}_0)$. If $\bar{x}_1$ is not better than $\bar{x}_0$, $\bar{x}_0$ is retained as the current base point. In this way, a monotone sequence of base points is generated. This process continues. As each new point is generated it is compared with the current base point. It is clear that a base point is simply the "best" value found, at any given stage of the search process. Some stopping rule must also be incorporated into the process.

The preceding description is sufficiently general to encompass almost any search method. Clearly, the strategy for selecting new candidate base points is of paramount importance. The particular strategy we shall incorporate into the preceding direct search method leads to the method known as "pattern search."

We shall describe pattern search as a method for minimizing a function, $f(\bar{x})$, where $\bar{x} = (x_1, x_2, \ldots, x_n)$ is an n-vector. What one does in any direct search algorithm is to vary the vector $\bar{x}$ until a minimum of $f(\bar{x})$ is obtained. Before defining how this is done in pattern search, a number of terms will be defined. The process of going from some given point to the next point is called a *move*. A move is considered to be a *success* if $f(\bar{x})$ is decreased; otherwise, it is considered to be a *failure*. In pattern search, we make two kinds of moves. The first type of move is called an exploratory move. The exploratory move consists of a restricted univariate search such that a move of very small length is taken in a single co-ordinate direction. Exploratory moves are made with respect to each component of $\bar{x}$. The information gained from the exploratory moves (success or failure) is then combined into a "pattern" which indicates a direction in which to move with a high probability of success. The pattern moves are larger in size than the exploratory moves. Pattern moves are made from base point to base point. After each successful

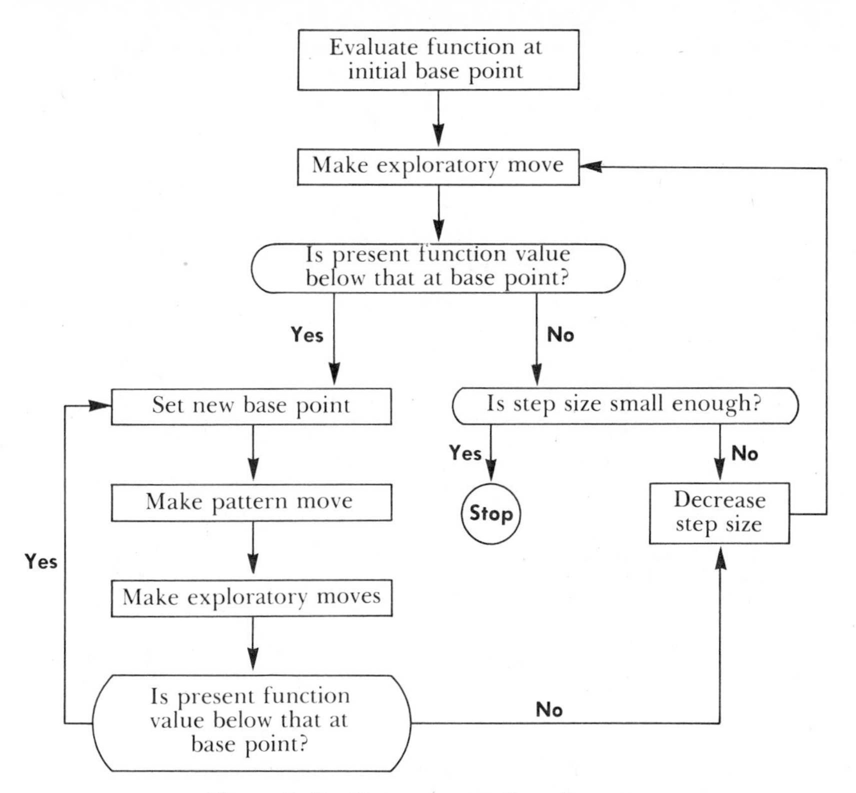

Figure 5-13 Pattern search flow diagram.

pattern move, further exploratory moves are made in an effort to improve the structure of the pattern.

From the general description given, it is clear that there are a large number of arbitrary combinations of details that need to be decided in order to fix a definite algorithm. The next algorithm given is the basic one due to Hooke and Jeeves. Its justification is its apparent success with a great many problems. One of the authors has used the pattern search algorithm on some fairly intractable problems arising in one particular area of technology with quite good results.

As we have discussed, the general approach of pattern search is to make a local "exploratory" search to seek out a promising direction and then to actually make a larger (global) "pattern" move. In the exploratory moves, a vector for a change is determined. In the pattern move, the change is actually made. Figure 5-13 presents the overall algorithm in terms of a flow diagram. Figure 5-14 gives the flow diagram for the exploratory moves. Together, Figures 5-13 and 5-14 define the pattern search method.

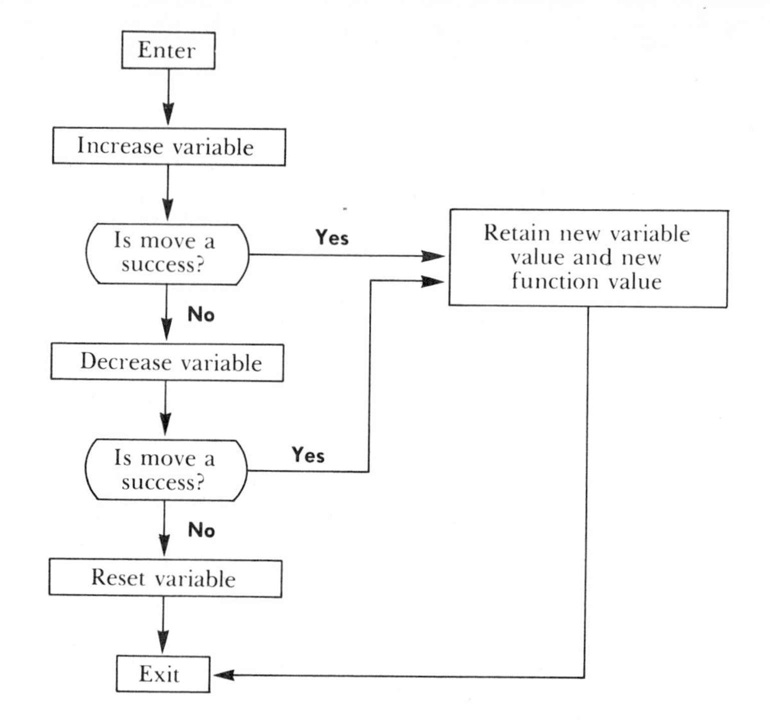

The above routine is performed for each variable, i.e., for each component of the vector $\bar{x} = (x_1, x_2, \ldots, x_n)$

Figure 5-14 Exploratory move flow diagram.

In Figure 5-13, the block which indicates "Make pattern move" simply changes the variable x_j by an amount Δx_j, i.e., $x_j + \Delta x_j$ becomes the new value of x_j. If the exploratory moves are successful, i.e., they result in a change in x_j by some amount δx_j, these increments are used to augment Δx_j. As an example of this process, suppose we were dealing with a function $z = f(x_1, x_2)$. If the exploratory moves were both successful then

$$x_1 + \delta x_1 \rightarrow x_1$$

$$x_2 + \delta x_2 \rightarrow x_2$$

At the same time, the increments Δx_1 and Δx_2 for the pattern moves would be modified in the same way:

$$\Delta x_1 + \delta x_1 \rightarrow \Delta x_1$$

$$\Delta x_2 + \delta x_2 \rightarrow \Delta x_2$$

Then a pattern move would be made:

$$x_1 + \Delta x_1 \to x_1$$
$$x_2 + \Delta x_2 \to x_2$$

Figure 5-15 indicates a typical pattern search for a function of two variables. The cross in Figure 5-15 indicates that a pattern move has been made which is not successful and therefore is rejected. The search continues from the previous base point and smaller pattern moves are made and a new pattern is built up as the search continues.

In Hooke and Jeeves[13] numerous examples of problem types and areas that pattern search has been applied to are given. One particular area that appears to be particularly amenable to this kind of treatment is that of nonlinear regression. In a nonlinear regression problem, one wishes to fit some data, say a set of observations, $\{x_k, y_k\}$ to a nonlinear form $y_k = g(x_k; a, b, c)$ where the function g involves the coefficients to be determined, i.e., a, b, c, in a nonlinear fashion. Such problems reduce to minimizing the sum of squares of the deviations of the residuals. This is a problem area where pattern search may be useful.

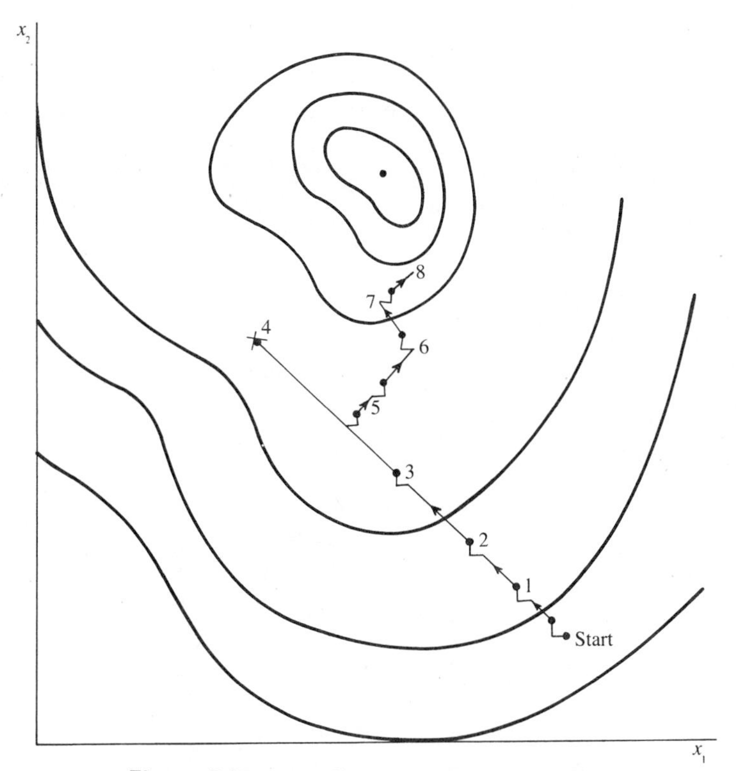

Figure 5-15 Two-dimensional pattern search.

EXERCISES

Exhaustive Search

1. Find the minimum of

$$f(x) = x^4 - 12x^3 + 15x^2 + 56x - 60$$

in the interval $0 \leq x \leq 15$ by means of exhaustive search.

Dichotomous Search

2. Find the minimum of $f(x) = 1 - 4x + 3x^2$ in the interval $0 \leq x \leq 1$ by means of dichotomous search.

Equal Interval Search

3. Solve the preceding problem using equal interval search.

4. Use equal interval search to find the maximum of

$$f(x) = x \sin x$$

in the interval [0, 10] where x is in radians.

5. Find the maximum value of $f(x) = Kxe^{-x^2}$ in the interval $0 \leq x \leq 2$, where K is a known constant, by means of Fibonaccian search.

Fibonacci Search

6. Solve exercise 2 using Fibonaccian search.

Grid Search

7. Consider the problem of locating a "center" or source with coordinates (x, y) in the Euclidean plane such that the sum of the distances from the source at (x, y) to a given set of points, (x_j, y_j), $j = 1, \ldots, n$ is to be minimized. This is formulated as follows:

$$\text{Min } z = \sum_{j=1}^{n} [(x_j - x)^2 + (y_j - y)^2]^{\frac{1}{2}}$$

We assume that the (x_j, y_j) are nonnegative and that (x, y) will therefore also be nonnegative. Find the point (x, y) which will minimize z by one of the multidimensional search methods. Specifically, solve the problem for the following set of data:

$$\begin{aligned}
(x_1, y_1) &= (5, 23) \\
(x_2, y_2) &= (9, 32) \\
(x_3, y_3) &= (15, 23) \\
(x_4, y_4) &= (21, 32) \\
(x_5, y_5) &= (26, 23) \\
(x_6, y_6) &= (31, 32) \\
(x_7, y_7) &= (16, 12)
\end{aligned}$$

(For a numerical method which is used, following an analysis of the problem based on classical optimization, which also can be used to solve this problem, see L. Cooper, "Location-Allocation Problems," *Operations Research 11*, 331–343 (1963).)

The Fletcher-Powell Method

8. Prove the convergence of the Fletcher-Powell method for a quadratic form. You may wish to consult Fletcher & Powell.[11]

9. For a certain chemical compound (ethyl formate) the following experimental data of vapor pressure (P) versus temperature (T) has been measured.

T °K	P, atm.
333.1	1.208
343.1	1.667
353.1	2.251
363.1	2.983
373.1	3.883
383.1	4.978
393.1	6.290
403.1	7.846
413.1	9.674

From certain theoretical considerations, it is considered likely that this data can be described or fitted to an equation of the form

$$P = e^{A - \frac{B}{T+C}}$$

Therefore, if we consider the data as being pairs of numbers, (T_j, P_j) we wish to minimize the sum of squares of the residuals in the curve fitting or we wish to find values of the three constants A, B and C such that

$$z = \sum_{j=1}^{n} \left(P_j - e^{A - \frac{B}{T_j + C}}\right)^2$$

is minimized. Attempt to solve this problem by Powell's method.

10. A function devised by Rosenbrock to test optimization methods is the following:

$$\text{Min } z = 100(x_2 - x_1^2)^2 + (1 - x_1)^2$$

The minimum is clearly at (1, 1) with $z = 0$. Attempt to solve this with some arbitrary starting point by Powell's method. This problem illustrates the utility of transformations in the variables, if they can be made. If we let $y_1 = x_1 - x_2^2$ and $y_2 = 1 - x_1$, you can investigate the effect on the method of steepest descent.

The Method of Steepest Descent

11. Solve exercise 9 by the method of steepest descent.

12. Solve exercise 10 by the method of steepest descent.

Pattern Search

13. Solve exercise 9 by pattern search.

14. Solve exercise 10 by pattern search.

General

15. Use one of the multivariate search methods to minimize:

$$z = 50(x_1 - x_2^3)^2 + (1 - x_2)^2$$

16. In the chapter on classical optimization we discussed the use of Lagrange multipliers for certain constrained problems of the form

$$\text{Min } z = f(\bar{x})$$

$$\text{Subject to: } g(\bar{x}) = b$$

in which one attempts to solve this problem, under certain conditions, by solving instead

$$\text{Min } F = f(\bar{x}) + \lambda[b - g(\bar{x})]$$

Can any of the multidimensional search methods be used to find $\bar{x}$ and λ which would minimize F?

17. How would you use one of the multivariate search methods to solve the following problem:

$$\text{Max } z = 3x_1e^{x_1} + 5x_2^3$$

$$\text{Subject to: } 2x_1 + x_2^2 = 150$$

18. How would you find the solution to the following simultaneous equations in terms of any of the methods discussed in this chapter?

$$3x_1 + 5x_2 = 21$$

$$x_1^2 + 2x_2^2 = 22$$

(Hint: First formulate an optimization problem which is to be solved.)

19. Describe how you would solve and then actually solve the following problem by one of the multivariate search methods:

$$\text{Max } z = 3x_1 + 4x_2$$

$$2x_1 + 5x_2 \leq 25$$

$$3x_1 + 4x_2 \geq 5$$

$$x_1, x_2 \geq 0$$

Are there any special problems involved?

20. The total cost of ordering and holding in inventory 15,000 units of a certain item over a certain period of time is

$$C_T = \frac{225{,}000}{x} + \frac{7x}{200} + 6000$$

What value of x, the number of items to order each time an order is placed, will make C_T a minimum if it is specified that $x \geq 3000$?

REFERENCES

1. Kiefer, J.: Optimum sequential search and approximation methods under minimum regularity assumptions. *J. Soc. Ind. & Appl. Math.*, *5*, No. 3 (1959).
2. Kiefer, J.: Sequential minimax search for a maximum. *Proc. Am. Math. Soc.*, *4* (1953).
3. Johnson, S. M.: Optimal Search for a Maximum is Fibonaccian. RAND Corp, Report P-856 (1956).

4. Bellman, R.: *Dynamic Programming*. Princeton University Press, Princeton (1957).
5. Mettrey, W.: Investigation of Monte Carlo solutions of nonlinear equations. M.S. thesis, Washington University, St. Louis, (1964).
6. Brooks, S. H.: A discussion of random methods for seeking maxima. *Oper. Res. 6* (1958).
7. Powell, M. J. D.: An efficient method for finding the minimum of a function of several variables without calculating derivatives. *Computer J.*, *7*, No. 2 (1964).
8. Davidon, W. C.: Variable Metric Method for Minimization. AEC Research and Development Report ANL-5990 (Rev.) (1959).
9. Cauchy, A. L.: Méthode générale pour la resolution de systémes d'equations simultanées. *Comptes Rendus*, *Ac. Sci.*, Paris, *25* (1847).
10. Hestenes, M. R., and Stiefel, E.: Methods of conjugate gradients for solving linear systems. *J. Res. Nat. Bur. Standards 49* (1952).
11. Fletcher, R., and Powell, M. J. D.: A rapidly convergent descent method for minimization. *Computer J.*, *6* (1963).
12. Fletcher, R., and Reeves, C. M.: Function minimization by conjugate gradients. *Computer J.*, *7* (1964).
13. Hooke, R., and Jeeves, T. A.: Direct search solution of numerical and statistical problems. *J. of Assn. Computing Machinery*, *8* (1961).
14. Isaacson, E., and Keller, H. B.: *Analysis of Numerical Methods*. John Wiley & Sons, New York (1966).

Chapter 6

LINEAR PROGRAMMING

6.1 INTRODUCTION

By far, the optimization technique in widest and most general use is known as linear programming. Evidence for its widespread use is shown very clearly by the fact that major digital computer manufacturers include computer codes for solving linear programming problems as a major inducement to purchase or rent some particular computer. The implication is usually that their particular computer code is vastly superior to others. This does attest to the importance of this area of optimization, even if the claims of the computer manufacturers are often inaccurate.

A linear programming problem is easily defined. It is the following:

$$\text{Max } z = \sum_{j=1}^{n} c_j x_j$$

Subject to

$$\sum_{j=1}^{n} a_{ij} x_j = b_i \qquad i = 1, 2, \ldots, m \tag{6-1}$$

$$x_j \geq 0 \qquad j = 1, 2, \ldots, n$$

How then does it differ from the kinds of general optimization problems, solution techniques for which were discussed in previous chapters? Previously, we considered problems of the general form:

$$\text{Max } z = f(\bar{x})$$

$$g_i(\bar{x}) = b_i \qquad i = 1, 2, \ldots, m \tag{6-2}$$

where only mild restrictions such as continuity or differentiability were placed on $f(\bar{x})$ and $g_i(\bar{x})$. Now, in (6-1) we have specified the *form* of f and the g_i once and for all. The coefficients, c_j, a_{ij}, and b_i are real scalars and may vary, but the form of the problem is fixed.

As the word "linear" in the name "linear programming" suggests, it can be seen in (6-1) that the objective function

$$f(\bar{x}) = \sum_{j=1}^{n} c_j x_j$$

is a *linear form* in the variables x_j, whose optimal values are to be determined. Similarly, the constraints in (6-1) are *linear* equations. Actually, the constraints of the physical problem that we are describing in any specific case, may be linear inequalities. However, as we shall see later, we will convert them to linear equalities before developing a solution method, or indeed, solving a problem. It should be borne in mind, however, that the "natural" description of a linear programming problem is frequently stated in terms of linear inequalities rather than equalities.

Let us now consider an example of a linear programming problem. Then we shall consider what are the implications of the restriction to linear relationships among the variables.

EXAMPLE

Suppose a manufacturer produces four different products which we will call A, B, C, and D. Suppose further that he has fixed amounts of three different resources, such as labor, raw materials, equipment, and the like, which can be combined in certain fixed and known ways to produce the four products. In short, he knows how much of each resource must be used to produce one unit of each product. Since he knows the variable cost of each of these resources, we assume further that he knows how much profit he makes for each unit of each product he sells. Suppose the actual numbers are as in Table 6.1 below. In the absence of any other physical or business constraints, the question the manufacturer wants to answer is how

Table 6.1 *Linear Programming Example*

	Product				Limitation
Resource	*A*	*B*	*C*	*D*	on Resource
1	5	2	3	1	300
2	1	2	1	2	200
3	1	0	1	0	100
Unit Profit	6	4	2	1	

much of each of the four products should he manufacture in order to maximize his profit? Even for a problem of this exceedingly small size (4 product variables and 3 resource constraints) this is not a trivial problem. The interdependence of the decisions required in order to use the resources "wisely" makes a trial and error or intuitive procedure (such as "making the most of the product with the highest unit profit") invalid in terms of profit maximization. Let us now formulate the problem in terms of a linear programming model.

Let x_j, $j = 1, 2, 3, 4$ represent the amounts of products A, B, C, and D that are to be manufactured. These are the numbers we wish to determine. Since we wish to maximize profit and the unit profits are given in the last line of Table 6.1, our objective function is

$$z = 6x_1 + 4x_2 + 2x_3 + x_4 \tag{6-3}$$

It is this quantity that we wish to maximize. However, there are certain restrictions on just how large x_1, x_2, x_3, and x_4 can be. Otherwise, to maximize z we would make them all as large as we please. The restrictions come about because of the limited amount of each of the resources that are available to us. For example, there are only 300 units of resource 1. For every unit of product A made we use 5 units of resource 1. For every unit of product B made we use 2 units of resource 1, and so forth. Hence, the quantities x_1, x_2, x_3, and x_4 will be limited by an inequality that states the total restriction on the use of resource 1. It is as follows:

$$5x_1 + 2x_2 + 3x_3 + x_4 \leq 300 \tag{6-4}$$

The relationship of the variables in (6-4) is written in the form of an inequality because it is possible that it may be more profitable, under certain circumstances, not to use all of resource 1. We allow for this possibility by using the inequality form. Similarly, the other constraints which relate to resources 2 and 3 are as follows:

$$\begin{aligned} x_1 + 2x_2 + x_3 + 2x_4 &\leq 200 \\ x_1 + x_3 &\leq 100 \end{aligned} \tag{6-5}$$

We have one further set of restrictions on the x_j, $j = 1, 2, 3, 4$, namely that they not be allowed to be negative numbers, i.e., they must be either zero or positive. Therefore we have

$$x_j \geq 0 \qquad j = 1, 2, 3, 4 \tag{6-6}$$

If we now put (6-3), (6-4), (6-5), and (6-6) together we have

$$\text{Max } z = 6x_1 + 4x_2 + 2x_3 + x_4$$

Subject to

$$\begin{aligned} 5x_1 + 2x_2 + 3x_3 + x_4 &\leq 300 \\ x_1 + 2x_2 + x_3 + 2x_4 &\leq 200 \\ x_1 + x_3 &\leq 100 \\ x_j \geq 0, \qquad j &= 1, 2, 3, 4 \end{aligned} \tag{6-7}$$

The solution to (6-7) will tell us how much of each product we shall make in order to maximize the profit of the operation.

The problem stated in (6-7) will serve to illustrate as well as define the characteristic "properties" of a linear programming problem. In a certain sense, this is trivial. We could easily dismiss the entire matter by merely saying that a linear programming problem is one in which the constraints are linear equations (or inequalities) and the objective function is a linear form. This is certainly an accurate mathematical description of (6-7) or the general formulation, (6-1). However, it does not tell us much about the implications of this restriction to linear equations and forms. Let us now consider this.

In the problem which is represented by equations (6-7) we know, for example, that it requires two units of resource 1 to make one unit of product B. The *assumption of linearity* is that it will then require four units of resource 1 to make two units of product B and, in general, it will require $2y$ units of resource 1 to make y units of product B. In other words, there is a *proportionality* or a simple multiplicative relationship between the units of resource requirements and the number of units of product B produced. This is an important property of linearity from the practical point of view. If, for example, it was the case for some manufacturing operation that 60 units of a given resource were required to make 30 units of some product but only 100 units of the resource were required to make 60 units of this product, then the proportionality assumption would not hold.†

If we refer again to the problem represented by equations (6-7), we may note that, for example, the amount of profit we obtain from selling one unit of product A is six profit units (whatever they may be) and that the amount of profit we would obtain from selling one unit of product B is four units of profit. The *assumption of linearity* is that the total profit from the two products is then 10 units. This is the *additivity* property of linear relationships and it states that returns or quantities or amounts or levels of activities, and the like, are additive in their effects. The additive property is obviously involved in the constraints of a linear programming model as well.

† Methods of dealing with such situations are given in Chapter 8.

These two properties, proportionality and additivity, are what essentially define a linear relationship. In order to use linear programming methodology, at least directly, the problem under study must satisfy these requirements. It is clear that many practical situations of interest may involve significant departures from linearity. For example, in many chemical and biological processes, exponential growth or decay laws hold. A well known example is the exponential decay law of radioactive atoms, according to which, if N_0 is the number of atoms present at time zero and N represents the number present at some subsequent time, t, i.e., $N = N(t)$, then N is given by

$$N = N_0 e^{-\lambda t} \tag{6-8}$$

where λ is a "decay constant" characteristic of the particular atomic species. We can see clearly that N is a *nonlinear* function of the variable t. If, however, in this case, it was necessary to include a relationship such as (6-8) in a linear programming model, we might proceed as follows. Let

$$v = \ln \frac{N}{N_0} \tag{6-9}$$

and now we can write

$$v + \lambda t = 0 \tag{6-10}$$

which is a linear equation. This kind of "trick" of using a transformation of variables can sometimes be used but not always. Nonlinearity is not so easily disposed of in general. For example, there is very little one could do with

$$x_1 \sin x_1 + \ln x_2 + x_2 x_3 = 30$$

in the usual context of a linear programming model.

6.2 GEOMETRIC INTERPRETATION OF LINEAR PROGRAMMING

In subsequent sections we shall discuss the theory of the "simplex" method of linear programming. It is this theory which leads to a computationally effective method for solving linear programming problems. Before we begin this, it is useful to consider some simple geometric representations of linear programming problems. These representations lead to some valuable insights concerning the solution to a linear programming problem which we shall verify and prove later. It should be borne in mind that while one shall easily obtain the solution to linear programming problems with two variables and perhaps be able to do so for three variables, beyond this number of variables the geometric approach would fail us completely.

Let us consider the following problem:

$$\text{Max } z = 2x_1 - 4x_2$$

Subject to

$$\begin{aligned} 3x_1 + 5x_2 &\geq 15 \\ 4x_1 + 9x_2 &\leq 36 \\ x_1, x_2 &\geq 0 \end{aligned} \tag{6-11}$$

Since, in (6-11), x_1, x_2 are restricted to be nonnegative we know we can confine our geometric representation of this problem to the nonnegative quadrant of two-dimensional Euclidean space. This is what we have done in Figure 6-1. An arrow pointing upward from the x_1 axis and one to the right from the x_2 axis can be seen, indicating that we are confined to the nonnegative quadrant of the x_1, x_2 plane. Let us now consider the constraints. If we first look at

$$3x_1 + 5x_2 \geq 15$$

we know from Chapter 3 that the equation $3x_1 + 5x_2 = 15$ will bound the allowable region in which (x_1, x_2) values may lie and that this region will be on one side or the other of this line. A simple way to test is to consider the origin, $(x_1, x_2) = (0, 0)$. It is clear that $3(0) + 5(0) = 0$ is not greater than or equal to 15. Hence the allowable region must be above the line $3x_1 + 5x_2 = 15$.

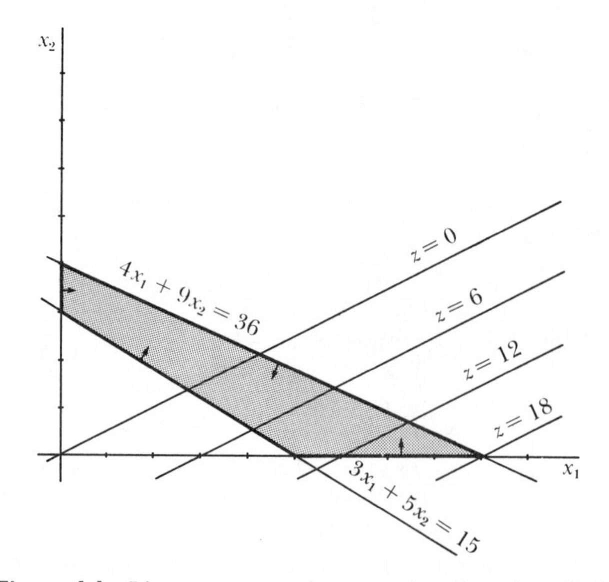

Figure 6-1 Linear programming example—Equations (6-11).

Accordingly, an arrow points upward from this line in Figure 6-1. A similar analysis will show that with respect to the inequality, $4x_1 + 9x_2 \leq 36$, the allowable region must lie below the line $4x_1 + 9x_2 = 36$. Hence, an arrow pointing downward is shown in Figure 6-1. The intersection of these two constraint lines and the two co-ordinate axes, $x_1 = 0$, $x_2 = 0$ gives rise to the shaded polygon which is shown in Figure 6-1. Any point inside this polygon or on its bounding lines is a feasible (allowable) solution to the problem given by the constraint equations of (6-11).

Having delineated the region in which the feasible solutions to our problem may lie, we must now consider the question of which of these solutions is the optimal solution. To answer this question we must examine the objective function, $z = 2x_1 - 4x_2$. For some particular value of z we can plot a contour line, i.e., a line along which the particular combinations of (x_1, x_2) values give this designated value of the objective function. A series of such constant z lines are shown in Figure 6-1. It can be seen that the direction of increasing z is toward the bottom of the shaded polygon of feasible solutions. It can also then be easily seen that the largest value of z corresponds to a single point. That is the point generated by the intersection of the x_1 axis $(x_2 = 0)$ and the constraint boundary, $4x_1 + 9x_2 = 36$. At this point $x_1 = 9$, $x_2 = 0$, $z = 18$. If we try to move into the polygon, we will decrease z. If we try to increase z, it will not be feasible, i.e., satisfy the constraints. Hence, the point $(x_1, x_2) = (9, 0)$ is the optimal solution and $z = 18$ is the maximum value of the objective function.

What is perhaps most remarkable about what we have observed in the solution of the problem depicted in Figure 6-1, is that all the salient features we have implicitly noted generalize to any number of variables and any number of constraints. Let us now explicitly state what they are.

1. The intersection of the quadrant boundaries and the constraint boundaries generated a convex polygon. This is also true in the n-dimensional case as was noted in Chapter 3 on convexity and n-dimensional geometry.

2. A solution to our linear programming problem was at a "corner" or extreme point of the polygon and not in the interior of the polygon of feasible solutions. This is also true in the general case of a convex polyhedron in an n-dimensional space. The solution is at an extreme point of the convex polyhedron. Hence, only these points need be examined for the optimal solution.

What we have noted here is stated in geometric language. What we shall do in our later discussion of the simplex method is prove these statements in the context of an algebraic formulation of the problem. It is not possible in the general case to use geometric methods to arrive at a solution as we did in the simple two-dimensional example given.

Before we leave the use of simple geometric examples, we shall examine a few more to illustrate points which will be dealt with later in the development of the simplex method.

In the example of equations (6-11) shown in Figure 6-1, there was a unique optimum solution. This, however, need not be the case. Consider the following problem:

$$\begin{aligned} \text{Max } z &= 2x_1 + 0.5x_2 \\ 6x_1 + 5x_2 &\leq 30 \\ 4x_1 + x_2 &\leq 12 \\ x_1, x_2 &\geq 0 \end{aligned} \tag{6-12}$$

Equations (6-12) are shown in Figure 6-2. It can be seen from arguments similar to those given previously, that the optimal value of the objective function is $z = 6$. It can also be seen that either of the extreme points designated A and B are optimal solutions. However, it is equally clear that any point on the line segment $\overline{AB}$ is also an optimal solution since it lies on the line $z = 6$. Hence, there are an infinite number of optimal solutions to the problem given by (6-12). In linear programming, when such a situation exists, we say that there are alternative optimal solutions, or simply alternative optima. The simplex algorithm finds a solution such as the one designated A or B, i.e., extreme point solutions. The algorithm can be used to find all such solutions, if it is necessary to do so.

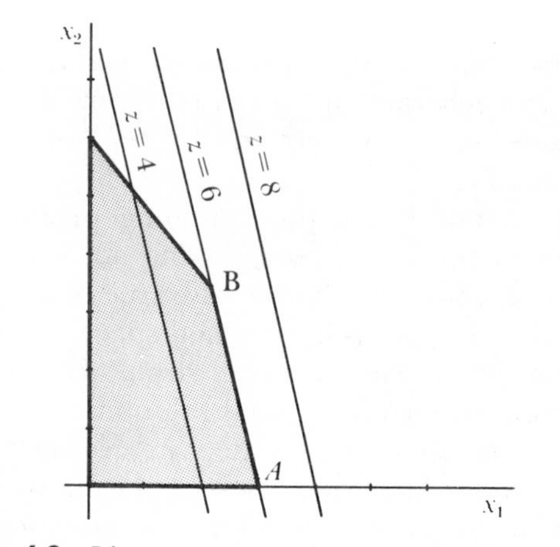

Figure 6-2 Linear programming example—Equations (6-12).

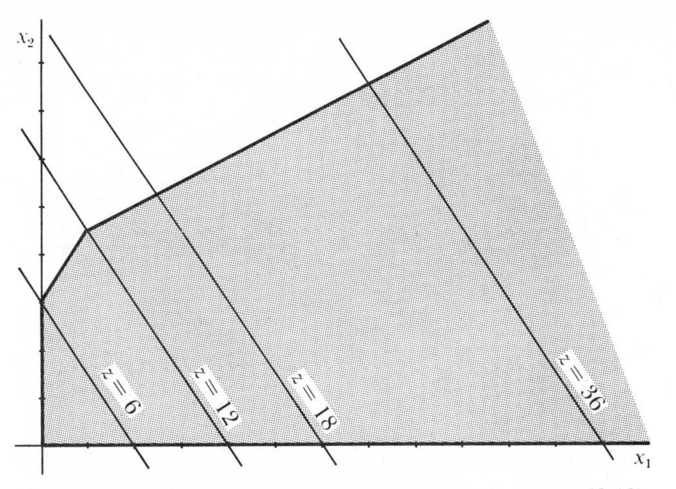

Figure 6-3 Linear programming example—Equations (6-13).

Let us now consider another example. Consider the following problem:

$$\begin{aligned} \text{Max } z &= 3x_1 + 2x_2 \\ -3x_1 + 2x_2 &\leq 6 \\ -4x_1 + 9x_2 &\leq 36 \\ x_1, x_2 &\geq 0 \end{aligned} \tag{6-13}$$

This problem is shown graphically in Figure 6-3. It can readily be seen that there is no largest finite value of z, i.e., it is always possible to find values of (x_1, x_2) such that z can be made arbitrarily large. When such a situation occurs the linear programming problem is said to have an *unbounded solution.*

In the example of Figure 6-3, not only does z become arbitrarily large but so do the values of (x_1, x_2). It is possible, however, for a problem to have an unbounded solution without all variables becoming arbitrarily large as z becomes arbitrarily large. Consider the following example:

$$\begin{aligned} \text{Max } z &= 2x_1 - 3x_2 \\ x_1 - x_2 &\geq 0 \\ x_2 &\leq 5 \\ x_1, x_2 &\geq 0 \end{aligned} \tag{6-14}$$

Equations (6-14) are represented graphically in Figure 6-4. We can see that the problem represented by (6-14) has an unbounded solution. This is true because as x_1 becomes arbitrarily large so does the value of the objective function. However, x_2 remains finite, i.e., $0 \leq x_2 \leq 5$.

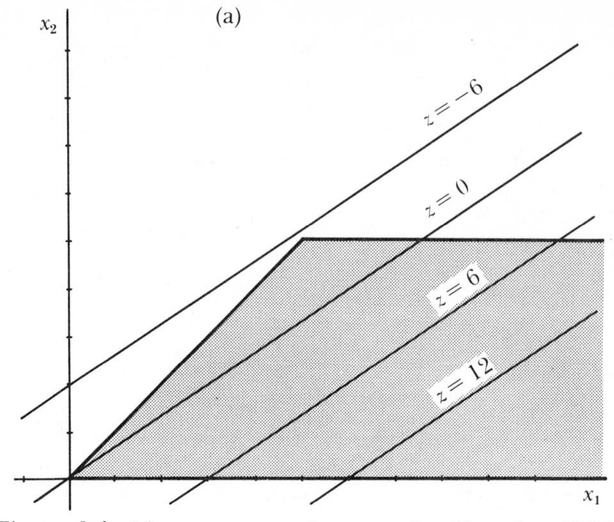

Figure 6-4 Linear programming example—Equations (6-14).

One last example of a kind of unboundedness is given in the following example:

$$\begin{aligned} \text{Max } z &= -4x_1 + 11x_2 \\ 3x_1 - 4x_2 &\geq -12 \\ 4x_1 - 11x_2 &\geq -44 \\ x_1, x_2 &\geq 0 \end{aligned} \tag{6-15}$$

The problem represented by equations (6-15) is depicted in Figure 6-5. In this problem we see that there is a finite maximum

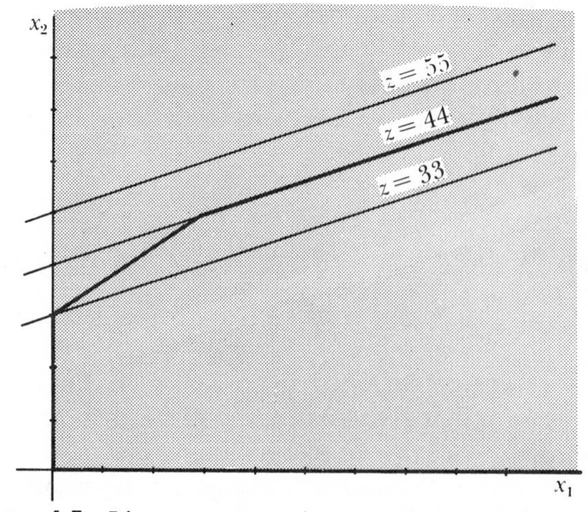

Figure 6-5 Linear programming example—Equations (6-15).

value that the objective can attain, viz., $z = 44$. Furthermore, we see that there is one optimal extreme point solution which occurs at the intersection of the constraint equations:

$$3x_1 - 4x_2 = -12$$
$$4x_1 - 11x_2 = -44$$

This point is $(x_1, x_2) = (\frac{268}{51}, \frac{88}{17})$. Any point along the upper boundary of the convex polygon is also an optimal solution. However, even though the maximum value of the objective function cannot exceed 44, values of (x_1, x_2) which give a value of $z = 44$ can become arbitrarily large. One can see this intuitively from Figure 6-5. It can also be seen from considering positive solutions to the equation

$$4x_1 - 11x_2 = -44$$

It is readily seen that $x_1 = \frac{11}{4}x_2 - 11$, and that as $x_2 \rightarrow \infty$, so does x_1. Yet, z remains at 44.

It should be emphasized that these last three examples, which relate to unboundedness either of the optimal value of the objective function or of the values of $\bar{x}$ which correspond to the optimal z or to both, are all pathological. They should not be encountered in any well-formulated problem whose variables represent actual physical entities of some kind. However, they may be encountered because of inadvertent errors during problem formulation or during mechanical manipulation of data. Hence, such situations must be recognized if they should occur, and corrective actions taken. It is an important feature of the computational method to be described later that there is a simple indication of unboundedness of the objective function so that this situation can be recognized and the computation terminated.

Another pathological situation which may arise because of errors during the stage of problem formulation relates to the nonexistence of solutions which satisfy either the constraints of the problem, $\sum_{j=1}^{n} a_{ij}x_j = b_i$, $i = 1, 2, \ldots, m$, or the nonnegativity restrictions $x_j \geq 0, j = 1, 2, \ldots, n$, or both. Consider the following problem:

$$\begin{aligned} \text{Max } z &= -3x_1 + 4x_2 \\ 2x_1 + 3x_2 &\leq 6 \\ 4x_1 + 5x_2 &\geq 20 \\ x_1, x_2 &\geq 0 \end{aligned} \qquad \textbf{(6-16)}$$

This problem is shown graphically in Figure 6-6. It can be seen that there are no solutions (x_1, x_2) such that $x_1, x_2 \geq 0$ which satisfy *both* constraints. This is clearly an impossible demand to make in any

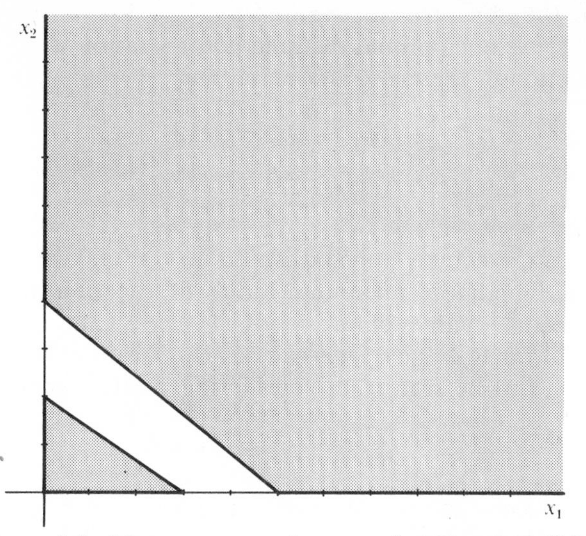

Figure 6-6 Linear programming example—Equations (6-16).

realistic problem. However, it is possible to phrase such demands in mathematical language, as we have illustrated in equations (6-16), without intending to do so.

Another possible situation that may arise is one in which there are solutions that satisfy the constraints but do not satisfy the non-negativity requirements. An example of such a problem is the

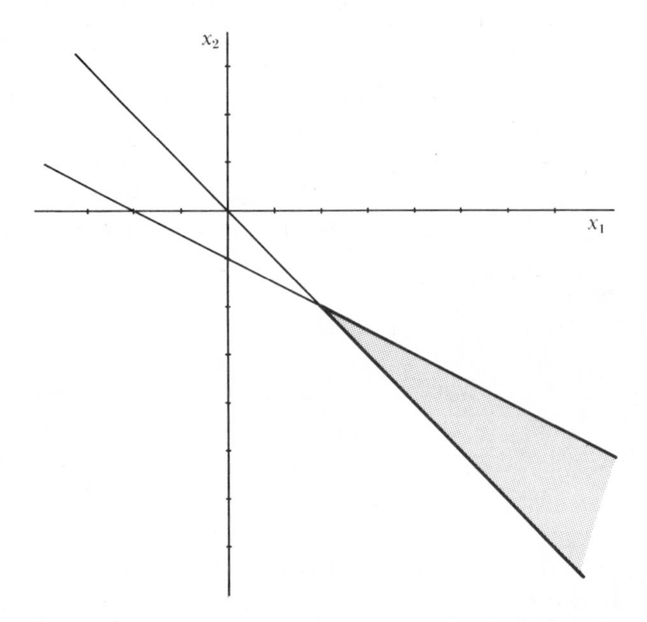

Figure 6-7 Linear programming example—Equations (6-17).

following:

$$\begin{aligned} \text{Max } z &= 2x_1 + x_2 \\ x_1 + x_2 &\leq 0 \\ x_1 + 2x_2 &\leq -2 \\ x_1, x_2 &\geq 0 \end{aligned} \tag{6-17}$$

The problem represented by equations (6-17) is shown in Figure 6-7. It can be readily seen in Figure 6-7 that, whereas there are values of (x_1, x_2) which satisfy the two inequalities of equations (6-17), there are no such values of (x_1, x_2) which also satisfy $x_1 \geq 0$, $x_2 \geq 0$. Hence, for this reason, there are no feasible solutions.

6.3 HOW NOT TO SOLVE A LINEAR PROGRAMMING PROBLEM

The seemingly facetious title of this section is meant to be taken seriously. One might well ask why, in this late date in the history of classical mathematical analysis, it should be considered at all unusual to solve a problem of the form

$$\begin{aligned} \text{Max } z &= \sum_{j=1}^{n} c_j x_j \\ \sum_{j=1}^{n} a_{ij} x_j &= b_i \qquad i = 1, 2, \ldots, m \\ x_j &\geq 0, \qquad j = 1, 2, \ldots, n \end{aligned} \tag{6-1}$$

In principle, with a few additional variables, we would be able to solve (6-1) by using the method of Lagrange multipliers described in Chapter 4. Let us consider how we might do this.

The theory of Lagrange multipliers, as we described it, applied to problems with equality constraints. As we noted in Chapter 4, inequality constraints such as $x_j \geq 0, j = 1, \ldots, n$ pose a problem. However, one way of handling such constraints is simply to replace them with

$$x_j - s_j^2 = 0 \qquad j = 1, 2, \ldots, n \tag{6-18}$$

This adds n additional variables, s_j. However, these are unrestricted variables. Now the variables x_j are constrained to be nonnegative by equations (6-18) and the constraints $x_j \geq 0$ need no longer be considered explicitly. The problem represented by equations (6-1) can now be represented, equivalently, by

$$\begin{aligned} \text{Max } z &= \sum_{j=1}^{n} c_j x_j \\ \sum_{j=1}^{n} a_{ij} x_j &= b_i \qquad i = 1, 2, \ldots, m \\ x_j - s_j^2 &= 0 \qquad j = 1, 2, \ldots, n \end{aligned} \tag{6-19}$$

We can then construct the Lagrangian function for the preceding problem:

$$F(\bar{x}, \bar{s}, \bar{\lambda}, \bar{\mu}) = \sum_{j=1}^{n} c_j x_j + \sum_{i=1}^{m} \lambda_i \left[b_i - \sum_{j=1}^{n} a_{ij} x_j \right] + \sum_{j=1}^{n} \mu_j [-x_j + s_j^2] \tag{6-20}$$

where $\bar{\lambda}$, $\bar{\mu}$ are m and n component vectors respectively of the Lagrange multipliers λ_i and μ_j and $\bar{s}$ is an n-vector of the s_j. We could now, in principle, differentiate the Lagrangian function and solve the resulting equations for the $m + 3n$ variables: λ_i, x_j, s_j, μ_j. Let us derive these equations to discover the nature of the problem encountered.

$$\frac{\partial F}{\partial x_j} = c_j - \sum_{i=1}^{n} \lambda_i a_{ij} - \mu_j = 0 \qquad j = 1, 2, \ldots, n$$

$$\frac{\partial F}{\partial u_j} = 2\mu_j s_j = 0 \qquad j = 1, 2, \ldots, n \tag{6-21}$$

$$\frac{\partial F}{\partial \lambda_i} = b_i - \sum_{j=1}^{n} a_{ij} x_j = 0 \qquad i = 1, 2, \ldots, m$$

$$\frac{\partial F}{\partial \mu_j} = -x_j + s_j^2 = 0 \qquad j = 1, 2, \ldots, n$$

If we examine equations (6-21) the nature of the difficulty associated with solving these equations becomes apparent. The set of equations

$$2\mu_j s_j = 0 \tag{6-22}$$

says that *either* $\mu_j = 0$, i.e., the constraint $x_j \geq 0$ is not limiting and, hence, $x_j > 0$ *or* $s_j = 0$ and $x_j = 0$. We cannot find an unconstrained maximum of F using these *necessary* conditions without testing every set of possibilities, i.e., finding all possible solutions of equations (6-21). This leads us to solve combinatorially large numbers of sets of equations. If there are n variables we would have to consider solving equations (6-21) for 2^n cases in accord with the condition that $\mu_j s_j = 0, j = 1, 2, \ldots, n$. Except for problems with only a relatively small number of variables, this is not a practical alternative.

The foregoing situation points out the need for an approach to solving linear programming problems which is quite different from any of the classical methods. Since problems involving hundreds and even thousands of variables exist, it is indeed fortunate that such a method has been found. The method that is used is called the *simplex method.* It is an iterative algebraic procedure which will obtain the global optimum to any well-posed linear programming problem in a finite number of steps. More important, however, than the fact that the simplex method is finite (the procedure we have been discussing is also finite) is the fact it is exceedingly efficient. In an insignificant number of iterations (compared to the total possible) it will find the global optimum. It was developed by George Dantzig in 1947 and since then has had a remarkable growth and has been widely adopted. In the following sections we shall develop the theory of the simplex method and indicate how it is used to solve linear programming problems.

6.4 THE SIMPLEX METHOD: BASIC THEOREMS

We consider the linear programming problem in the form given by equations (6-1), i.e.,

$$\text{Max } z = \sum_{j=1}^{n} c_j x_j$$

$$\sum_{j=1}^{n} a_{ij} x_j = b_i \qquad i = 1, 2, \ldots, m \tag{6-1}$$

$$x_j \geq 0 \qquad j = 1, 2, \ldots, n$$

We shall also frequently write (6-1) for convenience in matrix notation as

$$\text{Max } z = \bar{c}'\bar{x}$$

$$A\bar{x} = \bar{b} \tag{6-23}$$

$$\bar{x} \geq \bar{0}$$

It should be understood that even if the original statement of the linear programming problem contains inequalities, these can be converted by the addition of either nonnegative *slack* or *surplus* variables to equalities as follows. The inequality

$$\sum_{j=1}^{K} a_{ij} x_j \leq b_i$$

can be written as

$$\sum_{j=1}^{K} a_{ij} x_j + x_{K+i} = b_i \tag{6-24}$$

$$x_{K+i} \geq 0$$

where x_{K+i} is referred to as a *slack variable.* Similarly the inequality

$$\sum_{j=1}^{K} a_{ij} x_j \geq b_i$$

can be rewritten as

$$\sum_{j=1}^{K} a_{ij} x_j - x_{K+i} = b_i \tag{6-25}$$

$$x_{K+i} \geq 0$$

where x_{K+i} is called a *surplus variable.* In either case, they are simply additional nonnegative variables which have been added to the problem. The addition of these slack or surplus variables in no way changes the set of possible solutions to the original problem. Our starting point will always be of the form (6-1) or (6-23), a problem involving m equality constraints and n nonnegative variables.

We now state some definitions that will be used throughout our discussion of linear programming:

Feasible Solution:† A feasible solution to the linear programming problem is a vector $\bar{x} = (x_1, x_2, \ldots, x_n)$ which satisfies all the constraints of the problem.

Basic Feasible Solution: A basic feasible solution to the linear programming problem is a feasible solution with no more than m positive x_j. These positive x correspond to linearly independent columns of the matrix A.

Nondegenerate Basic Feasible Solution: A nondegenerate basic feasible solution is a basic feasible solution with exactly m positive x_j. Hence, a *degenerate* basic feasible solution has fewer than m positive x_j.

We shall now present the basic theoretical results which allow the construction of a computational algorithm for solving the linear programming problem. Where appropriate, we shall refer to results established earlier.

Theorem 6.1. The set of all feasible solutions to a linear programming problem is a convex set.

Proof: This was proved in Chapter 3, since the solutions must satisfy a set of linear constraints.

Theorem 6.2. The objective function of a linear programming problem assumes its maximum at an extreme point of the convex set of all feasible solutions.

Proof: The objective function of a linear programming problem may be considered to be a convex function since it is linear. By Theorem 6.1, the set of all feasible solutions generate a convex set. Hence, we are maximizing a convex function over a convex set. By the theorem proven in Section 3.11 of Chapter 4, such a maximum occurs at an extreme point of the convex set.

For the remaining theorems, we shall designate the n columns of the matrix A as $\bar{a}_1, \bar{a}_2, \ldots, \bar{a}_n$. We shall require sets of k linearly

---

† Some writers use the term feasible to refer to satisfying only the nonnegativity restrictions. For example, the widely used text by Hadley[1] adopts this usage.

independent vectors (columns of A). Without loss of generality, let us assume that these are the vectors $\bar{a}_1, \bar{a}_2, \ldots, \bar{a}_k$. It is clear that a renumbering of the vectors can accomplish this in all cases. Further, let X represent the convex set of feasible solutions.

Theorem 6.3. If there exists a set of $k \leq m$ linearly independent columns of A, viz. $\bar{a}_1, \bar{a}_2, \ldots, \bar{a}_k$ such that

$$x_1\bar{a}_1 + x_2\bar{a}_2 + \ldots + x_k\bar{a}_k = \bar{b}$$

with all $x_j \geq 0$, then the point $\bar{x} = (x_1, x_2, \ldots, x_k, 0, \ldots, 0)$ is an extreme point of the convex set of feasible solutions, X.

Proof: If $\bar{x}$ is not an extreme point then since it is feasible, it can be written as a convex combination of some two other points in X, i.e.,

$$\bar{x} = \lambda\bar{x}_1 + (1 - \lambda)\bar{x}_2 \qquad 0 < \lambda < 1 \tag{6-26}$$

Since all the x_j, the components of $\bar{x}$, are nonnegative and since the last $n - k$ components are zero, and since $0 < \lambda < 1$, it follows that the last $n - k$ components of $\bar{x}_1$ and $\bar{x}_2$ must also equal zero. Therefore,

$$\begin{aligned} \bar{x}_1 &= (x_1^1, x_2^1, \ldots, x_k^1, 0, \ldots, 0) \\ \bar{x}_2 &= (x_1^2\, x_2^2, \ldots, x_k^2, 0, \ldots, 0) \end{aligned} \tag{6-27}$$

We know that $\bar{x}$, $\bar{x}_1$, and $\bar{x}_2$ are all feasible solutions. Therefore, since (6-27) is true, we may write

$$\begin{aligned} x_1\bar{a}_1 + x_2\bar{a}_2 + \ldots + x_k\bar{a}_k &= \bar{b} \\ x_1^1\bar{a}_1 + x_2^1\bar{a}_2 + \ldots + x_k^1\bar{a}_k &= \bar{b} \\ x_1^2\bar{a}_1 + x_2^2\bar{a}_2 + \ldots + x_k^2\bar{a}_k &= \bar{b} \end{aligned} \tag{6-28}$$

If we subtract the second from the first of the preceding equations and also subtract the third from the second of these equations, we obtain

$$(x_1 - x_1^1)\bar{a}_1 + (x_2 - x_2^1)\bar{a}_2 + \ldots + (x_k - x_k^1)\bar{a}_k = \bar{0}$$
$$(x_1^1 - x_1^2)\bar{a}_1 + (x_2^1 - x_2^2)\bar{a}_2 + \ldots + (x_k^1 - x_k^2)\bar{a}_k = \bar{0}$$

However, by hypothesis, the columns of A, $\bar{a}_1, \ldots, \bar{a}_k$ are a set of linearly independent vectors. This implies that

$$x_j - x_j^1 = 0 \qquad j = 1, 2, \ldots, k$$
$$x_j^1 - x_j^2 = 0 \qquad j = 1, 2, \ldots, k$$

since all coefficients must vanish identically. Therefore, we have obtained the result that

$$x_j = x_j^1 = x_j^2 \qquad j = 1, 2, \ldots, k$$

and so our assumption that $\bar{x}$ could be expressed as a convex combination of some two other distinct points, $\bar{x}_1$ and $\bar{x}_2$, is not true. Therefore, by definition, since $\bar{x}$ cannot be expressed as a convex combination of two distinct points in X, it must be an extreme point of X.

Theorem 6.4. Let $\bar{x} = (x_1, x_2, \ldots, x_n)$ be an extreme point of X. Then the columns of A associated with positive x_j are linearly independent, with at most m of the x_j being positive.

Proof: Again we shall assume, without loss of generality, that the first k of the x_j are nonzero. Therefore we can write that

$$x_1\bar{a}_1 + x_2\bar{a}_2 + \ldots + x_k\bar{a}_k = \bar{b} \tag{6-29}$$

We shall prove the theorem by contradiction. To this end, let us assume that the vectors $\bar{a}_j$, $j = 1, 2, \ldots, k$ are linearly *dependent.* Therefore, there exists a linear combination of these vectors such that

$$\alpha_1\bar{a}_1 + \alpha_2\bar{a}_2 + \ldots + \alpha_k\bar{a}_k = \bar{0} \tag{6-30}$$

with at least one $\alpha_j \neq 0$. Let us now multiply (6-30) by some $\beta > 0$ and both add the result to (6-29) and subtract the result from (6-29). We therefore obtain two equations as follows:

$$\begin{aligned} \sum_{j=1}^{k} (x_j + \beta\alpha_j)\bar{a}_j &= \bar{b} \\ \sum_{j=1}^{k} (x_j - \beta\alpha_j)\bar{a}_j &= \bar{b} \end{aligned} \tag{6-31}$$

We now have two vectors as follows:

$$\begin{aligned} \bar{x}_1 &= (x_1 + \beta\alpha_1, x_2 + \beta\alpha_2, \ldots, x_k + \beta\alpha_k, 0, \ldots, 0) \\ \bar{x}_2 &= (x_1 - \beta\alpha_1, x_2 - \beta\alpha_2, \ldots, x_k - \beta\alpha_k, 0, \ldots, 0) \end{aligned} \tag{6-32}$$

The two vectors given by (6-32) may or may not be solutions, depending upon the magnitude of β. However, if $\beta > 0$ is chosen sufficiently small so that *all* of the first k components of both $\bar{x}_1$ and $\bar{x}_2$ are positive, which clearly can be done, then $\bar{x}_1$ and $\bar{x}_2$ will be feasible solutions and will therefore be in X. However, from the definition of $\bar{x}$ and (6-32) it is clear that

$$\bar{x} = \tfrac{1}{2}\bar{x}_1 + \tfrac{1}{2}\bar{x}_2$$

which contradicts the assumption that $\bar{x}$ is an extreme point. Hence, the assumption of linear dependence for $\bar{a}_1, \bar{a}_2, \ldots, \bar{a}_k$ has led to a contradiction. Therefore, the vectors $\bar{a}_1, \bar{a}_2, \ldots, \bar{a}_k$ are linearly independent.

Thus far we have spoken of k vectors which we have shown to be linearly independent. However, it was shown in Chapter 2 that every set of $n + 1$ vectors in an n-dimensional space is a linearly dependent set. Hence, by the foregoing result and the fact that the $\bar{a}_j$ are m-component vectors, we cannot have *more* than m positive x_j. This concludes the proof of the theorem.

If there is to be at least one solution to the set of equations $A\bar{x} = \bar{b}$, then as was shown in Chapter 2, if the rank of A equals m, there must be at least one set of m vectors from the set $\bar{a}_1, \ldots, \bar{a}_n$ that is linearly independent. Hence, this result with the preceding theorem allows us to assume that in all cases the set of vectors $\bar{a}_j$, $j = 1, 2, \ldots, n$ of the linear programming problem always contains a set of m linearly independent vectors.

From the foregoing theorems we can summarize the following pertinent results with respect to the solution of linear programming problems. We recall that the set X of feasible solutions is defined as $X = \{\bar{x} \mid A\bar{x} = \bar{b}, \bar{x} \geq \bar{0}\}$. We then have:

1. Every basic feasible solution to the linear programming problem corresponds to an extreme point of the convex set of feasible solutions, X.
2. Every extreme point of X has m linearly independent vectors from the set $\bar{a}_j, j = 1, 2, \ldots, n$ (the columns of A) associated with it.
3. There is some extreme point of X at which the objective function $z = \bar{c}'\bar{x}$ takes on its maximum value.

The preceding conclusions indicate that whereas the convex set X has an infinite number of solutions $\bar{x}$, we need only examine a *finite* subset of these, namely, those feasible solutions which correspond to extreme points of the convex set, X. These are solutions generated by sets of m linearly independent vectors from the n vectors $\bar{a}_1, \bar{a}_2, \ldots, \bar{a}_n$, the columns of the matrix A. Despite the fact that the set of solutions to be investigated is finite, it would be quite impossible to examine all of them for large m and n. An upper bound to the number of possible solutions is $\binom{n}{m}$, the maximum possible number of sets of m linearly independent vectors from a total set of n. It will be recalled that

$$\binom{n}{m} = \frac{n!}{m!\,(n - m)!}$$

and that, e.g., for $n = 300$, $m = 100$ (a not excessively large linear programming problem), we could not possibly examine this number of extreme point solutions.

What is obviously required is a computational procedure which will examine only a minute fraction of the total number of extreme point solutions and which will arrive at the optimal solution. Such a method is the simplex method which will be developed and described in the following section.

6.5 THE SIMPLEX METHOD: THEORY

In the previous section we developed some basic theorems about the nature of the solution to a linear programming problem. We will now develop some results that refer specifically to the "simplex method" for solving linear programming problems. In order to do this we need to adopt some standard notation, assumptions, and definitions.

We assume that our problem is stated in the form of (6-23)

$$\begin{aligned} \text{Max } z &= \bar{c}'\bar{x} \\ A\bar{x} &= \bar{b} \\ \bar{x} &\geq \bar{0} \end{aligned} \qquad \textbf{(6-23)}$$

where A is $m \times n$, $\bar{c}$, $\bar{x}$ are n-vectors, and $\bar{b}$ is an n-vector. We shall denote the jth column of A as $\bar{a}_j, j = 1, 2, \ldots, n$. A matrix B will consist of m linearly independent columns of A and therefore, it will constitute a *basis* for E^m. B will therefore be called a *basis matrix*. From the properties of a basis we know that any other vector in E^m can be written as a linear combination of the vectors in B. Therefore, any column of the A matrix can be written as a linear combination of the columns in B, i.e.,

$$\bar{a}_j = y_{1j}\bar{a}_1 + y_{2j}\bar{a}_2 + \ldots + y_{mj}\bar{a}_m = \sum_{i=1}^{m} y_{ij}a_i \qquad \textbf{(6-33)}$$

We have assumed for convenience that the basis B consists of $\bar{a}_1, \bar{a}_2, \ldots, \bar{a}_m$, i.e., the first m columns of A. This can be assumed since the numbering of the vectors can be made arbitrarily.

If we define an m-component vector $\bar{y}_j$ with components y_{ij}

$$\bar{y}_j = \begin{bmatrix} y_{1j} \\ y_{2j} \\ \cdot \\ \cdot \\ \cdot \\ y_{mj} \end{bmatrix}$$

then we can write (6-33) as

$$\bar{a}_j = B\bar{y}_j \qquad \textbf{(6-34)}$$

Since B is a basis for E^m and hence is a nonsingular matrix, we may also write

$$\bar{y}_j = B^{-1}\bar{a}_j \qquad \textbf{(6-35)}$$

If we consider the matrix A partitioned into the basis matrix B and a nonbasic matrix, N, then $A = (B, N)$ and we may write

$$A\bar{x} = (B, N)\bar{x} = (B, N)\begin{bmatrix} \bar{x}_B \\ \bar{x}_N \end{bmatrix} = \bar{b} \tag{6-36}$$

where $\bar{x}$ has been partitioned into $\bar{x}_B$ and $\bar{x}_N$ just as A has been partitioned. From (6-36) we have

$$B\bar{x}_B + N\bar{x}_N = \bar{b}$$

However, by the definition of a *basic solution* corresponding to the basis matrix B, we know that $\bar{x}_N = \bar{0}$. Therefore,

$$\begin{aligned} B\bar{x}_B &= \bar{b} \\ \bar{x}_B &= B^{-1}\bar{b} \end{aligned} \tag{6-37}$$

is a *basic solution* to $A\bar{x} = \bar{b}$. The variables in $\bar{x}_B$ are called *basic variables*.

If we now partition the vector $\bar{c}'$ as we have partitioned A and $\bar{x}$, we have:

$$\bar{c}' = (\bar{c}_{B'}, \bar{c}'_N) \quad \text{where} \quad \bar{c}'_B = (c_{B1}, c_{B2}, \ldots, c_{Bm})$$

and

$$z = \bar{c}'\bar{x} = (\bar{c}'_{B'}, \bar{c}'_N)\begin{bmatrix} \bar{x}_B \\ \bar{x}_N \end{bmatrix} = \bar{c}'_B\bar{x}_B + \bar{c}'_N\bar{x}_N = \bar{c}'_B\bar{x}_B \tag{6-38}$$

One last definition that we shall require is a scalar quantity associated with each $\bar{y}_j$ vector as follows:

$$z_j = \sum_{i=1}^{m} y_{ij}c_{Bi} = \bar{c}'_B\bar{y}_j \tag{6-39}$$

With the preceding notation and definitions we may proceed with developing the simplex theory. We assume we have some basic feasible solution. We must either show that this solution is the optimal basic feasible solution or determine an improved basic feasible solution. Let us first attend to the latter.

If we have a basic feasible solution to $A\bar{x} = \bar{b}$, we know that it can be written

$$\sum_{i=1}^{m} x_{Bi}\bar{a}_i = \bar{b} \tag{6-40}$$

because we assumed the first m columns of A were a basis, B. If we are to seek a new basic feasible solution, then we shall have to remove one or more vectors $\bar{a}_i$ from B, and replace them with some other vectors $\bar{a}_j$ from N. There are a great many ways this could be done. In the simplex method of Dantzig, only one vector at a time is removed from the basis matrix and replaced, in order to generate a new basis matrix and its associated basic feasible solution. If one thinks of this process geometrically, what we are doing is as follows. Each basic feasible solution corresponds to an extreme point of the convex set of feasible solutions. By replacing one vector at a time we move from one extreme point to an *adjacent* extreme point of the convex set. Let us now consider how to find an improved adjacent extreme point, or equivalently, an improved basic feasible solution by replacing one basis vector with another.

We assume we have a basic feasible solution of the form (6-40). If we consider the vectors of A which are not in the basis, as we know from (6-33), they can be expressed in terms of the basis vectors:

$$\bar{a}_j = \sum_{i=1}^{m} y_{ij}\bar{a}_i \qquad j = m + 1, \ldots, n \tag{6-41}$$

We are here again using the convention that the $\bar{a}_i$ are in the basis and the $\bar{a}_j$ are not. If we consider some vector, $\bar{a}_j$, which is to enter the basis, then let us single out from the basis vectors a vector $\bar{a}_r$ such that $y_{rj} \neq 0$. Using (6-41) we may write

$$\bar{a}_j = y_{rj}\bar{a}_r + \sum_{\substack{i=1 \\ i\neq r}}^{m} y_{ij}\bar{a}_i \tag{6-42}$$

If we solve (6-42) for $\bar{a}_r$, we have

$$\bar{a}_r = \frac{1}{y_{rj}}\,\bar{a}_j - \sum_{\substack{i=1 \\ i\neq r}}^{m} \frac{y_{ij}}{y_{rj}}\,\bar{a}_i \tag{6-43}$$

If we now substitute $\bar{a}_r$ from (6-43) into our basic solution given by (6-40) we have

$$\sum_{\substack{i=1 \\ i\neq r}}^{m} x_{Bi}\bar{a}_i + \frac{x_{Br}}{y_{rj}}\,\bar{a}_j - x_{Br}\sum_{\substack{i=1 \\ i\neq r}}^{m} \frac{y_{ij}}{y_{rj}}\,\bar{a}_i = \bar{b} \tag{6-44}$$

Upon rearrangement, from (6-44) we obtain

$$\sum_{\substack{i=1 \\ i\neq r}}^{m} \left(x_{Bi} - x_{Br}\frac{y_{ij}}{y_{rj}}\right)\bar{a}_i + \frac{x_{Br}}{y_{rj}}\,\bar{a}_j = \bar{b} \tag{6-45}$$

which is a new basic solution. We have not insured, however, that it is feasible because although the constraints $A\bar{x} = \bar{b}$ are satisfied, the substitution we have just carried out does not insure that the new x_{Bi} are necessarily nonnegative. If we examine (6-45) we see that the new values of x_{Bi}, which we shall designate $\hat{x}_{Bi}$, have the form

$$\begin{aligned} \hat{x}_{Bi} &= x_{Bi} - x_{Br}\frac{y_{ij}}{y_{rj}} \qquad i \neq r \\ \hat{x}_{Br} &= \frac{x_{Br}}{y_{rj}} \end{aligned} \tag{6-46}$$

It is clear that the possibility exists that some of these could be less than zero. Hence, we must require that

$$\hat{x}_{Bi} = x_{Bi} - x_{Br}\frac{y_{ij}}{y_{rj}} \geq 0 \qquad i \neq r$$
$$\hat{x}_{Br} = \frac{x_{Br}}{y_{rj}} \geq 0 \tag{6-47}$$

If $x_{Br} > 0$, then we must require that $y_{rj} > 0$. If all the $y_{ij} \leq 0$, then the $\hat{x}_{Bi}$ will be nonnegative. However, if some of the $y_{ij} > 0$, then whether or not $x_{Bi} - x_{Br}\frac{y_{ij}}{y_{rj}}$ would be nonnegative would depend upon which particular $\bar{a}_r$ we had chosen to remove from the basis. Let us see how to choose the correct $\bar{a}_r$. If some $y_{ij} > 0$, we can divide the first equation of (6-47) by y_{ij} to obtain

$$\frac{x_{Bi}}{y_{ij}} - \frac{x_{Br}}{y_{rj}} \geq 0 \tag{6-48}$$

If we wish to guarantee that (6-48) will always hold, then we can simply choose the column r to be removed from the basis B as follows:

$$\frac{x_{Br}}{y_{rj}} = \min_i \left\{\frac{x_{Bi}}{y_{ij}}, y_{ij} > 0\right\} \tag{6-49}$$

If r is chosen according to (6-49) then (6-48) will always be true.

There are a number of special cases to consider which although they cause no problems should be mentioned. First, the minimum in (6-49) may not be unique. In this case one or more variables in the new basic solution will be zero and hence we will have a degenerate basic feasible solution. If we had started with a degenerate solution, which is perfectly possible, and x_{Br} was zero, then the new basic solution would also be degenerate since $\hat{x}_{Br} = \frac{x_{Br}}{y_{rj}} = 0$. However, we could have started with a degenerate solution and *not* have the new solution be degenerate. One can see this from examining (6-49). If none of the y_{ij} were greater than zero for $x_{Bi} = 0$ in the degenerate solution, then x_{Br} would not be determined from the degenerate x_{Bi} and hence the new solution might not be degenerate.

We have seen how to determine the vector $\bar{a}_r$ which is to leave the basis, B. We compute $\frac{x_{Br}}{y_{rj}}$ from (6-49) and this tells us which column r of the basis matrix B is to be removed. We then have a new basis matrix, $\hat{B}$ consisting of columns $\bar{a}_i$, $i \neq r$ and a new column $\bar{a}_j$. The new basic solution is given by $\hat{\bar{x}}_B = \hat{B}^{-1}\bar{b}$ and the new values of the basic variables are given by equations (6-46).

Our principal concern in the preceding was to insure that if a vector $\bar{a}_j$ came into the basis and a vector $\bar{a}_r$ was removed from the basis, the new solution would be a feasible solution. However, we said nothing about how to select the vector $\bar{a}_j$ which is to enter the basis in order to give us a new basic feasible solution. What we would like to be able to do is to find a new basic feasible solution which will give us an improved (larger) value of the objective function or have some indication that the current basic feasible solution is the optimal solution. Let us inquire into how to do this.

For the original basic feasible solution, the objective function is given by

$$z = \bar{c}'\bar{x} = \bar{c}'_B\bar{x}_B = \sum_{i=1}^{m} c_{Bi}x_{Bi} \tag{6-50}$$

Similarly, the new objective function z is given by

$$\hat{z} = \bar{c}'\bar{x} = \hat{\bar{c}}'_B\hat{\bar{x}}_B = \sum_{i=1}^{m} \hat{c}_{Bi}\hat{x}_{Bi} \tag{6-51}$$

According to our notation, since $\bar{a}_r$ has been removed from the basis and $\bar{a}_j$ has entered, we have

$$\begin{aligned} \hat{c}_{Bi} &= c_{Bi} \qquad i \neq r \\ \hat{c}_{Br} &= c_j \end{aligned} \tag{6-52}$$

Substituting from equations (6-46) and (6-52) into (6-51) we have

$$\hat{z} = \sum_{\substack{i=1 \\ i\neq r}}^{m} c_{Bi}\left(x_{Bi} - x_{Br}\frac{y_{ij}}{y_{rj}}\right) + \frac{x_{Br}}{y_{rj}}c_j \tag{6-53}$$

If we now note that the missing term in the summation of equation (6-53) is

$$c_{Br}\left(x_{Br} - x_{Br}\frac{y_{rj}}{y_{rj}}\right) = 0$$

then we can add this term to the summation to yield

$$\hat{z} = \sum_{i=1}^{m} c_{Bi}\left(x_{Bi} - x_{Br}\frac{y_{ij}}{y_{rj}}\right) + \frac{x_{Br}}{y_{rj}}c_j \tag{6-54}$$

We can rearrange equation (6-54) as follows:

$$\begin{aligned} \hat{z} &= \sum_{i=1}^{m} c_{Bi}x_{Bi} - \frac{x_{Br}}{y_{rj}}\sum_{i=1}^{m} c_{Bi}y_{ij} + \frac{x_{Br}}{y_{rj}}c_j \\ &= z - \frac{x_{Br}}{y_{rj}}z_j + \frac{x_{Br}}{y_{rj}}c_j \end{aligned}$$

Rearranging, we have

$$\hat{z} = z + \frac{x_{Br}}{y_{rj}}(c_j - z_j) \tag{6-55}$$

Equation (6-55) gives us the information that we have been seeking. We know that $\frac{x_{Br}}{y_{rj}} > 0$ in the absence of degeneracy. Therefore, if we select a vector such that $c_j - z_j > 0$ (or $z_j - c_j < 0$ as it is usually written), we will have made $\hat{z} > z$. Even in the presence of degeneracy, $\hat{z} \geq z$. If we choose the vector $\bar{a}_j$ to enter the basis such that $\frac{x_{Br}}{y_{rj}}(c_j - z_j)$ is the largest possible, for those $\bar{a}_j, j = m + 1, \ldots, n$, then we will have the greatest possible increase in z. In order to minimize the amount of calculation, in actual practice a vector $\bar{a}_j$ is selected such that $z_j - c_j$ is the smallest of the $z_j - c_j < 0$. This seems to be quite satisfactory and greatly reduces the amount of calculation required, since it can be calculated without reference to the vector leaving the basis.

We have one more major point to discuss and that is when to terminate the calculation. We have seen that, in the absence of degeneracy, as long as there is a vector $\bar{a}_j$ not in the basis with a value of $z_j - c_j < 0$, we will obtain a monotonic, nonzero increase in the objective function, i.e., $\hat{z} > z$. Hence, we can continue such a process until there are no vectors $\bar{a}_j$ with $z_j - c_j < 0$. This means that for all vectors $\bar{a}_j$, $z_j - c_j \geq 0$. Suppose this occurs at a point $\bar{x}_B$ with corresponding value of the objective function, z_0. We shall demonstrate that when this occurs, i.e., $z_j - c_j \geq 0, j = m + 1, \ldots, n$, z_0 is the maximum value of the objective function.

Let $x'_j \geq 0$, $j = 1, 2, \ldots, n$ be any feasible solution to $A\bar{x} = \bar{b}$. Therefore,

$$\sum_{j=1}^{n} x'_j \bar{a}_j = \bar{b} \qquad x'_j \geq 0 \qquad j = 1, 2, \ldots, n \tag{6-56}$$

The value of the objective function for this solution is

$$z' = c_1 x'_1 + c_2 x'_2 + \ldots + c_n x'_n \tag{6-57}$$

We know that we can express any vector $\bar{a}_j$ in A as a linear combination of the basis vectors:

$$\bar{a}_j = \sum_{i=1}^{m} y_{ij} \bar{a}_i \tag{6-58}$$

If we substitute from (6-58) into (6-56), we have

$$x'_1 \sum_{i=1}^{m} y_{i1} \bar{a}_i + x'_2 \sum_{i=1}^{m} y_{i2} \bar{a}_i + \ldots + x'_n \sum_{i=1}^{m} y_{in} \bar{a}_i = \bar{b}$$

which upon rearrangement becomes

$$\left[\sum_{j=1}^{n} x'_j y_{1j}\right] \bar{a}_1 + \left[\sum_{j=1}^{n} x'_j y_{2j}\right] \bar{a}_2 + \ldots + \left[\sum_{j=1}^{n} x'_j y_{mj}\right] \bar{a}_m = \bar{b} \tag{6-59}$$

Equation (6-59) expresses the vector $\bar{b}$ in terms of the basis vectors, $\bar{a}_i$. However, it will be recalled from Chapter 2 that the expression of a vector in terms of a set of basis vectors is unique. If we compare equations (6-56) and (6-59), it is clear that

$$x_{Bi} = \sum_{j=1}^{n} x'_j y_{ij} \qquad i = 1, 2, \ldots, m \tag{6-60}$$

Now let us examine the objective function, z'. We know that $z_j - c_j \geq 0$ for all $\bar{a}_j$ not in the basis. For those columns of A in the basis, we have

$$\bar{y}_j = \bar{y}_i = B^{-1}\bar{a}_i = \bar{e}_i \qquad j = i = 1, 2, \ldots, m \tag{6-61}$$

where $\bar{e}_i$ = is the *unit vector* with a "1" in the ith row and zeros elsewhere. From (6-61) we see that

$$z_j = \bar{c}'_B \bar{y}_j = \bar{c}'_B \bar{e}_i = c_j \qquad j = 1, 2, \ldots, m \tag{6-62}$$

Therefore $z_j - c_j = 0$ for $j \leq m$ and now it is true that $z_j - c_j \geq 0$ for all j, i.e., for every column of A. Since $z_j \geq c_j$, $j = 1, 2, \ldots, n$ we can substitute this into (6-57) to obtain

$$z_1 x'_1 + z_2 x'_2 + \ldots + z_n x'_n \geq z' \tag{6-63}$$

By definition, $z_j = \sum_{i=1}^{m} y_{ij} c_{Bi}$. If we substitute this definition into (6-63) we have

$$x'_1 \sum_{i=1}^{m} y_{i1} c_{Bi} + x'_2 \sum_{i=1}^{m} y_{i2} c_{Bi} + \ldots + x'_n \sum_{i=1}^{m} y_{in} c_{Bi} \geq z'$$

which upon rearrangement becomes

$$\left[\sum_{j=1}^{n} x'_j y_{1j}\right] c_{B1} + \left[\sum_{j=1}^{n} x'_j y_{2j}\right] c_{B2} + \ldots + \left[\sum_{j=1}^{n} x'_j y_{mj}\right] c_{Bm} \geq z' \tag{6-64}$$

Substituting from (6-60) into (6-64) we have

$$c_{B1} x_{B1} + c_{B2} x_{B2} + \ldots + c_{Bm} x_{Bm} = \bar{c}'_B \bar{x}_B = z_0 \geq z' \tag{6-65}$$

Equation (6-65) says that when we have reached a solution $\bar{x}_B$ for which all $z_j - c_j \geq 0$, its value of z_0 is at least as great as any other feasible solution. Therefore, z_0 is the maximum value of the objective function.

In our previous discussion of how to proceed from one basic feasible solution to another, we assumed we could always replace one vector in the basis with another, if we were not yet optimal, i.e., if one or more $z_j - c_j < 0$. However, if we examine the criterion for removal of a vector from the basis

$$\frac{x_{Br}}{y_{rj}} = \min_i \left\{\frac{x_{Bi}}{y_{ij}}, y_{ij} > 0\right\} \tag{6-49}$$

it will be seen that at least one y_{ij} must be greater than zero for $i = 1, 2, \ldots, m$. We shall now see that if it should happen that all the $y_{ij} \leq 0$, $i = 1, 2, \ldots, m$ when we try to insert some vector into the basis, this is an indication of an *unbounded solution.*

Let us assume that we have a basic feasible solution $\bar{x}_B$ and, as usual,

$$\sum_{i=1}^{m} x_{Bi}\bar{a}_i = \bar{b}$$

If we both add and subtract $\phi\bar{a}_k$ to the preceding we have

$$\sum_{i=1}^{m} x_{Bi}\bar{a}_i - \phi\bar{a}_k + \phi\bar{a}_k = \bar{b} \tag{6-66}$$

ϕ is any scalar and $\bar{a}_k$ is the vector about to enter the basis. We can, of course, express $\bar{a}_k$ as a linear combination of the basis vectors

$$\bar{a}_k = \sum_{i=1}^{m} y_{ik}\bar{a}_i \tag{6-67}$$

Substituting (6-67) into (6-66) we have

$$\sum_{i=1}^{m} x_{Bi}\bar{a}_i - \phi \sum_{i=1}^{m} y_{ik}\bar{a}_i + \phi\bar{a}_k = \bar{b}$$

which upon rearrangement yields

$$\sum_{i=1}^{m} (x_{Bi} - \phi y_{ik})\bar{a}_i + \phi\bar{a}_k = \bar{b} \tag{6-68}$$

The solution exhibited in equation (6-68) is a feasible but nonbasic solution since it has $m + 1$ nonzero variables. The objective function for this solution is

$$\hat{z} = \sum_{i=1}^{m} c_{Bi}(x_{Bi} - \phi y_{ik}) + c_k\phi$$

or

$$\hat{z} = z + \phi(c_k - z_k) \tag{6-69}$$

It can be seen from equations (6-68) and (6-69) that if $z_k - c_k < 0$ and all the $y_{ik} \leq 0$, then $x_{Bi} - \phi y_{ik} \geq 0$ for all $\phi \geq 0$ and $\hat{z}$ can be made arbitrarily large as ϕ increases. The linear programming problem then has an unbounded solution. As we have indicated previously, no properly formulated linear programming problem, purporting to deal with the real world, can have an unbounded maximum. Such an indication in the course of a calculation is a sign that an error has been made in the formulation of the original problem.

In the theory discussed we have assumed, in various places, that degeneracy did not exist. Let us examine this assumption now. We showed that when all $z_j - c_j \geq 0$ for all $\bar{a}_j$ not in the basis, we would have the optimal solution. However, whether or not we can reach that particular basic feasible solution may conceivably depend upon the presence of degeneracy. We can see this by examining equation (6-55):

$$\hat{z} = z + \frac{x_{Br}}{y_{rj}}(c_j - z_j) \tag{6-55}$$

If degeneracy is present, i.e., $x_{Br} = 0$, then $\hat{z} = z$. It turns out that when degeneracy occurs, one cannot be sure that some basis will not be repeated. The reason for this is that, in the absence of degeneracy, $\hat{z} > z$. If this is true, we can never return to a previous basis. However, if $\hat{z} = z$, which is possible when degeneracy is present, we may return to a previously examined basis. The possibility then exists for an endless cycling through the same chain of basic feasible solutions. If this occurs, the computation may not terminate even though there are only a finite number of bases. Such a process is called *cycling* and it can occur in the presence of degeneracy.

There exist several mathematical techniques which allow degenerate solutions to be avoided and hence prevent the occurrence of cycling. (See Hadley[1] and Dantzig.[2]) It turns out, however, that no practical problem has ever been observed to exhibit cycling, even though degeneracy is quite prevalent in linear programming problems. In fact, one has to go to considerable trouble to construct a linear programming problem which will exhibit cycling of bases when the ordinary simplex computational method is employed. Therefore, even though the presence of degeneracy causes a theoretical problem, which can be resolved, it is never a problem in practice.

In Figure 6-2, we illustrated an example of a linear programming problem in which there was more than one basic feasible solution which yielded the optimal value of the objective function. Indeed, there were an infinity of nonbasic feasible solutions which also yielded the optimal value of the objective function. Let us now show this in a general way.

Let us assume that a particular linear programming problem has p different basic feasible solutions which are also optimal. Let $\bar{x}_l$, $l = 1, 2, \ldots, p$, be the n-vectors representing each such solution. Now let us consider any convex combination of these solutions.

$$\bar{x} = \sum_{l=1}^{p} \mu_l \bar{x}_l \qquad \mu_l \geq 0 \qquad l = 1, 2, \ldots, p$$

where (6-70)

$$\sum_{l=1}^{p} \mu_l = 1$$

Since each $\bar{x}_l \geq 0$ and the $\mu_l \geq 0$, it is clear that $\bar{x} \geq 0$. Further, since $A\bar{x}_l = \bar{b}$ by assumption, then

$$A\bar{x} = A\sum_{l=1}^{p} \mu_l \bar{x}_l = \sum_{l=1}^{p} \mu_l A\bar{x}_l = \sum_{l=1}^{p} \mu_l \bar{b} = \bar{b}\sum_{l=1}^{p} \mu_l = \bar{b}$$

Therefore, $\bar{x}$ is a feasible solution, although it is clearly not necessarily basic. We also assumed that Max $z = \bar{c}'\bar{x}_l$. Therefore,

$$\bar{c}'\bar{x} = \bar{c}'\sum_{l=1}^{p} \mu_l \bar{x}_l = \sum_{l=1}^{p} \mu_l \bar{c}'\bar{x}_l = \sum_{l=1}^{p} \mu_l(\text{Max } z) = \text{Max } z$$

Therefore, $\bar{x}$ is also an optimal solution. What we have shown is that any convex combination of p different optimal basic feasible solutions will also be an optimal solution.

It is a relatively simple matter to generate some or all of the alternative optimal basic feasible solutions, if one should wish to do so. Suppose one has an optimal basic feasible solution to a linear programming problem. Since this is the case, $z_j - c_j \geq 0$, $j = 1, 2, \ldots, n$. Suppose further that there exists a vector $\bar{a}_k$, not in the basis, such that $z_k - c_k = 0$ and that $y_{ik} > 0$ for at least one $i = 1, 2, \ldots, m$. Then, since

$$\frac{x_{Br}}{y_{rk}} = \text{Min}\left\{\frac{x_{Bi}}{y_{ik}}, y_{ik} > 0\right\}$$

and

$$\hat{z} = z + \frac{x_{Br}}{y_{rk}}(c_k - z_k)$$

if $\bar{a}_k$ enters the basis $\hat{z} = z$. Therefore, we have an alternate optimal solution. Hence, we can always find alternate optimal solutions, if we should wish to do so, by this method.

In the following section we shall develop the simplex computational algorithm in its tableau form. Before we do this, however, let us illustrate the iterative process developed in the foregoing theory by means of a simple example. Let us consider an example we examined previously in equations (6-11) and graphically in Figure 6-1:

$$\begin{aligned} \text{Max } z &= 2x_1 - 4x_2 \\ 3x_1 + 5x_2 &\geq 15 \\ 4x_1 + 9x_2 &\leq 36 \\ x_1, x_2 &\geq 0 \end{aligned} \qquad \textbf{(6-11)}$$

First we convert the constraints of (6-11) into equalities by adding slack or surplus variables:

$$\begin{aligned} 3x_1 + 5x_2 - x_3 &= 15 \\ 4x_1 + 9x_2 + x_4 &= 36 \\ x_1, x_2, x_3, x_4 &\geq 0 \end{aligned}$$

Let us now define the A matrix and its constituent vectors, $\bar{a}_j$, $j = 1, 2, 3, 4$:

$$A = \begin{bmatrix} 3 & 5 & -1 & 0 \\ 4 & 9 & 0 & 1 \end{bmatrix} \qquad \bar{b} = \begin{bmatrix} 15 \\ 36 \end{bmatrix}$$

$$\bar{a}_1 = \begin{bmatrix} 3 \\ 4 \end{bmatrix} \qquad \bar{a}_2 = \begin{bmatrix} 5 \\ 9 \end{bmatrix} \qquad \bar{a}_3 = \begin{bmatrix} -1 \\ 0 \end{bmatrix} \qquad \bar{a}_4 = \begin{bmatrix} 0 \\ 1 \end{bmatrix}$$

We take as our initial basis, $\bar{a}_1$ and $\bar{a}_4$. This means that

$$x_1\bar{a}_1 + x_4\bar{a}_4 = \bar{b}$$

$$x_1\begin{bmatrix} 3 \\ 4 \end{bmatrix} + x_4\begin{bmatrix} 0 \\ 1 \end{bmatrix} = \begin{bmatrix} 15 \\ 36 \end{bmatrix}$$

Therefore,

$$x_1 = 5 \qquad x_4 = 16 \qquad x_2 = x_3 = 0$$

Our initial basic feasible solution is $\bar{x}_B = \begin{bmatrix} 5 \\ 16 \end{bmatrix}$. Now we shall express the vectors not in the basis, i.e., $\bar{a}_2$, $\bar{a}_3$ in terms of $\bar{a}_1$ and $\bar{a}_4$.

$$\bar{a}_2 = y_{12}\bar{a}_1 + y_{42}\bar{a}_4$$

$$\bar{a}_3 = y_{13}\bar{a}_1 + y_{43}\bar{a}_4$$

$$\begin{bmatrix} 5 \\ 9 \end{bmatrix} = y_{12}\begin{bmatrix} 3 \\ 4 \end{bmatrix} + y_{42}\begin{bmatrix} 0 \\ 1 \end{bmatrix}$$

This yields

$$\bar{y}_2 = [y_{12}, y_{42}] = [\tfrac{5}{3}, \tfrac{7}{3}]$$

Similarly,

$$\begin{bmatrix} -1 \\ 0 \end{bmatrix} = y_{13}\begin{bmatrix} 3 \\ 4 \end{bmatrix} + y_{43}\begin{bmatrix} 0 \\ 1 \end{bmatrix}$$

which yields

$$\bar{y}_3 = [y_{13}, y_{43}] = [-\tfrac{1}{3}, \tfrac{4}{3}]$$

Let us now compute z_j for $\bar{a}_2$ and $\bar{a}_3$.

$$z_j = \bar{c}'_B\bar{y}_j \qquad \bar{c}'_B = (2, 0) \qquad c_2 = -4$$

$$c_3 = 0$$

$$z_2 = (2, 0)\begin{bmatrix} \frac{5}{3} \\ \frac{7}{3} \end{bmatrix} = \tfrac{10}{3}$$

$$z_3 = (2, 0)\begin{bmatrix} -\frac{1}{3} \\ \frac{4}{3} \end{bmatrix} = -\tfrac{2}{3}$$

$$z_2 - c_2 = \tfrac{10}{3} + 4 = \tfrac{22}{3}$$

$$z_3 - c_3 = -\tfrac{2}{3} - 0 = -\tfrac{2}{3}$$

Since only $z_3 - c_3 < 0$, $\bar{a}_3$ will be chosen to enter the basis. Let us now determine which vector must be removed. The rule is

$$\operatorname*{Min}_i \left\{\frac{x_{Bi}}{y_{i3}}, y_{i3} > 0\right\}$$

We have only two choices $\frac{x_1}{y_{13}}$ or $\frac{x_4}{y_{43}}$. However, since $y_{13} = -\frac{1}{3} < 0$, we can only consider $\frac{x_4}{y_{43}}$. Hence, in this simple case, $\bar{a}_4$ leaves the basis.

The original value of the objective function was

$$z = \bar{c}_B'\bar{x}_B = (2, 0)\begin{bmatrix} 5 \\ 16 \end{bmatrix} = 10$$

According to our formula

$$\hat{z} = z + \frac{x_4}{y_{43}}(c_3 - z_3) = 10 + \frac{16}{\frac{4}{3}}\left(\frac{2}{3}\right) = 18$$

and as we see $\hat{z} > z$. We can check this by direct computation of $\hat{z}$. The new value of $\bar{c}_B$ is $\hat{\bar{c}}_B' = (2, 0)$. Now let us compute the new value of $\bar{x}_B$:

$$x_1\bar{a}_1 + x_3\bar{a}_3 = \bar{b}$$

$$x_1\begin{bmatrix} 3 \\ 4 \end{bmatrix} + x_3\begin{bmatrix} -1 \\ 0 \end{bmatrix} = \begin{bmatrix} 15 \\ 36 \end{bmatrix}$$

Therefore,

$$x_1 = 9 \qquad x_3 = 12 \qquad \hat{\bar{x}}_B = \begin{bmatrix} 9 \\ 12 \end{bmatrix}$$

$$\hat{z} = \hat{\bar{c}}_B'\hat{\bar{x}}_B = (2, 0)\begin{bmatrix} 9 \\ 12 \end{bmatrix} = 18$$

Let us now repeat our previous work and express the vectors not in the basis, $\bar{a}_2$, $\bar{a}_4$ in terms of $\bar{a}_1$ and $\bar{a}_3$, our basic vectors.

$$\bar{a}_2 = y_{12}\bar{a}_1 + y_{32}\bar{a}_3$$

$$\bar{a}_4 = y_{14}\bar{a}_1 + y_{34}\bar{a}_3$$

$$\begin{bmatrix} 5 \\ 9 \end{bmatrix} = y_{12}\begin{bmatrix} 3 \\ 4 \end{bmatrix} + y_{32}\begin{bmatrix} -1 \\ 0 \end{bmatrix}$$

This yields

$$\bar{y}_2 = [y_{12}, y_{32}] = [\tfrac{9}{4}, \tfrac{7}{4}]$$

Similarly,

$$\begin{bmatrix} 0 \\ 1 \end{bmatrix} = y_{14}\begin{bmatrix} 3 \\ 4 \end{bmatrix} + y_{34}\begin{bmatrix} -1 \\ 0 \end{bmatrix}$$

$$\bar{y}_4 = [y_{14}, y_{34}] = [\tfrac{1}{4}, \tfrac{3}{4}]$$

Next, we compute z_j for $\bar{a}_2$ and $\bar{a}_3$, noting that $\bar{c}'_B = (2, 0)$, $c_2 = -4$, $c_4 = 0$

$$z_2 = \bar{c}'_B \bar{y}_2 = (2, 0)\begin{bmatrix} \frac{9}{4} \\ \frac{7}{4} \end{bmatrix} = \tfrac{9}{2}$$

$$z_4 = \bar{c}'_B \bar{y}_4 = (2, 0)\begin{bmatrix} \frac{1}{4} \\ \frac{3}{4} \end{bmatrix} = \tfrac{1}{2}$$

$$z_2 - c_2 = \tfrac{9}{2} + 4 = \tfrac{17}{2} > 0$$

$$z_4 - c_4 = \tfrac{1}{2} - 0 = \tfrac{1}{2} > 0$$

Since all $z_j - c_j \geq 0$ for vectors not in the basis, we have the optimal solution to the linear programming problem, i.e., $x_1 = 9$, $x_2 = 0$, $x_3 = 12$, $x_4 = 0$, and $z = 18$.

6.6 THE SIMPLEX METHOD: COMPUTATIONAL TECHNIQUES

In this section we shall state the actual computational algorithm used to solve linear programming problems. There are many variations in practice but they can all be explained in terms of the rules we give. We shall also give examples of the solution of simple problems.

Before we describe the algorithm and the "tableau format," we need to consider how to simply transform the components of $\bar{y}_j$, the coefficients expressing nonbasic vectors as linear combinations of basis vectors. We will require such a transformation each time we enter a new vector into the basis and simultaneously remove one of the previous basis vectors.

Let us assume that $\bar{a}_k$ is to enter the basis and $\bar{a}_r$ is to be removed from the basis. We know that for any vector we can write

$$\bar{a}_j = \sum_{i=1}^{m} y_{ij}\bar{a}_i \qquad \textbf{(6-58)}$$

We can rewrite (6-58) in terms of certain vectors $\bar{a}_r$ and $\bar{a}_k$ as follows:

$$\bar{a}_k = y_{rk}\bar{a}_r + \sum_{\substack{i=1 \\ i \neq r}}^{m} y_{ik}\bar{a}_i \qquad \textbf{(6-71)}$$

If we now solve (6-71) for $\bar{a}_r$, since $y_{rk} \neq 0$, we have

$$\bar{a}_r = \frac{1}{y_{rk}}\,\bar{a}_k - \sum_{\substack{i=1 \\ i \neq r}}^{m} \frac{y_{rk}}{y_{ik}}\,\bar{a}_i \qquad \textbf{(6-72)}$$

If we rewrite (6-58) as

$$\bar{a}_j = y_{rj}\bar{a}_r + \sum_{\substack{i=1 \\ i \neq r}}^{m} y_{ij}\bar{a}_i \qquad \textbf{(6-73)}$$

and now substitute (6-72) into (6-73), we have

$$\bar{a}_j = \sum_{\substack{i=1 \\ i \neq r}}^{m} \left(y_{ij} - \frac{y_{rj} y_{ik}}{y_{rk}} \right) \bar{a}_i + \frac{y_{rj}}{y_{rk}} \bar{a}_k \tag{6-74}$$

Equation (6-58) expressed any vector $\bar{a}_j$ in terms of the current basis vectors, $\bar{a}_i$, $i = 1, 2, \ldots, m$. Then we inserted $\bar{a}_k$ and removed $\bar{a}_r$. Therefore, equation (6-74) expresses any vector $\bar{a}_j$ in terms of the new set of basis vectors, $\bar{a}_i$, $i = 1, 2, \ldots, m$, $i \neq r$, and $\bar{a}_k$. If we rewrite equation (6-74) as

$$\bar{a}_j = \sum_{i=1}^{m} \hat{y}_{ij} \hat{\bar{a}}_i \tag{6-75}$$

Then, if we define $\hat{\bar{a}}_i = \bar{a}_i$, $i \neq r$, and $\hat{\bar{a}}_r = \bar{a}_k$, we have

$$\begin{aligned} \hat{y}_{ij} &= y_{ij} - \frac{y_{rj} y_{ik}}{y_{rk}} \qquad i \neq r \\ \hat{y}_{rj} &= \frac{y_{rj}}{y_{rk}} \end{aligned} \tag{6-76}$$

Equations (6-76) are computational equations which allow us to calculate a new set of y_{ij} from the old, after determining which vectors are to enter and leave the basis.

We must also transform the $z_j - c_j$. It is more efficient to do this than to recalculate $\hat{z}_j - c_j$ from its definition. First we note that by definition

$$\hat{z}_j - c_j = \sum_{i=1}^{m} \hat{c}_{Bi} \hat{y}_{ij} - c_j \tag{6-77}$$

where $\hat{c}_{Bi} = c_{Bi}$, $i \neq r$, and $\hat{c}_{Br} = c_k$. If we now substitute from equations (6-76) into (6-77) we have

$$\hat{z}_j - c_j = \sum_{\substack{i=1 \\ i \neq r}}^{m} c_{Bi} \left(y_{ij} - \frac{y_{rj} y_{ik}}{y_{rk}} \right) + \frac{y_{rj}}{y_{rk}} c_k - c_j \tag{6-78}$$

If we also note that the missing term in the summation is

$$c_{Br} \left(y_{rj} - \frac{y_{rj} y_{rk}}{y_{rk}} \right) = 0 \tag{6-79}$$

we can combine (6-78) and (6-79) to obtain

$$\begin{aligned} \hat{z}_j - c_j &= \sum_{i=1}^{m} c_{Bi} \left(y_{ij} - \frac{y_{rj} y_{ik}}{y_{rk}} \right) + \frac{y_{rj}}{y_{rk}} c_k - c_j \\ &= \sum_{i=1}^{m} c_{Bi} y_{ij} - \frac{y_{rj}}{y_{rk}} \sum_{i=1}^{m} c_{Bi} y_{ik} + \frac{y_{rj}}{y_{rk}} c_k - c_j \\ &= \sum_{i=1}^{m} c_{Bi} y_{ij} - c_j - \frac{y_{rj}}{y_{rk}} \left[\sum_{i=1}^{m} c_{Bi} y_{ik} - c_k \right] \\ &= z_j - c_j - \frac{y_{rj}}{y_{rk}} (z_k - c_k) \end{aligned}$$

Therefore,

$$\hat{z}_j - c_j = z_j - c_j - \frac{y_{rj}}{y_{rk}}(z_k - c_k) \qquad \textbf{(6-80)}$$

In the previous discussions, we have always assumed that we had a basic feasible solution and, if it was not an optimal solution, we would seek another basic feasible solution with an increased value of the objective function. What must now be considered is how to obtain an *initial* basic feasible solution.

The simplest case is when the original linear programming problem has the form

$$\begin{aligned} \text{Max } z &= \bar{c}'\bar{x} \\ D\bar{x}_D &\leq \bar{b} \\ \bar{x} &\geq \bar{0} \end{aligned} \qquad \textbf{(6-81)}$$

For this case, assuming $\bar{b} \geq \bar{0}$, we may add a slack variable to each constraint so that we now have

$$D\bar{x}_D + I\bar{x}_s = \bar{b} \qquad \textbf{(6-82)}$$

where $\bar{x}_s$ is the vector of slack variables, one for each constraint, and I is an $m \times m$ identity matrix. Therefore, $A = (D, I)$ in $A\bar{x} = \bar{b}$ and $\bar{x} = \begin{bmatrix} \bar{x}_D \\ \bar{x}_s \end{bmatrix}$. It is easy to see why this is such a convenient case. If we set $\bar{x}_D = \bar{0}$, we have $I\bar{x}_s = \bar{b}$ or $\bar{x}_s = \bar{b}$ as the initial basic feasible solution. Furthermore, since the basis matrix $B = I$, $B^{-1} = I^{-1} = I$ and

$$\bar{y}_j = B^{-1}\bar{a}_j = I\bar{a}_j = \bar{a}_j \qquad \textbf{(6-83)}$$

Since the slack variables have no c_j associated with them, $\bar{c}_B = \bar{0}$ and $z = \bar{c}_B'\bar{x}_B = 0$. The $z_j - c_j$ are also simply calculated as

$$z_j - c_j = \bar{c}_B'\bar{y}_j - c_j = -c_j \qquad \textbf{(6-84)}$$

If the constraints have the form given in (6-82), then the presence of an identity basis is obvious. However, if by inspection we can identify which columns of the matrix A in $A\bar{x} = \bar{b}$, lead to an $m \times m$ identity submatrix, we can just as easily do what is described in equations (6-81) to (6-84). However, there will be situations in which this is not possible. For the situation in which an identity submatrix is not present, we need a perfectly general method of proceeding. We will now describe such a method. First, however, a cautionary note. The method we are about to describe is probably the simplest one to employ for small problems to be done by a student or one familiarizing himself with the essential features of computational linear programming. In fact, it was used in the early days of linear programming calculations. However, for large problems, which the scientist, engineer, businessman, and people in other fields will be involved in solving, standard digital computer programs will be used. These are of great complexity and they use so-called "two-phase" or "composite" algorithms. The spirit of these methods is similar to what we shall describe in that they start with a partial or full set of "artificial" variables, as we shall. However, the details of these methods are quite different and a complete description of them is beyond the scope of this introductory work.

The method of *artificial variables* for finding an initial basic feasible solution that we shall describe is due to Dantzig. We assume that our constraints are in the form $A\bar{x} = \bar{b}$ and that $\bar{b} \geq \bar{0}$. We now add a set of m *artificial* variables $\bar{x}_a$ to the constraints to derive a new set of constraints:

$$A\bar{x} + I\bar{x}_a = \bar{b} \tag{6-85}$$

The similarity of equation (6-85) to (6-82) is obvious. It is, therefore, a simple matter to find an initial basic feasible solution to equations (6-85). We simply let $\bar{x} = \bar{0}$ and $\bar{x}_a = \bar{b}$. However, this solution is *not* a feasible solution to $A\bar{x} = \bar{b}$. At this point it is not obvious that we have gained anything. We might make the following observation, however. We could add the artificial variables $\bar{x}_a$ to the problem, as we have described, and then employ the simplex method of iterating from one basic feasible solution to another. If we do this in such a way that we reach a point where $\bar{x}_a = \bar{0}$, we will then have a basic feasible solution to $A\bar{x} = \bar{b}$, our original problem. We could then drop the artificial variables completely and proceed with simplex iterations to find the optimal solution. This is precisely what is done.

Clearly, a way must be found to force $\bar{x}_a = \bar{0}$ at some point. The way to do this is quite simple. If we assign to these artificial variables values of c_j, which we shall call c_{ai} and which are extremely small in a maximization problem (or extremely large in a minimization problem), we should ultimately be able to drive the columns corresponding to these variables out of the basis. In short, we provide heavy penalties, in terms of the objective function, for continuing to keep these variables as basic.

If we are maximizing, then $c_{ai} = -M$, $M > 0$ and if we are minimizing, $c_{ai} = M$, $M > 0$, where M is an extremely large number. As we shall see it is not necessary to specify its value in a hand computation. It is a simple matter to show that once an artificial variable leaves the basis, it can never re-enter.

It should be noted that it is not always necessary to add a full complement of artificial vectors. If one or more of the vectors $\bar{a}_j$ are unit vectors of the form $\bar{e}_i$, then they can be used, along with whatever $\bar{e}_i$ are added to make up a full identity matrix, I. This way one may be able to add fewer artificial variables and hence reduce the number of simplex iterations. This will be illustrated in the problems.

To summarize, instead of the original problem,

$$\begin{aligned} \text{Max } z &= \bar{c}'\bar{x} \\ A\bar{x} &= \bar{b} \\ \bar{x} &\geq \bar{0} \end{aligned} \tag{6-86}$$

we solve instead

$$\begin{aligned} \text{Max } z &= \bar{c}'\bar{x} - M\bar{1}\bar{x}_a \\ A\bar{x} + I\bar{x}_a &= \bar{b} \\ \bar{x}, \bar{x}_a &\geq \bar{0} \end{aligned} \tag{6-87}$$

We first drive all x_{ai} to zero and then drop the artificial variables and continue the simplex iterative process until we reach the optimal basic feasible solution.

If one uses the foregoing method, or one of the other methods which make use of artificial variables, some consideration must be given to certain possible anomalous cases. If when the optimality criterion is satisfied, i.e., $z_j - c_j \geq 0$, $j = 1, 2, \ldots, n$, there are no artificial vectors in the basis, then it is clear that we have found the optimal basic feasible solution. On the other hand, if there are one or more artificial vectors (vectors corresponding to artificial variables) in the basis at a positive level, then it is clear that there is no basic feasible solution to $A\bar{x} = \bar{b}$, $\bar{x} \geq \bar{0}$. If there were, the artificial variables would have been driven to zero. There is one other case, however. There might be one or more artificial vectors in the basis, but at a zero level. In this case, we clearly have a feasible solution to the original problem. However, this is an indication of a degenerate basic feasible solution and, possibly, of redundancy in the constraints. This is discussed in detail in Hadley.[1] The reader is also asked to examine these two cases in the problems at the end of this chapter.

Let us now describe the tableau format. Previously, for simplicity we considered that vectors $\bar{a}_1, \ldots, \bar{a}_m$ constituted a basis. In order to have a more general notation let us now designate whichever subset of m vectors $\bar{a}_j, j = 1, 2, \ldots, n$, which are in the basis as $B = (\bar{b}_1, \bar{b}_2, \ldots, \bar{b}_m)$. Let us also designate the artificial vectors as $\bar{a}_{n+1}, \bar{a}_{n+2}, \ldots, \bar{a}_{n+m}$ if there is a full set. There may be fewer than m vectors, as we have discussed previously. What we mean by a "simplex tableau" is a convenient means of representation of all the necessary information at each iteration.† This consists of

1. The vectors in the basis
2. The current basic feasible solution, $\bar{x}_B$

† The tableau can also be regarded as a representation of the detached coefficients of the variables in the constraints and objective function when one represents the nonbasic variables in terms of the basic variables (or vice-versa as some authors prefer). We have chosen not to emphasize this aspect of the tableau format.

3. The current value of the objective function, z

4. The representation of all the vectors in terms of the basis vectors, $\bar{y}_i$

5. The current values of $z_j - c_j$ for all j

6. The prices in the basis, $\bar{c}_B$, as well as the price associated with each variable

A convenient way to display all this information, the *simplex tableau*, is shown in Table 6.2. We can see that all the information listed is displayed in the tableau in Table 6.2.

Let us now review the simplex method with reference to the tableau. First, we examine all the $z_j - c_j$ in the bottom row. If all the $z_j - c_j \geq 0$, and this is a maximization problem, then the current solution is optimal. If one or more of the $z_j - c_j < 0$, then the current solution $\bar{x}_B$ is not optimal. We select one of the columns with $z_j - c_j < 0$ (usually the one with the most negative value) to enter the basis. We have designated previously, the column to enter the basis as column k. If there should be a tie, choose any one of the columns. If we now examine the y_{ik}, the components of $\bar{y}_k$, and all $y_{ik} \leq 0$ (we merely examine the $\bar{y}_j$ values in column k), then we know there is an unbounded solution. If at least one $y_{ik} > 0$, we can insert $\bar{a}_k$ into the basis. We determine the vector to be removed from

$$\frac{x_{Br}}{y_{rk}} = \min_i \left\{\frac{x_{Bi}}{y_{ik}}, y_{ik} > 0\right\} \qquad \textbf{(6-88)}$$

This expression indicates which column, column r, is to be removed from the basis. Column r is to be replaced by column k in the basis. In order to calculate the new tableau from the current tableau we use the formulas developed earlier in this section. The $\bar{y}_j$ are transformed according to equations (6-76):

$$\hat{y}_{ij} = y_{ij} - \frac{y_{rj}y_{ik}}{y_{rk}} \qquad i \neq r$$

$$\hat{y}_{rj} = \frac{y_{rj}}{y_{rk}} \qquad \textbf{(6-76)}$$

Table 6.2 *Simplex Method Tableau*

		c_j	c_1	c_2	$\cdots$	c_j	$\cdots$	c_n	$-M$	$-M$	$\cdots$	$-M$
$\bar{c}_B$	BASIS VECTORS	x_B	$\bar{a}_1$	$\bar{a}_2$	$\cdots$	$\bar{a}_j$	$\cdots$	$\bar{a}_n$	$\bar{a}_{n+1}$	$\bar{a}_{n+2}$	$\cdots$	$\bar{a}_{n+m}$
c_{B1}	$\bar{b}_1$	x_{B1}	y_{11}	y_{12}	$\cdots$	y_{1j}	$\cdots$	y_{1n}	$y_{1,n+1}$	$y_{1,n+2}$	$\cdots$	$y_{1,n+m}$
c_{B2}	$\bar{b}_2$	x_{B2}	y_{21}	y_{22}	$\cdots$	y_{2j}	$\cdots$	y_{2n}	$y_{2,n+1}$	$y_{2,n+2}$	$\cdots$	$y_{2,n+m}$
$\vdots$	$\vdots$	$\vdots$	$\vdots$	$\vdots$	$\vdots$	$\vdots$	$\vdots$	$\vdots$	$\vdots$	$\vdots$	$\vdots$	$\vdots$
c_{Bm}	$\bar{b}_m$	x_{Bm}	y_{m1}	y_{m2}	$\cdots$	y_{mj}	$\cdots$	y_{mn}	$y_{m,n+1}$	$y_{m,n+2}$	$\cdots$	$y_{m,n+m}$
		z	$z_1 - c_1$	$z_2 - c_2$	$\cdots$	$z_j - c_j$	$\cdots$	$z_n - c_n$	$z_{n+1} - c_{n+1}$	$z_{n+2} - c_{n+2}$	$\cdots$	$z_{n+m} - c_{n+m}$

The x_{Bi} and z can also be transformed using the preceding equations, as a little algebra will show, i.e.,

$$\hat{x}_{Bi} = x_{Bi} - \frac{y_{ik}x_{Br}}{y_{rk}} \qquad i \neq r$$

$$\hat{x}_{Br} = \frac{x_{Br}}{y_{rk}} \tag{6-89}$$

$$\hat{z} = z - \frac{x_{Br}}{y_{rk}}(z_k - c_k)$$

The $z_j - c_j$ can be transformed according to equations (6-80):

$$\hat{z}_j - c_j = z_j - c_j - \frac{y_{rj}}{y_{rk}}(z_k - c_k) \tag{6-80}$$

We can combine all the transformation formulas into one set by means of the following notation. If we let $y_{i0} = x_{Bi}, i = 1, 2, \ldots, m$; $y_{m+1,0} = z$ and finally $y_{m+1,j} = z_j - c_j, j = 1, 2, \ldots, n$, then we may write as transformation formulas for the entire tableau

$$\hat{y}_{ij} = y_{ij} - y_{ik}\frac{y_{rj}}{y_{rk}} \qquad \begin{array}{l} j = 0, 1, \ldots, n \\ i = 1, 2, \ldots, m+1 \\ i \neq r \end{array} \tag{6-90}$$

$$\hat{y}_{rj} = \frac{y_{rj}}{y_{rk}} \qquad j = 0, 1, \ldots, n$$

In practice, these equations are easier to use than appear at first glance. The process will become quite clear by means of the following examples.

EXAMPLE 6.1

$$\begin{aligned} \text{Max } z &= 4x_1 + 3x_2 \\ 2x_1 + 3x_2 &\leq 18 \\ 4x_1 + 2x_2 &\leq 10 \\ x_1, x_2 &\geq 0 \end{aligned} \tag{6-91}$$

First we add slack variables to the constraints to obtain

$$\begin{aligned} 2x_1 + 3x_2 + x_3 &= 18 \\ 4x_1 + 2x_2 + x_4 &= 10 \end{aligned} \tag{6-92}$$

Our vectors $\bar{a}_j$ and $\bar{b}$ are as follows:

$$\bar{a}_1 = \begin{bmatrix} 2 \\ 4 \end{bmatrix} \quad \bar{a}_2 = \begin{bmatrix} 3 \\ 2 \end{bmatrix} \quad \bar{a}_3 = \begin{bmatrix} 1 \\ 0 \end{bmatrix} \quad \bar{a}_4 = \begin{bmatrix} 0 \\ 1 \end{bmatrix} \quad \bar{b} = \begin{bmatrix} 18 \\ 10 \end{bmatrix}$$

Table 6.3 *Tableau 1 for Example 6.1*

	c_j		4	3	0	0
$\bar{c}_B$	BASIS VECTORS	$\bar{x}_B$	$\bar{a}_1$	$\bar{a}_2$	$\bar{a}_3$	$\bar{a}_4$
0	$\bar{a}_3$	18	2	3	1	0
0	$\bar{a}_4$	10	④	2	0	1
		0	−4	−3	0	0

Since $(\bar{a}_3, \bar{a}_4) = \begin{bmatrix} 1 & 0 \\ 0 & 1 \end{bmatrix} = I$, we have an initial basic feasible solution with $\bar{a}_3$ and $\bar{a}_4$ in the basis. Since x_3 and x_4 do not appear in the objective function, their prices are zero and therefore $\bar{c}'_B = (0, 0)$. In addition $\bar{x}_B = [18, 10]$, $z = \bar{c}'_B\bar{x}_B = 0$ and

$$z_j - c_j = \bar{c}'_B\bar{y}_j - c_j = -c_j$$

The first tableau is shown in Table 6.3. Since $z_1 - c_1 < 0$ and $z_2 - c_2 < 0$, we do not have the optimal solution. If we use the criterion for selecting the vector to enter the basis

$$z_k - c_k = \operatorname*{Min}_j\,(z_j - c_j < 0) \tag{6-93}$$

Then $z_1 - c_1 = \text{Min}\,(z_1 - c_1, z_2 - c_2) = \text{Min}\,(-4, -3) = -4$. Therefore, $\bar{a}_1$ enters the basis. Now we need to determine which vector leaves the basis. This is given by

$$\frac{x_{Br}}{y_{r1}} = \operatorname*{Min}_i \left\{\frac{x_{Bi}}{y_{i1}}, y_{i1} > 0\right\}$$

$$= \text{Min}\left(\frac{18}{2}, \frac{10}{4}\right) = \frac{10}{4} = \frac{x_{B2}}{y_{21}}$$

Therefore, $\bar{a}_4$ leaves the basis, since it is the second "column" in the basis.

The element $y_{rk} = y_{21} = 4$ is circled in Tableau 1 for convenience. First we transform the row leaving the basis as follows:

$$\hat{x}_{B2} = \hat{y}_{20} = \frac{y_{20}}{y_{21}} = \frac{10}{4} = \frac{5}{2}$$

$$\hat{y}_{21} = \frac{y_{21}}{y_{21}} = 1$$

$$\hat{y}_{22} = \frac{y_{22}}{y_{21}} = \frac{2}{4} = \frac{1}{2}$$

$$\hat{y}_{23} = \frac{y_{23}}{y_{21}} = \frac{0}{4} = 0$$

$$\hat{y}_{24} = \frac{y_{24}}{y_{21}} = \frac{1}{4}$$

To transform row 1, we compute

$$\frac{y_{ik}}{y_{rk}} = \frac{y_{11}}{y_{21}} = \frac{2}{4} = \frac{1}{2}$$

Now we use

$$\hat{y}_{ij} = y_{ij} - \frac{y_{11}}{y_{21}} y_{2j}$$

$$\hat{y}_{10} = y_{10} - \tfrac{1}{2} y_{21} = 18 - \tfrac{1}{2}(10) = 13$$

$$\hat{y}_{11} = y_{11} - \tfrac{1}{2} y_{21} = 2 - \tfrac{1}{2}(4) = 0$$

$$\hat{y}_{12} = y_{12} - \tfrac{1}{2} y_{22} = 3 - \tfrac{1}{2}(2) = 2$$

$$\hat{y}_{13} = y_{13} - \tfrac{1}{2} y_{23} = 1 - \tfrac{1}{2}(0) = 1$$

$$\hat{y}_{14} = y_{14} - \tfrac{1}{2} y_{24} = 0 - \tfrac{1}{2}(1) = -\tfrac{1}{2}$$

We similarly transform the $z_j - c_j$ row by computing

$$\frac{y_{31}}{y_{21}} = \frac{-4}{4} = -1$$

Therefore, $\hat{y}_{3j} = y_{3j} - \dfrac{y_{31}}{y_{21}} y_{2j} = y_{3j} + y_{2j}$

$$\hat{y}_{30} = \hat{z} = 0 + 10 = 10$$

$$\hat{y}_{31} = \hat{z}_1 - c_1 = -4 + 4 = 0$$

$$\hat{y}_{32} = \hat{z}_2 - c_2 = -3 + 2 = -1$$

$$\hat{y}_{33} = \hat{z}_3 - c_3 = 0 + 0 = 0$$

$$\hat{y}_{34} = \hat{z}_4 - c_4 = 0 + 1 = 1$$

We can now represent all this information in the new Tableau 2 given in Table 6.4. From Tableau 2 we see that only $z_2 - c_2 < 0$. Therefore it will enter the basis. We compute

$$\frac{x_{Br}}{y_{r2}} = \text{Min}\left(\frac{13}{2}, \frac{\frac{5}{2}}{\frac{1}{2}}\right) = 5$$

Table 6.4 *Tableau 2 for Example 6.1*

$\bar{c}_B$	Basis Vectors c_j	$\bar{x}_B$	$\bar{a}_1$ (4)	$\bar{a}_2$ (3)	$\bar{a}_3$ (0)	$\bar{a}_4$ (0)
0	$\bar{a}_3$	13	0	2	1	$-\frac{1}{2}$
4	$\bar{a}_1$	$\frac{5}{2}$	1	$\textcircled{\frac{1}{2}}$	0	$\frac{1}{4}$
		10	0	−1	0	1

Therefore, $\bar{a}_1$ leaves the basis (even though it came in on the previous iteration). $y_{rk} = y_{22} = \frac{1}{2}$. First we transform the second row:

$$\hat{x}_{B2} = \hat{y}_{20} = \frac{y_{20}}{y_{22}} = \frac{\frac{5}{2}}{\frac{1}{2}} = 5$$

$$\hat{y}_{21} = \frac{y_{21}}{y_{22}} = \frac{1}{\frac{1}{2}} = 2$$

$$\hat{y}_{22} = \frac{y_{22}}{y_{22}} = \frac{\frac{1}{2}}{\frac{1}{2}} = 1$$

$$\hat{y}_{23} = \frac{y_{23}}{y_{22}} = \frac{0}{\frac{1}{2}} = 0$$

$$\hat{y}_{24} = \frac{y_{24}}{y_{22}} = \frac{\frac{1}{4}}{\frac{1}{2}} = \frac{1}{2}$$

To transform row 1, we compute $\dfrac{y_{12}}{y_{22}} = \dfrac{2}{\frac{1}{2}} = 4$

$$\hat{y}_{10} = y_{10} - 4y_{20} = 13 - 4(\tfrac{5}{2}) = 13 - 10 = 3$$
$$\hat{y}_{11} = y_{11} - 4y_{21} = 0 - 4(1) = 0 - 4 = -4$$
$$\hat{y}_{12} = y_{12} - 4y_{22} = 2 - 4(\tfrac{1}{2}) = 2 - 2 = 0$$
$$\hat{y}_{13} = y_{13} - 4y_{23} = 1 - 4(0) = 1 - 0 = 1$$
$$\hat{y}_{14} = y_{14} - 4y_{24} = -\tfrac{1}{2} - 4(\tfrac{1}{4}) = -\tfrac{1}{2} - 1 = -\tfrac{3}{2}$$

To transform row 3, we compute $\dfrac{y_{32}}{y_{22}} = \dfrac{-1}{\frac{1}{2}} = -2$

$$\hat{y}_{30} = y_{30} + 2y_{20} = 10 + 2(\tfrac{5}{2}) = 15 = z$$
$$\hat{y}_{31} = y_{31} + 2y_{21} = 0 + 2(1) = 2 = z_1 - c_1$$
$$\hat{y}_{32} = y_{32} + 2y_{22} = -1 + 2(\tfrac{1}{2}) = 0 = z_2 - c_2$$
$$\hat{y}_{33} = y_{33} + 2y_{23} = 0 + 2(0) = 0 = z_3 - c_3$$
$$\hat{y}_{34} = y_{34} + 2y_{24} = 1 + 2(\tfrac{1}{4}) = \tfrac{3}{2} = z_4 - c_4$$

We show this information in Tableau 3, Table 6.5. It can be seen that $z_j - c_j \geq 0$, $j = 1, 2, 3, 4$. Therefore, we have the optimal solution.

EXAMPLE 6.2

$$\begin{aligned} \text{Max } z &= 3x_1 + 8x_2 \\ 3x_1 + 4x_2 &\leq 20 \\ x_1 + 3x_2 &\geq 6 \\ x_1, x_2 &\geq 0 \end{aligned} \qquad \textbf{(6-94)}$$

Table 6.5 *Tableau 3 for Example 6.1*

$\bar{c}_B$	Basis Vectors	$\bar{x}_B$	c_j: 4 $\bar{a}_1$	3 $\bar{a}_2$	0 $\bar{a}_3$	0 $\bar{a}_4$
0	$\bar{a}_3$	3	-4	0	1	$-\frac{3}{2}$
3	$\bar{a}_2$	5	2	1	0	$\frac{1}{2}$
		15	2	0	0	$\frac{3}{2}$

First we add slack and surplus variables to the constraints of (6-94) to obtain the following.

$$3x_1 + 4x_2 + x_3 = 20$$
$$x_1 + 3x_2 - x_4 = 6 \tag{6-95}$$

The vectors of our problem are as follows:

$$\bar{a}_1 = \begin{bmatrix} 3 \\ 1 \end{bmatrix} \quad \bar{a}_2 = \begin{bmatrix} 4 \\ 3 \end{bmatrix} \quad \bar{a}_3 = \begin{bmatrix} 1 \\ 0 \end{bmatrix} \quad \bar{a}_4 = \begin{bmatrix} 0 \\ -1 \end{bmatrix} \quad \bar{b} = \begin{bmatrix} 20 \\ 6 \end{bmatrix}$$

Since we do not have an identity submatrix contained within the columns of A we shall use an artificial basis to obtain an initial basic feasible solution after we add artificial variables, x_{a1}, x_{a2}. Hence, the constraints we are actually beginning with at this point are

$$3x_1 + 4x_2 + x_3 + x_{a1} = 20 \tag{6-96}$$
$$x_1 + 3x_2 - x_4 + x_{a2} = 6$$

We attach a price of $-M$ to both artificial variables. We are now in a position to construct the first solution and tableau. The initial basis is $B = I = (\bar{a}_5, \bar{a}_6)$ where $\bar{a}_5 = \begin{bmatrix} 1 \\ 0 \end{bmatrix}$, $\bar{a}_6 = \begin{bmatrix} 0 \\ 1 \end{bmatrix}$. Note that we need not have added $\bar{a}_5$ since we already had $\bar{a}_3 = \begin{bmatrix} 1 \\ 0 \end{bmatrix}$. We have added a *complete* artificial basis only for illustrative purposes. $\bar{c}'_B = (-M, -M)$ since initially the artificial variables are basic. Also, it is clear that $\bar{x}_B = [20, 6]$ and

$$z = (-M, -M)\begin{bmatrix} 20 \\ 6 \end{bmatrix} = -26M$$

The $z_j - c_j$ are calculated as follows:

$$z_j - c_j = \bar{c}'_B\bar{y}_j - c_j = \bar{c}'_B B^{-1}\bar{a}_j - c_j = \bar{c}'_B I\bar{a}_j - c_j = \bar{c}'_B\bar{a}_j - c_j$$

Table 6.6 *Tableau 1 for Example 6.2*

	c_j		3	8	0	0	$-M$	$-M$
$\bar{c}_B$	Basis Vectors	$\bar{x}_B$	$\bar{a}_1$	$\bar{a}_2$	$\bar{a}_3$	$\bar{a}_4$	$\bar{a}_5$	$\bar{a}_6$
$-M$	$\bar{a}_5$	20	3	4	1	0	1	0
$-M$	$\bar{a}_6$	6	1	③	0	-1	0	1
			$-4M$	$-7M$				
		$-26M$	-3	-0	$-M$	M	0	0

Therefore,

$$z_1 - c_1 = (-M, -M)\begin{bmatrix}3\\1\end{bmatrix} - 3 = -4M - 3$$

$$z_2 - c_2 = (-M, -M)\begin{bmatrix}4\\3\end{bmatrix} - 8 = -7M - 8$$

$$z_3 - c_3 = (-M, -M)\begin{bmatrix}1\\0\end{bmatrix} - 0 = -M$$

$$z_4 - c_4 = (-M, -M)\begin{bmatrix}0\\-1\end{bmatrix} - 0 = M$$

$$z_5 - c_5 = (-M, -M)\begin{bmatrix}1\\0\end{bmatrix} + M = 0$$

$$z_6 - c_6 = (-M, -M)\begin{bmatrix}0\\1\end{bmatrix} + M = 0$$

The first tableau is shown in Table 6.6. The most negative $z_j - c_j$ is $-7M - 8$. Therefore $\bar{a}_2$ will be chosen to enter the basis. To determine the vector to leave the basis

$$\frac{x_{Br}}{y_{rk}} = \text{Min}\left\{\frac{20}{4}, \frac{6}{3}\right\} = \frac{6}{4} = \frac{x_{B2}}{y_{22}}$$

Therefore $\bar{a}_6$ leaves the basis. We now transform all the quantities in Table 6.6, just as in the previous example and we obtain Tableau 2 in Table 6.7. If we examine the $z_j - c_j$ of Tableau 2 in Table 6.7, we see that the most negative is that of $\bar{a}_1$. Hence $\bar{a}_1$ is chosen to enter the basis. The vector to leave is $\bar{a}_1$ since

$$\frac{x_{B2}}{y_{21}} = \text{Min}\left\{\frac{12}{\frac{5}{3}}, \frac{2}{\frac{1}{3}}\right\} = 6$$

We now transform Tableau 2 to obtain Tableau 3 in Table 6.8. The most negative $z_j - c_j$ in Tableau 3 is that of $\bar{a}_4$. Hence, $\bar{a}_4$ will

Table 6.7 *Tableau 2 for Example 6.2*

	c_j		3	8	0	0	$-M$	$-M$
$\bar{c}_B$	Basis Vectors	$\bar{x}_B$	$\bar{a}_1$	$\bar{a}_2$	$\bar{a}_3$	$\bar{a}_4$	$\bar{a}_5$	$\bar{a}_6$
$-M$	$\bar{a}_5$	12	$\frac{5}{3}$	0	1	$\frac{4}{3}$	1	$-\frac{4}{3}$
8	$\bar{a}_2$	2	$\frac{1}{3}$ (circled)	1	0	$-\frac{1}{3}$	0	$\frac{1}{3}$
		$-12M$ $+16$	$-\frac{5}{3}M$ $-\frac{1}{3}$	0	$-M$	$-\frac{4}{3}M$ $-\frac{8}{3}$	0	$\frac{7}{3}M$ $+\frac{8}{3}$

Table 6.8 *Tableau 3 for Example 6.2*

	c_j		3	8	0	0	$-M$	$-M$
$\bar{c}_B$	Basis Vectors	$\bar{x}_B$	$\bar{a}_1$	$\bar{a}_2$	$\bar{a}_3$	$\bar{a}_4$	$\bar{a}_5$	$\bar{a}_6$
$-M$	$\bar{a}_5$	2	0	-5	1	3 (circled)	1	-3
3	$\bar{a}_1$	6	1	3	0	-1	0	1
		$-2M$ $+18$	0	$5M$ $+1$	$-M$	$-3M$ -3	0	$4M$ $+3$

enter the basis. The vector to leave is clearly $\bar{a}_5$ since the only positive y_{i4} is y_{14}. If we now transform Tableau 3, we obtain Tableau 4 in Table 6.9. We see from Tableau 4 that not all $z_j - c_j \geq 0$. We do not calculate $z_j - c_j$ for $\bar{a}_5$ and $\bar{a}_6$ since, having driven them out of the basis, they need never re-enter. $z_2 - c_2$ is -4 and therefore $\bar{a}_2$ can enter the basis and $\bar{a}_1$ is removed.

Table 6.9 *Tableau 4 for Example 6.2*

	c_j		3	8	0	0	$-M$	$-M$
$\bar{c}_B$	Basis Vectors	$\bar{x}_B$	$\bar{a}_1$	$\bar{a}_2$	$\bar{a}_3$	$\bar{a}_4$	$\bar{a}_5$	$\bar{a}_6$
0	$\bar{a}_4$	$\frac{2}{3}$	0	$-\frac{5}{3}$	$\frac{1}{3}$	1	$\frac{1}{3}$	-1
3	$\bar{a}_1$	$\frac{20}{3}$	1	$\frac{4}{3}$ (circled)	$\frac{1}{3}$	0	$\frac{1}{3}$	0
		20	0	-4	1	0		

Table 6.10 *Tableau 5 for Example 6.2*

	c_j		3	8	0	0
$\bar{c}_B$	Basis Vectors	$\bar{x}_B$	$\bar{x}_1$	$\bar{a}_2$	$\bar{a}_3$	$\bar{a}_4$
0	$\bar{a}_4$	9	$\frac{5}{4}$	0	$\frac{3}{4}$	1
8	$\bar{a}_2$	5	$\frac{3}{4}$	1	$\frac{1}{4}$	0
		40	3	0	2	0

Table 6.10 gives the next tableau. We have not transformed the columns corresponding to $\bar{a}_5$ and $\bar{a}_6$ since they are no longer needed.

It can be seen that we now have the optimal solution to the problem.

EXERCISES

Use of Classical Methods

1. Solve the following linear programming problem by the method of Section 6.3:

$$\text{Max } z = 2x_1 + x_2 + 3x_3 + 5x_4$$
$$x_1 + 2x_2 - x_3 + 3x_4 = 10$$
$$2x_1 - 3x_2 + 5x_3 - 6x_4 = -5$$
$$x_1, x_2, x_3, x_4 \geq 0$$

Basic Solutions

2. Find the optimal solution to the previous problem by computing all basic solutions and selecting the one that maximizes the objective function.

The Simplex Method—Theory

3. Prove that if a vector is removed from the basis at some iteration of the simplex method, then it cannot re-enter the basis at the next iteration.

The Simplex Method—Computation

4. Solve the following problem by the simplex method, without using the artificial variables:

$$\text{Max } z = 2x_1 + 3x_2 + x_3 + 6x_4$$
$$7x_1 + 2x_2 + 3x_3 + x_4 \leq 6$$
$$2x_1 + 5x_2 + x_3 + 3x_4 \leq 10$$
$$x_1 + 3x_2 + 5x_3 + 2x_4 \leq 8$$
$$x_1, x_2, x_3, x_4 \geq 0$$

5. Solve the following problem by the simplex method without using artificial variables:

$$\text{Max } z = 3x_1 - 4x_2 + 5x_3 + x_4$$
$$-3x_1 - 2x_2 + 6x_3 - 9x_4 \geq 0$$
$$2x_1 + 4x_2 + 8x_3 - 5x_4 \geq -3$$
$$3x_1 - x_2 + 2x_3 + 4x_4 \leq 15$$
$$x_j \geq 0, \qquad \text{all } j$$

6. Solve the following problem graphically and by the simplex method:

$$\text{Max } z = 3x_1 + x_2$$
$$2x_1 + 4x_2 \leq 21$$
$$5x_1 + 3x_2 \leq 18$$
$$x_1, x_2 \geq 0$$

7. Solve the following problem by the simplex method using artificial variables:

$$\text{Min } z = 2x_1 + 3x_2 + x_3 + 4x_4$$
$$6x_1 - 2x_2 + x_3 - 4x_4 \geq 3$$
$$3x_1 + x_2 - 4x_3 - 3x_4 \geq 6$$
$$-2x_1 + 4x_2 + 3x_3 + 6x_4 \geq 8$$
$$x_j \geq 0, \qquad \text{all } j$$

8. Given the following tableau as an intermediate stage in the solution of a minimization problem:

$\bar{c}_B$	Basis Vectors	$\bar{x}_B$	$\bar{a}_1$	$\bar{a}_2$	$\bar{a}_3$	$\bar{a}_4$	$\bar{a}_5$	$\bar{a}_6$
−1	$\bar{a}_1$	4	1	$\frac{2}{3}$	0	0	$\frac{4}{3}$	0
−3	$\bar{a}_4$	2	0	$-\frac{7}{3}$	3	1	$-\frac{2}{3}$	0
1	$\bar{a}_6$	2	0	$-\frac{2}{3}$	−2	0	$\frac{2}{3}$	1
		−8	0	$\frac{8}{3}$	−11	0	$\frac{4}{3}$	0

If the inverse of the *current* basis is

$$B^{-1} = \frac{1}{3}\begin{bmatrix} 1 & 1 & -1 \\ 1 & -2 & 2 \\ -1 & 2 & 1 \end{bmatrix}$$

find the original problem.

9. Solve the following linear programming problem:

$$\text{Max } z = 3x_1 + 2x_2 + x_3$$
$$2x_1 + 5x_2 - 3x_3 - x_4 = 20$$
$$4x_1 - 2x_2 + 3x_3 - x_4 = 10$$
$$x_1, x_2, x_3, x_4, \geq 0$$

What can you say about the solution?

General

10. Devise a method to solve a set of m simultaneous equations in m variables by means of the simplex method. Keep in mind that the variables, in general, will be unrestricted. What should be done about this?

11. How could one use linear programming to solve the following problem:

$$\text{Max } z = 3x_1 - x_2 + x_4$$
$$x_1 - x_2 - x_3 \leq 3$$
$$2x_1 - x_2 + x_4 = 5$$
$$|x_1 + x_3| \leq 3$$
$$|x_2| \geq 1$$

12. Use the simplex method to invert the following matrix:

$$S = \begin{bmatrix} 4 & 3 & 6 \\ 1 & 2 & 3 \\ 8 & 2 & 1 \end{bmatrix}$$

13. A manufacturer of metal products makes three products A, B, and C, all of which require machining, polishing and assembling of component parts. The amounts of these operations required for one unit of A are 3, 1 and 2 hours, respectively. Similarly, they are 2, 1 and 1 hours for product B and 4, 1 and 2 hours for product C. There are 300 hours of total machining time available, 100 hours of polishing time, and 200 hours of assembling time. The unit profits that can be made from the sale of products A, B and C are \$3, \$2 and \$4 respectively. How many units of each product should be produced so as to maximize total profit?

14. If you had solved a linear programming problem and you were asked to exhibit the next best solution for comparison with the optimal solution, how would you find it? Can there be more than one? Do this for the linear programming problem in exercise 4.

15. Consider the problem of inserting two vectors into the basis at each iteration instead of a single vector. Give an algorithm for doing this. Does it seem to be a reasonable thing to do?

16. Consider the following problem:

$$\text{Maximize } z = x_2$$
$$\text{Subject to: } 2x_1 + 3x_2 \leq 9$$
$$|x_1 - 2| \leq 1$$
$$x_1, x_2 \geq 0$$

   (a) Solve this problem graphically.
   (b) How can the problem be reformulated so that it could be solved by the simplex method (do not construct the simplex tableau).

17. Consider the following problem:

$$\text{Maximize } z = x_2$$
$$\text{Subject to: } 2x_1 + 3x_2 \leq 9$$
$$|x_1 - 2| \geq 1$$
$$x_1, x_2 \geq 0$$

   (a) Illustrate the constraint set graphically and find the optimal solution.
   (b) How might the simplex method be employed to solve such a problem?

REFERENCES

1. Hadley, G.: *Linear Programming*. Addison-Wesley, Reading, Massachusetts (1962).
2. Dantzig, G. B.: *Linear Programming and Its Extensions*. Princeton U. Press, Princeton (1963).
3. Charnes, A., and Cooper, W. W.: *Management Models and Industrial Applications of Linear Programming*. John Wiley & Sons, New York (1961).
4. Gass, S. I.: *Linear Programming: Methods and Applications*. Third Edition, McGraw-Hill, New York (1969).

Chapter 7

LINEAR PROGRAMMING: ADDITIONAL TOPICS

7.1 THE THEORY OF DUALITY IN LINEAR PROGRAMMING

Associated with every linear programming problem, there is another type called the *dual linear programming problem* which bears an important and special relationship to the first problem. It will be the purpose of this section to present this relationship and make clear what are the significant aspects of this connection between the two problems.

Let us consider the original linear programming problem, which we shall now refer to as the *primal* problem to be given in the following form:

$$\begin{aligned} \text{Max } z &= \bar{c}'\bar{x} \\ D\bar{x} &\leq \bar{d} \\ \bar{x} &\geq \bar{0} \end{aligned} \qquad \textbf{(7-1)}$$

where D is a matrix, $D = [d_{ij}]$, $i = 1, 2, \ldots, s$; $j = 1, 2, \ldots, t$ and

$$\bar{d} = \begin{bmatrix} d_1 \\ d_2 \\ \cdot \\ \cdot \\ \cdot \\ d_s \end{bmatrix} \qquad \bar{x} = \begin{bmatrix} x_1 \\ x_2 \\ \cdot \\ \cdot \\ \cdot \\ x_t \end{bmatrix} \qquad \bar{c}' = (c_1, c_2, \ldots, c_t)$$

We now define the dual problem as follows:

$$\begin{aligned} \text{Min } v &= \bar{d}'\bar{w} \\ D'\bar{w} &\geq \bar{c}' \\ \bar{w} &\geq \bar{0} \end{aligned} \qquad \textbf{(7-2)}$$

When the primal is stated as in (7-1), the problem in (7-2) is usually called the *symmetric dual*. The reason for this term is that if the constraints of (7-1) were written as equalities, then an unsymmetric dual representation results. This will be discussed later. The dimensions of the dual problem (7-2) are different from those of the primal (7-1). The primal problem had s constraints and t variables, whereas the dual problem has t constraints and s variables. We shall see later that this property can be used for computational purposes.

A convenient schematic representation of the relationship between the two problems is as follows:

$$
\begin{array}{c}
[x_1, x_2, \ldots, x_t] \\[6pt]
\begin{bmatrix} w_1 \\ w_2 \\ \cdot \\ \cdot \\ \cdot \\ w_s \end{bmatrix}
\begin{bmatrix} d_{11} & d_{12} & \cdots & d_{1t} \\ d_{21} & d_{22} & \cdots & d_{2t} \\ \cdot & \cdot & & \cdot \\ \cdot & \cdot & & \cdot \\ \cdot & \cdot & & \cdot \\ d_{s1} & d_{s2} & & d_{st} \end{bmatrix}
\leq
\begin{bmatrix} d_1 \\ d_2 \\ \cdot \\ \cdot \\ \cdot \\ d_s \end{bmatrix} \\[6pt]
\geq \\[6pt]
[c_1, c_2, \ldots, c_t]
\end{array}
\qquad (7\text{-}3)
$$

In the representation given in (7-3) the columns of the matrix D are the activity vectors for the primal problem, while the rows of D are the activity vectors for the dual problem. Although the preceding representation is used by several authors, perhaps a more suggestive scheme is as follows:

$$
\begin{array}{ccccccc}
d_{11}(x_1)(w_1) & + & d_{12}(x_2)(w_1) & + \ldots + & d_{1t}(x_t)(w_1) & \leq & d_1 \\
+ & & + & & + & & \\
d_{21}(x_1)(w_2) & + & d_{22}(x_2)(w_2) & + \ldots + & d_{2t}(x_t)(w_2) & \leq & d_2 \\
\cdot & & \cdot & & \cdot & & \\
\cdot & & \cdot & & \cdot & & \\
\cdot & & \cdot & & \cdot & & \\
d_{s1}(x_1)(w_s) & + & d_{s2}(x_2)(w_s) & + \ldots + & d_{st}(x_t)(w_s) & \leq & d_s \\
\geq & & \geq & & \geq & & \\
c_1 & & c_2 & \ldots & c_t & &
\end{array}
\qquad (7\text{-}4)
$$

In (7-4) the x_j are associated with the rows and the w_i are associated with the columns.

To illustrate the preceding concepts, consider the following primal problem:

$$\text{Max } z = 3x_1 + 2x_2$$
$$-2x_1 + 7x_2 \leq -24$$
$$3x_1 + 4x_2 \leq 31$$
$$x_1 - 5x_2 \leq 20$$
$$x_1, x_2 \geq 0$$

The dual of this problem is then given by

$$\text{Min } v = -24w_1 + 31w_2 + 20w_3$$
$$-2w_1 + 3w_2 + w_3 \geq 3$$
$$7w_1 + 4w_2 - 5w_3 \geq 2$$
$$w_1, w_2, w_3 \geq 0$$

The first important result we need to establish is in the following simple theorem.

Theorem 7.1. The dual of the dual is the primal.

Proof: Suppose we are given the primal and dual problems as in (7-1) and (7-2). Consider the dual formulation

$$\text{Min } v = \bar{d}'\bar{w}$$
$$D'\bar{w} \geq \bar{c} \qquad \textbf{(7-2)}$$
$$\bar{w} \geq \bar{0}$$

Let us rewrite this dual as

$$\text{Max } v_1 = \text{Max } (-v) = (-\bar{d}')\bar{w}$$
$$(-D')\bar{w} \leq -\bar{c} \qquad \textbf{(7-5)}$$
$$\bar{w} \geq \bar{0}$$

Since $-\text{Max } v_1 = \text{Min } v$, we can see that (7-5) is equivalent to the dual (7-2). Furthermore, (7-5) is in the form of the primal problem. If we treat (7-5) as a primal problem and write down its dual, we have

$$\text{Min } z_1 = -\bar{c}'\bar{x}$$
$$-D\bar{x} \geq -\bar{d} \qquad \textbf{(7-6)}$$
$$\bar{x} \geq \bar{0}$$

or rewriting (7-6) and noting that

$$\text{Min } z_1 = -\bar{c}'\bar{x} = -\text{Max } z = \bar{c}'\bar{x}$$

we have

$$\begin{aligned} \text{Max } z &= \bar{c}'\bar{x} \\ D\bar{x} &\leq \bar{d} \\ \bar{x} &\geq \bar{0} \end{aligned} \tag{7-1}$$

which gives us (7-1) again. Hence, we have shown that the dual of the dual is the primal.

The foregoing result indicates that deciding which problem is the primal one and which is the dual one is arbitrary. We could just as well have considered a problem of the form

$$\begin{aligned} \text{Min } z &= \bar{c}'\bar{x} \\ D\bar{x} &\geq \bar{d} \\ \bar{x} &\geq \bar{0} \end{aligned} \tag{7-7}$$

as the primal problem. Then the dual would be of the form

$$\begin{aligned} \text{Max } v &= \bar{d}'\bar{w} \\ D'\bar{w} &\leq \bar{c} \\ \bar{w} &\geq \bar{0} \end{aligned} \tag{7-8}$$

In other words, there is complete symmetry between primal-dual pairs.

Until now we have not indicated why we should be interested in these two problems called the primal and the dual. It turns out that the feasible solutions and the optimal solutions to both problems are related in a very special way. The following theorems will indicate the relationships.

Theorem 7.2. If $\bar{x}$ is any feasible solution to (7-1) and $\bar{w}$ is any feasible solution to (7-2) then $\bar{c}'\bar{x} \leq \bar{d}'\bar{w}$, or equivalently $z \leq v$.

Proof: By hypothesis, $\bar{x}$ is any feasible solution to (7-1). Therefore, it satisfies

$$D\bar{x} \leq \bar{d} \tag{7-9}$$

Since $\bar{w} \geq 0$ we can multiply (7-9) by $\bar{w}'$ to obtain

$$\bar{w}'D\bar{x} \leq \bar{w}'\bar{d} = \bar{d}'\bar{w} \tag{7-10}$$

Similarly, since $\bar{w}$ is any feasible solution to (7-2) we have

$$D'\bar{w} \geq \bar{c} \tag{7-11}$$

and since $\bar{x} \geq 0$ we have

$$\bar{x}'D'\bar{w} \geq \bar{x}'\bar{c}$$

or

$$\bar{w}'D\bar{x} \geq \bar{c}'\bar{x} \tag{7-12}$$

From (7-10) and (7-12) we have

$$\bar{c}'\bar{x} \leq \bar{w}'D\bar{x} \leq \bar{d}'\bar{w}$$

and therefore

$$\bar{c}'\bar{x} \leq \bar{d}'\bar{w} \quad \text{or} \quad z \leq v$$

Theorem 7.2 indicates the very important result that given any pair of feasible solutions to the primal and dual problems, where the primal is a maximization problem and the dual is a minimization problem, the value of the primal objective function is always less than or equal to the value of the dual objective function.

Theorem 7.3. If $\bar{x}^*$ is a feasible solution to (7-1) and $\bar{w}^*$ is a feasible solution to (7-2) such that $\bar{c}'\bar{x}^* = \bar{d}'\bar{w}^*$, then $\bar{x}^*$ is an optimal solution to (7-1) and $\bar{w}^*$ is an optimal solution to (7-2).

Proof: By hypothesis, $\bar{c}'\bar{x}^* = \bar{d}'\bar{w}^*$. By Theorem 7.2, we know that for any feasible solution, $\bar{x}$, $\bar{c}'\bar{x} \leq \bar{d}'\bar{w}^* = \bar{c}'\bar{x}^*$. Therefore, $\bar{x}^*$ is optimal. Similarly, for any feasible solution, $\bar{w}$, $\bar{d}'\bar{w} \geq \bar{c}'\bar{x}^* = \bar{d}'\bar{w}^*$, and so, $\bar{w}^*$ is also an optimal solution.

Theorem 7.3 gives the very important result that when one has two solutions, one to the primal and one to the dual, such that the objective functions are equal, both solutions are optimal. One last theorem will conclude the delineation of the major characteristics of duality in linear programming.

Theorem 7.4. If one of the pair of primal and dual problems, (7-1) and (7-2), has an optimal solution, then the other also has an optimal solution.

Proof: We shall prove this theorem by exhibiting a solution to the dual problem which is obtained by suitable modification of the optimal solution and optimality conditions of the primal problem. If we were to solve the primal problem

$$\begin{aligned} \text{Max } z &= \bar{c}'\bar{x} \\ D\bar{x} &\leq \bar{d} \\ \bar{x} &\geq \bar{0} \end{aligned} \tag{7-1}$$

by the simplex method, we would convert (7-1) to

$$\begin{aligned} \text{Max } z &= \bar{c}'\bar{x} \\ D\bar{x} + I\bar{x}_s &= \bar{d} \\ \bar{x}, \bar{x}_s &\geq \bar{0} \end{aligned} \tag{7-13}$$

where $\bar{x}_s$ is the vector of slack variables. Using our usual notation, $\bar{x}_B$ is the optimal basic feasible solution, B is the basis matrix, and $\bar{c}_B$ is the vector of prices in the basis. Since we have an optimal solution $\bar{x}_B$ for (7-13) it is true that for this solution $z_j - c_j \geq 0$ for all j. Therefore,

$$z_j \geq c_j \qquad j = 1, 2, \ldots, t + s \tag{7-14}$$

Since $z_j = \bar{c}'_B \bar{y}_j = \bar{c}'_B B^{-1} \bar{d}_j$, where the $\bar{d}_j$ are the columns of D, we have

$$\bar{c}'_B B^{-1} \bar{d}_j \geq c_j \qquad j = 1, 2, \ldots, t + s \tag{7-15}$$

If we write (7-15) in matrix form, we have

$$\bar{c}'_B B^{-1} D \geq \bar{c}' \tag{7-16}$$

If we now define $\bar{w}' = \bar{c}'_B B^{-1}$, we have from (7-16)

$$\bar{w}' D \geq \bar{c}'$$

or equivalently

$$D' \bar{w} \geq \bar{c} \tag{7-17}$$

It can be seen from (7-17) that the $\bar{w}$ so defined satisfies the constraints of (7-2), the dual problem, as is indicated by equation (7-17). Hence, we have constructed a solution from the optimal solution to the primal problem which satisfies the dual constraints. We must now show that, in addition to satisfying the constraints (7-17), $\bar{w} \geq \bar{0}$. This can be readily seen, if we recall that for an optimal solution to the primal, all $z_j - c_j \geq 0$. In particular, this is also true for the slack vectors associated with the variables in the vector $\bar{x}_s$. We then have

$$z_j - c_j \geq 0 \qquad j = t + 1, \ldots, t + s \tag{7-18}$$

Again we have

$$\bar{c}'_B B^{-1} \bar{e}_i - c_j \geq 0 \qquad j = t + 1, \ldots, t + s \tag{7-19}$$

However, c_j for $j = t + 1, \ldots, t + s$ are all zero. Therefore, writing (7-19) in matrix form

$$\bar{c}'_B B^{-1} I \geq 0 \tag{7-20}$$

or

$$\bar{w}' = \bar{c}'_B B^{-1} \geq \bar{0} \tag{7-21}$$

We have now shown in (7-17) and (7-21) that $\bar{w}' = \bar{c}'_B B^{-1}$ is a feasible solution to the dual problem. To show that it is the optimal solution to the dual we note that

$$v = \bar{d}' \bar{w} = \bar{w}' \bar{d} = \bar{c}'_B B^{-1} \bar{d} = \bar{c}'_B \bar{x}_B = \max z$$

Therefore, $\bar{w}$ is the optimal solution, by Theorem 7.3.

Theorem 7.4 shows us how to find the solution to the dual from the solution to the primal, if we should wish it, since $\bar{w}' = \bar{c}'_B B^{-1}$. These values will be found in the simplex tableau under the columns corresponding to the slack variables.

Let us now consider the primal problem in a more usual form, viz., the form it takes when ready to apply the simplex method.

$$\begin{aligned} \text{Max } z &= \bar{c}'\bar{x} \\ D\bar{x} &= \bar{d} \\ \bar{x} &\geq \bar{0} \end{aligned} \tag{7-22}$$

In order to see what the dual of (7-22) is, let us make (7-22) look like (7-1). We can do this easily since $D\bar{x} = \bar{d}$ can be replaced with the pair of inequalities, $D\bar{x} \leq \bar{d}$ and $D\bar{x} \geq \bar{d}$, or what is equivalent, $D\bar{x} \leq d$ and $-D\bar{x} \leq -\bar{d}$. We now have as a problem equivalent to (7-22):

$$\begin{aligned} \text{Max } z &= \bar{c}'\bar{x} \\ D\bar{x} &\leq \bar{d} \\ -D\bar{x} &\leq -\bar{d} \\ \bar{x} &\geq \bar{0} \end{aligned} \tag{7-23}$$

Equation (7-23) is in the same form as (7-1). We shall subdivide the dual variables into two sets $\bar{w}_1$ corresponding to the constraints $D\bar{x} \leq \bar{d}$ and $\bar{w}_2$ corresponding to $-D\bar{x} \leq -\bar{d}$. Then the dual of (7-23) is

$$\begin{aligned} \text{Min } v &= \bar{d}'\bar{w}_1 - \bar{d}'\bar{w}_2 \\ D'\bar{w}_1 - D'\bar{w}_2 &\geq \bar{c} \\ \bar{w}_1, \bar{w}_2 &\geq 0 \end{aligned} \tag{7-24}$$

If we now define $\bar{w} = \bar{w}_1 - \bar{w}_2$, we can rewrite (7-24) more compactly as

$$\begin{aligned} \text{Min } v &= \bar{d}'\bar{w} \\ D'\bar{w} &\geq \bar{c} \\ \bar{w} &\text{ unrestricted} \end{aligned} \tag{7-25}$$

$\bar{w}$ is unrestricted since $\bar{w}_1$ and $\bar{w}_2$ are nonnegative. Hence, their difference could be positive, negative, or zero. Therefore, (7-25) is the dual of (7-22). We see that the consequence of writing the primal problem with equations instead of inequalities is to make the dual variables unrestricted in sign.

There is one other major result relating to primal-dual pairs and that is concerned with unboundedness. We have shown in the previous theorems that if the primal has an optimal solution then so does the dual. However, what if the primal has an unbounded solution? How does this affect the dual? It can be seen after a moment's reflection, that if the primal has an unbounded solution the dual would have to have either no solution at all or an unbounded solution. This is obviously the case because if the dual had a finite solution, since the dual of the dual is the primal, the primal would have a finite solution (see Theorems 7.1 and 7.3). More precisely, we can show the following.

Theorem 7.5. If the primal problem has an unbounded solution, the dual problem has no feasible solution.

Proof: Consider the primal-dual pairs (7-1) and (7-2). Theorem 7.2 asserts that for any pair of feasible solutions, $\bar{x}$ and $\bar{w}$

$$\bar{c}'\bar{x} \leq \bar{d}'\bar{w}$$

In particular, this would be true for

$$\max z = \bar{c}'\bar{x}$$

However, our hypothesis is that $\max z = \infty$. Therefore,

$$\infty = \max z = \bar{c}'\bar{x} \leq \bar{d}'\bar{w} \tag{7-26}$$

According to (7-26), the vector $\bar{w}$ in order to be feasible would require components without bound. Clearly, there is no feasible $\bar{w}$ whose components are all finite.

There are many other aspects of duality that are of importance. More details can be obtained in Hadley[2] or some of the other references on linear programming. One aspect worth mentioning, from a computational point of view, is the fact that if one has a linear programming problem

$$\text{Max } z = \bar{c}'\bar{x}$$
$$A\bar{x} = \bar{b}$$
$$\bar{x} \geq \bar{0}$$

such that $m > n$, it is more economical to solve the dual problem

$$\text{Min } v = \bar{b}'\bar{w}$$
$$A'\bar{w} \geq \bar{c}$$
$$\bar{w} \text{ unrestricted}$$

since the basis will now be of size $n < m$. There are application areas where this approach is of value. Since the objective functions are equal at optimality, the dual of the dual is the primal and dual variables can be computed from primal variables, one can obtain the solution to the original problem with considerably less computational effort.

7.2 OTHER SIMPLEX ALGORITHMS

In the following discussion we shall briefly describe a number of other simplex-related algorithms for solving linear programming problems. We shall not specify them in as much detail as in the development of the previous material on the simplex method. The purpose here is to present the motivation and content of the methods rather than all the computational details. Many of the more purely computational aspects of some of these topics are given in Orchard-Hays.[1]

The Revised Simplex Method

This method is completely equivalent theoretically to the simplex method. The criteria which are used to determine which vector is to enter and which vector is to leave the basis are exactly the same as in the simplex method. The use of the term "revised" refers only to what quantities are stored in a tableau, and how one tableau is transformed into the next. The motivation for the development of the revised simplex method is to minimize the amount of storage in a digital computer required to contain the necessary information for the solution of large problems of importance to business, government, and other users. There is a commonly held misconception that the revised simplex method requires less computational effort. (See Simmonard,[3] Chung,[4] Gass.[5]) This is not precisely true in the way in which it is generally implied. There are advantages of precision and storage which make the revised simplex method very attractive and the one in almost exclusive use, but the decrease, if any, in the amount of computation is not a major factor. For a clear discussion of this matter one can consult Hadley.[2]

The revised simplex method proceeds from the basic observation that if one had available B^{-1}, the inverse of the columns of A in $A\bar{x} = \bar{b}$ that are in the basis, one could construct all quantities of interest since

$$\begin{aligned} \bar{x}_B &= B^{-1}\bar{b} \\ z &= \bar{c}'_B\bar{x}_B \\ z_j - c_j &= \bar{c}'_B B^{-1}\bar{a}_j - c_j \end{aligned} \tag{7-27}$$

Hence, if one could find an easy way to transform B^{-1} from iteration to iteration, all other quantities of interest are simply calculated. This requires the storage of an $m \times m$ matrix in contrast to ordinary

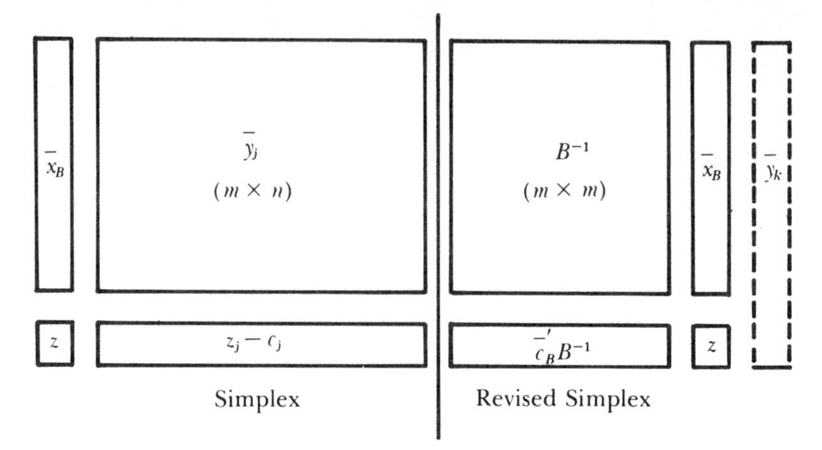

Figure 7-1 Simplex and revised simplex schema.

simplex where an $m \times n$ matrix of $\bar{y}_j$ are stored. Schematically, the two methods are contrasted as shown in Figure 7-1.

Since high speed core memory is the limiting factor in digital computer storage for linear programming problems, the revised simplex storage obviously requires much less. It does require, however, the storage of the original $\bar{a}_j$ on magnetic tape, drum, or disc, because as is indicated in (7-27), the $z_j - c_j$ must be calculated using the $\bar{a}_j$, since they are not transformed at each iteration. Only the product of $\bar{c}'_B B^{-1}$ which is used to calculate z_j is transformed along with B^{-1} and the other quantities. The last column in the revised simplex tableau, $\bar{y}_k$, is calculated *after* the vector to enter the basis is determined by examining the $z_j - c_j$. It is not directly transformed at each iteration.

The same basic transformation formulas are used to transform B^{-1} in revised simplex as are used in the ordinary simplex method. A representation of the inverse matrix called the "product form of the inverse" is used to minimize the amount of computation and to derive certain other benefits. A detailed discussion of some of these points is given in Orchard-Hays[1] and Hadley.[2]

The Dual Simplex Algorithm

The dual simplex method originated with C. E. Lemke. (See Lemke[6].) In the course of applying the simplex method to the dual of a linear programming problem, he realized that a different algorithm from the ordinary primal simplex algorithm we have been discussing could be devised. The algorithm has certain special uses, e.g., when adding an additional constraint to a linear programming problem for which one has already obtained the optimal solution. However, it has not been widely used as a general linear programming algorithm because it can be time-consuming and it is somewhat difficult to obtain an initial basic feasible solution for many problems.

The basic philosophy of the dual simplex algorithm as contrasted with the primal simplex algorithm is easily and simply stated. In primal simplex we begin with a basic feasible solution for which not all the $z_j - c_j$ are nonnegative. If they were, we would, of course, have the optimal basic feasible solution. We then make changes in the basis, one vector at a time, maintaining nonnegativity of the variables until all $z_j - c_j \geq 0$ and hence the optimal solution has been obtained. By contrast, in the dual simplex algorithm, we begin with a basic but *not feasible* solution for which all $z_j - c_j \geq 0$. We then make changes in the basis, one vector at a time, maintaining all the $z_j - c_j \geq 0$, until we have a feasible solution. At this point, we have the optimal feasible solution, since we have maintained the $z_j - c_j \geq 0$, at each iteration. In brief, in primal simplex we maintain feasibility while changing the basis until we satisfy the optimality criterion. In dual simplex we maintain the satisfaction of the optimality criterion without being feasible, while changing the basis until we obtain our first feasible solution, at which point we are done.

Let us consider the given linear programming problem (our primal) as

$$\begin{aligned} \text{Max } z &= \bar{c}'\bar{x} \\ A\bar{x} &= \bar{b} \\ \bar{x} &\geq \bar{0} \end{aligned} \tag{7-28}$$

From our discussion of duality we know that the dual of (7-28) is

$$\begin{aligned} \text{Min } v &= \bar{b}'\bar{w} \\ A'\bar{w} &\geq \bar{c} \\ \bar{w} &\text{ unrestricted} \end{aligned} \tag{7-29}$$

As in the previous sections, A is an $m \times n$ matrix, $B = (\bar{b}_1, \ldots, \bar{b}_m)$ is the basis matrix under discussion. We note that the constraints of (7-29) can be written in equation form as

$$\bar{w}'\bar{a}_j \geq c_j \qquad j = 1, 2, \ldots, n \tag{7-30}$$

Suppose that the basis, B is such that

$$\bar{w}' = \bar{c}_B' B^{-1}$$

is a solution to the dual problem, i.e.,

$$\begin{aligned} \bar{w}'\bar{b}_i &= c_{Bi} \qquad i = 1, 2, \ldots, m \\ \bar{w}'\bar{a}_j &\geq c_j \qquad j \text{ not in basis} \end{aligned} \tag{7-31}$$

Since $\bar{w}' = \bar{c}'_B B^{-1}$ we can see from (7-31) that

$$\bar{w}'\bar{a}_j - c_j \geq 0 \qquad j = 1, 2, \ldots, n$$

or

$$\bar{c}'_B B^{-1}\bar{a}_j - c_j \geq 0 \qquad j = 1, 2, \ldots, n$$

or

$$z_j - c_j \geq 0 \qquad j = 1, 2, \ldots, n \tag{7-32}$$

Equation (7-32) states that for such a solution to the dual as we have assumed, the optimality criterion for the primal problem is satisfied. If it were also true that

$$\bar{x}_B = B^{-1}\bar{b} \geq \bar{0}$$

then we would also have the optimal solution to both primal and dual problems since

$$z = \bar{c}'\bar{x} = \bar{c}'_B\bar{x}_B = \bar{c}'_B B^{-1}\bar{b} = \bar{w}'\bar{b} = v \tag{7-33}$$

From the foregoing discussion, we must conclude that if we have such a solution $\bar{w}' = \bar{c}'_B B^{-1}$ and it is *not* optimal, since it satisfies the optimality criterion, it must not be feasible, i.e., one or more of the x_{Bi} is negative. Let us consider this situation. Let us designate the rows of B^{-1} as $\bar{\beta}^j$. Now consider some $x_{Bi} < 0$. In particular, let $x_{Br} < 0$ where

$$x_{Br} = \bar{\beta}^r\bar{b} < 0 \tag{7-34}$$

Suppose we derive a new solution vector, $\hat{\bar{w}}'$ from $\bar{w}' = \bar{c}'_B B^{-1}$ as follows:

$$\hat{\bar{w}}' = \bar{w}' - \theta\bar{\beta}^r \tag{7-35}$$

From this it follows that

$$\hat{v} = \hat{\bar{w}}'\bar{b} = \bar{w}'\bar{b} - \theta\bar{\beta}^r\bar{b} = v - \theta x_{Br} \tag{7-36}$$

Since $x_{Br} < 0$, if $\theta < 0$, then $\hat{v} < v$ and therefore $\hat{\bar{w}}$ will result in a reduced value of v. If, in addition, $\hat{\bar{w}}$ also satisfies the constraints of (7-29) we have a new solution to the dual problem (7-29) with $\hat{v} < v$. Let us now see what must be done with respect to the constraints.

$$\hat{\bar{w}}'\bar{b}_i = \bar{w}'\bar{b}_i - \theta\bar{\beta}^r\bar{b}_i \tag{7-37}$$

Observe that $\bar{w}'\bar{b}_i = c_{Bi}$ and that since $B^{-1}B = I$ that $\bar{\beta}^r\bar{b}_i = \delta_{ir}$ where δ_{ir} is the *Kronecker delta*, $\delta_{ir} = \begin{cases} 1, & i = r \\ 0, & i \neq r \end{cases}$. Therefore,

$$\hat{\bar{w}}'\bar{b}_i = c_{Bi} - \theta\delta_{ir} \tag{7-38}$$

An equivalent way of writing (7-38) is

$$\begin{aligned} \hat{\bar{w}}'\bar{b}_i &= c_{Bi} \qquad i \neq r \\ \hat{\bar{w}}'\bar{b}_r &= c_{Br} - \theta \geq c_{Br} \text{ if } \theta \leq 0 \end{aligned} \tag{7-39}$$

For all $\bar{a}_j$ which are not in the primal basis we have

$$\hat{\bar{w}}'\bar{a}_j = \bar{w}'\bar{a}_j - \theta\bar{\beta}^r\bar{a}_j \tag{7-40}$$

First, we dispose of the case of dual unboundedness. If $\bar{\beta}^r\bar{a}_j \geq 0$ for every $\bar{a}_j$ not in the basis, then for $\theta \leq 0$,

$$\bar{w}'\bar{a}_j \geq c_j \tag{7-41}$$

For this case, θ can be made an arbitrarily large negative number and we will always have a solution to the dual. Since

$$\hat{v} = v - \theta x_{Br'}$$

from (7-36), we see that $\hat{v}$ can be made arbitrarily small and therefore the dual has an unbounded solution. From Section 7.1 we know that in this case the primal has no feasible solution.

Therefore, we assume that there is at least one $\bar{a}_j$ for which $\bar{\beta}^r\bar{a}_j < 0$. We recall that

$$\bar{y}_j = B^{-1}\bar{a}_j$$

or

$$y_{rj} = \bar{\beta}^r\bar{a}_j \tag{7-42}$$

If we now consider $y_{rj} < 0$, in order to satisfy

$$\hat{\bar{w}}'\bar{a}_j = \bar{w}'\bar{a}_j - \theta\bar{\beta}^r\bar{a}_j \geq c_j$$

it is necessary that

$$\bar{w}'\bar{a}_j - c_j \geq \theta y_{rj} \tag{74-3}$$

Since $y_{rj} < 0$, this imposes the restriction on θ that

$$\theta \geq \frac{\bar{w}'\bar{a}_j - c_j}{y_{rj}} \tag{7-44}$$

Since we wish to decrease v as much as possible, the best we can possibly do is choose θ so that

$$\theta = \operatorname*{Max}_j \left\{ \frac{\bar{w}'\bar{a}_j - c_j}{y_{rj}}, y_{rj} < 0 \right\} \tag{7-45}$$

However, (7-45) is more simply stated in terms of primal notation. We noted earlier that $\bar{w}' = \bar{c}_B B^{-1}$ and $z_j = \bar{c}_B B^{-1}\bar{a}_j$. Therefore, $\bar{w}'\bar{a}_j - c_j = z_j - c_j \geq 0$. Hence,

$$\theta = \frac{z_k - c_k}{y_{rk}} = \operatorname*{Max}_j \left\{ \frac{z_j - c_j}{y_{rj}}, y_{rj} < 0 \right\} \tag{7-46}$$

We now summarize the preceding. If one or more $x_{Bi} < 0$, say x_{Br}, then we have found a new solution to the dual given by (7-39) which satisfies the dual constraints and results in a reduced value of the dual objective function. If we take $\bar{a}_k$, as determined from (7-46), together with the $\bar{b}_i$ $(i \neq r)$ we have a new primal basis $\hat{B}$ in which column r of B has been replaced by $\bar{a}_k$. If $\hat{\bar{x}}_B = \hat{B}^{-1}\bar{b}$ still contains negative components we can repeat the preceding process.

While we had recourse to the dual problem in the development of the foregoing argument, the dual simplex algorithm is a method for solving a directly stated linear programming problem and not its dual. It can be stated in primal terms as follows:

Dual Simplex Algorithm

1. We must begin with a basic solution to the primal with $z_j - c_j \geq 0$ for all j and not feasible in the sense that one or more of the $x_{Bi} < 0$.

2. We determine the vector to be removed from the basis by

$$x_{Br} = \operatorname*{Min}_{i} \{x_{Bi}, x_{Bi} < 0\} \tag{7-47}$$

This means that $\bar{b}_r$ is removed from B, and x_{Br} becomes zero.

3. We determine the vector to enter the basis by

$$\theta = \frac{z_k - c_k}{y_{rk}} = \operatorname*{Max}_{j} \left\{\frac{z_j - c_j}{y_{rj}}, y_{rj} < 0\right\} \tag{7-48}$$

4. All quantities of interest are transformed by the usual transformation formulas discussed in the previous section on the revised simplex method.

5. When a basic solution is reached such that $\bar{x}_B \geq \bar{0}$, we have obtained the optimal solution.

It should be noted that in the dual simplex method, the order of determining the vectors to enter and leave the basis is reversed from that of the primal simplex method. This is a result of deriving the rules and, hence, the "path to optimality" from that of the dual problem. In other words, we are attempting, loosely speaking, to minimize v rather than maximize z. Of course, when $v = z$, we have the optimal solution since $z = \bar{c}'\bar{x} \leq \bar{b}'\bar{w} = v$ for all solutions and equality occurs for the optimal solution. The rule used to determine the vector to be *removed* is somewhat arbitrary, just as the rule to determine the vector to enter was arbitrary in primal simplex.

The chief problem in using the dual simplex algorithm is finding an initial basic solution with all $z_j - c_j \geq 0$. Generally speaking, this is not easy to do. Hence, the algorithm is seldom used as a general algorithm for solving linear programming problems. However, it has some special uses which will be discussed in a subsequent section of this chapter.

Other Linear Programming Algorithms

Several other linear programming algorithms have been devised. It is outside the scope of this text to discuss all of them. A few words about two of them are in order, however.

A method known as the *primal-dual algorithm* was proposed in 1956. (See Dantzig, Ford, and Fulkerson.[7]) Its motivation is an attempt to avoid the often lengthy first phase of the usual simplex calculation, during which the artificial variables are driven to zero, in order to obtain an initial basic feasible solution. The reason this first phase is undesirable is that during this phase of the calculation, essentially nothing is being accomplished (except perhaps by chance) in improving the value of the objective function and hence in approaching the optimal solution. This is true because the criterion which is used to determine which vectors shall enter the basis during this first phase is concerned only with forcing the artificial variables to become zero. This has nothing whatever to do with the original objective function of the linear programming problem. Only after the artificial variables become zero do we proceed with simplex iterations to achieve the optimal solution. The primal-dual algorithm is devised so that achieving a feasible solution and improving the value of the objective function can be carried out simultaneously. The details of this method are discussed in Hadley.[2] Experience with the algorithm indicates that it is probably superior to ordinary simplex (see Mueller and Cooper[8]) but it is not known how it compares with the so-called composite simplex algorithm which we discuss next.

The *composite simplex algorithm* is not a different algorithm from the basic primal simplex algorithm in the sense that the dual simplex and primal-dual algorithms are. These latter have different theoretical bases for determining how to move from one basic feasible solution to another. The composite simplex algorithm (see Orchard-Hays[1]), on the other hand, is a variant of the revised simplex algorithm and thus is a variety of primal simplex. What the composite simplex algorithm does is avoid explicitly introducing artificial variables, but as long as some variable is negative which should not be, it utilizes a form of the objective function which will ultimately force these variables to become positive. In this way, it attempts to achieve the same results as the primal-dual algorithm, viz., to become feasible and also move toward optimality all in one set of iterations. Almost all working computer codes in recent years have been of the "composite" variety. There are many variations and embellishments of these basic ideas that are incorporated in such computer codes. Nevertheless, no one has yet clearly established, either on theoretical or empirical grounds, whether or not the composite algorithm is vastly superior to the ordinary two-phase methods, as its adherents claim (see Marble[9]).

7.3 THE DECOMPOSITION PRINCIPLE

The "decomposition principle" of Dantzig and Wolfe[10] was devised to make it possible to solve very large linear programs more efficiently by decomposing a larger problem into a series of smaller problems. The use of the principle presupposes that the linear programming problem has the structure as described in the following paragraph. For many areas of application this is the case.

We assume that the large linear programming problems in which we are interested have the following structure:

$$\text{Max } z = \sum_{j=1}^{p} \bar{c}_j' \bar{x}_j$$

$$\begin{bmatrix} A_1 & A_2 & A_3 & \cdots & A_p \\ A_{p+1} & \bar{0} & \bar{0} & \cdots & \bar{0} \\ \bar{0} & A_{p+2} & \bar{0} & \cdots & \bar{0} \\ \bar{0} & \bar{0} & A_{p+3} & \cdots & \bar{0} \\ \cdot & \cdot & \cdot & & \cdot \\ \cdot & \cdot & \cdot & & \cdot \\ \cdot & \cdot & \cdot & & \cdot \\ \bar{0} & \bar{0} & \bar{0} & & A_{2p} \end{bmatrix} \begin{bmatrix} \bar{x}_1 \\ \bar{x}_2 \\ \\ \cdot \\ \cdot \\ \cdot \\ \cdot \\ \bar{x}_p \end{bmatrix} = \begin{bmatrix} \bar{b}_0 \\ \bar{b}_1 \\ \bar{b}_2 \\ \cdot \\ \cdot \\ \cdot \\ \cdot \\ \bar{b}_p \end{bmatrix} \tag{7-49}$$

$$\bar{x}_j \geq 0 \qquad j = 1, 2, \ldots, p$$

In the problem of (7-49), each matrix A_j is of order $m_0 \times n_j$, $j = 1, 2, \ldots, p$, and each matrix A_{p+j} is of order $m_j \times n_j$, $j = 1, 2, \ldots, p$; the $\bar{x}_j$ and $\bar{c}_j$ each have n_j components and the $\bar{b}_j$ are vectors with m_j components.

It can be seen that many problems could have their constraint rows rearranged so that the form of (7-49) could be obtained. Without the decomposition principle, which we shall describe later, it would be necessary to solve (7-49) as a single problem for which a basis of size $\sum_{j=0}^{p} m_j$ is necessary. With the decomposition method we shall instead solve a set of p problems each of which has a basis of size m_j.

We first note that if the set of points $\bar{x}_j$ is strictly bounded and satisfies

$$A_{p+j}\bar{x}_j = \bar{b}_j \qquad j = 1, 2, \ldots, p \qquad \textbf{(7-50)}$$
$$\bar{x}_j \geq \bar{0}$$

then the set of points $\bar{x}_j$ is a closed convex set with a finite number of extreme points. If so, any point in this convex set can be represented as a convex combination of its extreme points. Let us designate these extreme points as $\bar{x}^*_{jk}$, $k = 1, 2, \ldots, k_j$. From this it follows that any feasible solution to (7-50) can be written as

$$\bar{x}_j = \sum_{k=1}^{k_j} \rho_{jk}\bar{x}_{jk} \qquad j = 1, 2, \ldots, p$$
$$\sum_{k=1}^{k_j} \rho_{jk} = 1 \qquad \textbf{(7-51)}$$
$$\rho_{jk} \geq 0 \qquad k = 1, 2, \ldots, k_j$$

Conversely, any $\bar{x}_j$ of the form (7-51) will satisfy (7-50).

If we substitute from (7-51) into our original problem (7-49), we can now express this as

$$\text{Max } z = \sum_{j=1}^{p} \sum_{k=1}^{k_j} \rho_{jk}\bar{c}_j\bar{x}^*_{jk}$$
$$\sum_{j=1}^{p} \sum_{k=1}^{k_j} \rho_{jk}A_j\bar{x}^*_{jk} = \bar{b}_0 \qquad \textbf{(7-52)}$$
$$\sum_{k=1}^{k_j} \rho_{jk} = 1 \qquad j = 1, 2, \ldots, p$$
$$\rho_{jk} \geq 0 \qquad k = 1, 2, \ldots, k_j$$

It will be noted that, assuming the extreme points were known, (7-52) is a linear programming problem in variables ρ_{jk}. Suppose we designate the values of ρ_{jk} in an optimal solution to (7-52) as ρ^*_{jk}, $j = 1, 2, \ldots, p$; $k = 1, 2, \ldots, k_j$. If this is the case then the optimal solution to (7-49) would be

$$\bar{x}_j = \sum_{k=1}^{k_j} \rho^*_{jk}\bar{x}^*_{jk} \qquad j = 1, 2, \ldots, p \qquad \textbf{(7-53)}$$

If we define

$$\bar{g}_{jk} = A_j\bar{x}^*_{jk} \qquad f_{jk} = \bar{c}_j{}'\bar{x}^*_{jk} \qquad \begin{matrix} j = 1, 2, \ldots, p \\ k = 1, 2, \ldots, k_j \end{matrix} \qquad \textbf{(7-54)}$$

then we can write (7-52) as

$$\text{Max } z = \sum_{j=1}^{p} \sum_{k=1}^{k_j} f_{jk}\rho_{jk}$$

$$\sum_{j=1}^{p} \sum_{k=1}^{k_j} \rho_{jk}\bar{g}_{jk} = \bar{b}_0$$

$$\sum_{k=1}^{k_j} \rho_{jk} = 1 \qquad j = 1, 2, \ldots, p \tag{7-55}$$

$$\rho_{jk} \geq 0 \qquad \begin{array}{l} j = 1, 2, \ldots, p \\ k = 1, 2, \ldots, k_j \end{array}$$

The advantage of the formulation given in (7-55) over the original problem statement in (7-49) is that instead of having a basis of size $\sum_{j=1}^{p} m_j$ it has a basis of size $m_0 + p$. It has many more variables, however. Its principal disadvantage is that it requires a knowledge of all the extreme points $\bar{x}_{jk}^*$ since these are necessary to define the coefficients of (7-55) as is indicated in (7-54). If it were necessary to generate all the extreme points of each of the convex sets

$$\begin{array}{r} A_{p+j}\bar{x}_j = b_j \\ \bar{x}_j \geq 0 \end{array} \qquad j = 1, 2, \ldots, p \tag{7-50}$$

then this approach would be quite hopeless, since the number of extreme points is prohibitively large from the point of view of computation. Fortunately, it is not necessary to do so. All we need to do is to generate the $\bar{g}_{jk}$ and f_{jk} as they are required, i.e., a particular extreme point at a time. Let us now examine how this is done.

Let us first combine the $m_0 + p$ constraints of (7-55) into one set as

$$\sum_{j=1}^{p} \sum_{k=1}^{k_j} \rho_{jk}\bar{q}_{jk} = \bar{b} \tag{7-56}$$

where $\bar{q}'_{jk} = [\bar{g}'_{jk}, \bar{e}_j]$, $\bar{b}' = [\bar{b}'_0, \bar{1}']$, where $\bar{1}$ is the *sum vector* having p components each equal to 1. Let B be any basis matrix of order $m_0 + p$ for the system of constraints given in (7-56). A basic feasible solution to (7-56) will be denoted $\bar{\rho}_B$, where $\bar{\rho}_B = B^{-1}\bar{b}$. We now let $\bar{s}' = \bar{f}'_B B^{-1}$ and partition $\bar{s}' = (\bar{s}'_1, \bar{s}'_2)$ where $\bar{s}_1$ consists of the first m_0 components of $\bar{s}$, and $\bar{s}_2$ consists of the last p components of $\bar{s}$. Let us now calculate $z_{jk} - f_{jk}$:

$$z_{jk} - f_{jk} = \bar{f}'_B B^{-1}\bar{q}_{jk} - f_{jk} = \bar{s}'\bar{q}_{jk} - f_{jk} = (\bar{s}'_1, \bar{s}'_2)\bar{q}_{jk} - f_{jk} \tag{7-57}$$

However, $\bar{q}_{jk} = [\bar{g}_{jk}, \bar{e}_j]$. Substituting into (7-57) gives

$$z_{jk} - f_{jk} = (\bar{s}_1', \bar{s}_2')[\bar{g}_{jk}, \bar{e}_j] - f_{jk} = \bar{s}_1'\bar{g}_{jk} + s_{j2} - f_{jk}'$$
$$= (\bar{s}_1'A_j - \bar{c}_j')\bar{x}_{jk}^* + s_{j2} \; ; \tag{7-58}$$

s_{j2} is the jth component of $\bar{s}_2'$.

We need the $z_{jk} - f_{jk}$ to determine whether or not the particular basic feasible solution we are dealing with, $\bar{\rho}_B$ is optimal. Therefore, we must examine the $z_{jk} - f_{jk}$ for all possible j, k. One way to do this is to seek the minimum of the $z_{jk} - f_{jk}$ over all j, k. If this minimum is nonnegative, then the solution is optimal. If it is negative, we must continue our iterative calculation. Let us formulate this approach and see where it leads.

$$\underset{\substack{j=1,\ldots,p\\k=1,\ldots,k_j}}{\text{Min}} (z_{jk} - f_{jk})$$

$$= \text{Min}\left\{\underset{k}{\text{Min}}\,(z_{1k} - f_{1k}), \underset{k}{\text{Min}}\,(z_{2k} - f_{2k}), \ldots, \underset{k}{\text{Min}}\,(z_{pk} - f_{pk})\right\} \tag{7-59}$$

Let us now consider (7-58) again. We see that

$$z_{jk} - f_{jk} = (\bar{s}_1'A_j - \bar{c}_j')\bar{x}_{jk}^* + s_{j2}$$

Hence, for a fixed j, the $\underset{k}{\text{Min}}\,(z_{jk} - f_{jk})$ occurs at some extreme point of the convex set of feasible solutions to

$$A_{p+j}\bar{x}_j = \bar{b}_j$$
$$\bar{x}_j \geq \bar{0} \tag{7-60}$$

This gives us the key to the method for avoiding the enumeration of all the extreme points. Since an extreme point corresponds to a basic feasible solution of (7-60), the minimum value over $k = 1, 2, \ldots, k_j$ for a fixed value of j is s_{j2} plus the optimal solution to the problem

$$\text{Min } z_j = (\bar{s}_1'A_j - \bar{c}_j')\bar{x}_j$$
$$A_{p+j}\bar{x}_j = \bar{b}_j \tag{7-61}$$
$$\bar{x}_j \geq \bar{0}$$

It can be seen that the optimal solution to (7-61) gives us the extreme point, $\bar{x}_{jk}^*$, that we require since the value of $z_{jk} - f_{jk}$ calculated from this extreme point is a minimum over all values of k. We can then use this value of $\bar{x}_{jk}^*$ to calculate $\bar{g}_{jk}$ and f_{jk} and therefore, $\bar{q}_{jk}$.

To summarize the procedure, we see that to find the minimum $z_{jk} - f_{jk}$ for all j, k we solve p linear programming problems of the form (7-61). If z_j^* is the optimal value of z_j for each of these problems, we then compute $z_j^* + s_{j2}$ and finally

$$\underset{\substack{j=1,\ldots,p \\ k=1,\ldots,k_j}}{\text{Min}} (z_{jk} - f_{jk}) = \underset{j}{\text{Min}} (z_j^* + s_{j2}) = z_t^* + s_{t2} \qquad \textbf{(7-62)}$$

We now let $\bar{x}_{tu}^*$ be an optimal extreme point solution of (7-61) for $= t$. Using our definitions we obtain

$$\bar{g}_{tu} = A_t\bar{x}_{tu}^*, \qquad \bar{q}'_{tu} = [\bar{g}'_{tu}, \bar{e}'_t], \qquad f_t = \bar{c}'_t\bar{x}_{tu}^*$$

and $\bar{q}_{tu}$ enters the basis at the next iteration. We now return to our original main problem (7-55) and transform to get a new basis, $\hat{B}^{-1}$ as well as $\hat{\bar{s}}$, $\hat{\bar{\rho}}_B$, and continue. At each step we solve p new linear programming problems, until the optimal solution is achieved.

The advantage of this procedure is perhaps not obvious. We solve p problems at each iteration of basis size m_j instead of solving one problem at each iteration of basis size $\sum_{j=0}^{p} m_j$. For suitably large problems this is the only practical procedure that can be followed.

One additional point should be mentioned. When we get to the stage of solving the set of p problems (7-61), if any of these have special structures, special simplified algorithms such as transportation problem algorithms or dynamic programming algorithms may be used. (These are discussed later in this text.) This will also speed up the calculation.

7.4 POSTOPTIMAL ANALYSIS

By postoptimal analysis of linear programming problems we refer to additional calculations that may be performed to ascertain how the optimal solution would change if specified changes occurred in some of the structural parameters of the problem. The parameters we refer to are the requirements, prices, or even perhaps some of the elements of $A = [a_{ij}]$.

We may be interested in changing some of these parameters because some of them are arbitrarily set (or perhaps not precisely known) and it is natural to wonder what effect an arbitrary change (or an amount of uncertainty) will have on the optimal solution. Another reason for postoptimal analysis is simply human error. After a problem is solved we may find clerical or other errors in one or two parameters and we then wish to re-solve the problem.

Obviously, when the preceding situations exist, one could start from scratch and solve in the usual way the linear programming problem with the new values of the parameters. If this was all we ever did there would be no need for this discussion. It turns out in many cases that it is not necessary to begin all over again. One can "continue" in various ways from the optimal solution to the original problem. We shall look at a few representative situations of this type.

Changes in Prices

Let us assume we have an optimal basic feasible solution to

$$\begin{aligned} \text{Max } z &= \bar{c}'\bar{x} \\ A\bar{x} &= \bar{b} \\ \bar{x} &\geq \bar{0} \end{aligned} \qquad \textbf{(7-63)}$$

In the usual notation, $\bar{x}_B = B^{-1}\bar{b}$ and $z = \bar{c}'_B\bar{x}_B$. Suppose c_l, i.e., c_j for $j = l$, is changed to $c_l + \alpha$, where α is an arbitrary scalar. Let the new values of c_j be c_j^+. Then

$$\begin{aligned} c_j^+ &= c_j \qquad j \neq l \\ c_l^+ &= c_l + \alpha \end{aligned} \qquad \textbf{(7-64)}$$

If c_l is not in the basis, then the $z_j - c_j$ are

$$\begin{aligned} z_j^+ - c_j^+ &= \bar{c}'_B\bar{y}_j - c_j \geq 0 \qquad j \neq l \\ z_l^+ - c_l^+ &= \bar{c}'_B\bar{y}_j - (c_l + \alpha) \end{aligned}$$

If $c_l + \alpha \leq \bar{c}'_B\bar{y}_j$ then $\bar{x}_B = B^{-1}\bar{b}$ is still the optimal solution. If not, then $z_l - c_l \leq 0$ and additional simplex iterations can be made to obtain the optimal solution. If c_l is in the basis, then the $z_j - c_j$ are

$$z_j^+ - c_j^+ = \bar{c}_B^{+\prime}\bar{y}_j - c_j^+ \qquad j = 1, 2, \ldots, n$$

and all the $z_j - c_j$ have to be checked and additional simplex iterations may be required.

More generally, suppose $\bar{c}^+ = \bar{c} + \bar{\alpha}$ then

$$\begin{aligned} z_j^+ - c_j^+ &= \bar{c}_B^{+\prime}\bar{y}_j - c_j^+ = (\bar{c}'_B + \bar{\alpha}'_B)\bar{y}_j - (c_j + \alpha_j) \\ &= \bar{c}'_B\bar{y}_j - c_j + \bar{\alpha}'_B\bar{y}_j - \alpha_j \\ &= z_j - c_j + \bar{\alpha}'_B\bar{y}_j - \alpha_j \end{aligned} \qquad \textbf{(7-65)}$$

Therefore,

$$z_j^+ - c_j^+ = z_j - c_j + \bar{\alpha}'_B\bar{y}_j - \alpha_j$$

If one or more of the $z_j^+ - c_j^+$ are <0, then one must perform additional simplex iterations.

Changes in Requirements

If the requirements vector, $\bar{b}$ is changed to $\bar{b}^+$ by $\bar{b}^+ = \bar{b} + \bar{\gamma}$, then the new basic solution, $\bar{x}_B^+$ is given by

$$\bar{x}_B^+ = B^{-1}\bar{b}^+ = B^{-1}(\bar{b} + \bar{\gamma}) = B^{-1}\bar{b} + B^{-1}\bar{\gamma}$$

Therefore, **(7-66)**

$$\bar{x}_B^+ = \bar{x}_B + B^{-1}\bar{\gamma}$$

If $\bar{x}_B^+ \geq \bar{0}$, we are still optimal, since the $z_j - c_j$ are unchanged. However, if one or more of the $x_{Bi}^+ < 0$ because the corresponding element of $B^{-1}\bar{\gamma}$ is a negative number whose magnitude exceeds that of x_{Bi}, then we no longer have a feasible solution. This is an ideal situation in which to apply the dual simplex algorithm, since the $z_j - c_j \geq 0$ and one or more of the $x_{Bi} < 0$. If the change in $\bar{b}$ is not too drastic, a small number of dual simplex iterations will find the optimal solution.

Changes in Constraint Coefficients

If a coefficient a_{il} is changed to a_{il}^+ it will affect some column of A, which we designate column $\bar{a}_l$. If $\bar{a}_l$ is not in B, then $\bar{x}_B = B^{-1}\bar{b}$ is unchanged. Let us now examine the $z_j - c_j$.

For $j \neq l$, we have

$$z_j^+ - c_j = \bar{c}_B' B^{-1}\bar{a}_j^+ - c_j$$

Therefore

$$z_j^+ - c_j = \bar{c}_B' B^{-1}\bar{a}_j - c_j \qquad j \neq l$$
$$z_l^+ - c_l = \bar{c}_B' B^{-1}\bar{a}_l^+ - c_j \qquad \textbf{(7-67)}$$

If $z_l^+ - c_l \geq 0$, then the former optimal solution is still optimal. If $z_l^+ - c_l < 0$, then additional primal simplex iterations may be taken to obtain the optimal solution.

If $\bar{a}_l$ is in the basis, B, then the situation is a great deal more complicated. We now have a new matrix, B_+. However, this matrix may not be a basis, since the substitution of $\bar{a}_l^+$ for $\bar{a}_l$ might destroy the linear independence of the $\bar{a}_j$, $j \neq l$ and $\bar{a}_l^+$. Furthermore, if B_+ is a basis, $\bar{x}_{B_+} = B_+^{-1}\bar{b}$ may no longer be feasible and if it is feasible, it may no longer be optimal. Hence, this kind of change is considerably more complex than when $\bar{a}_l$ was not in the basis. In short, we must check for each of these possibilities. If B_+ is a basis and $\bar{x}_{B_+} \geq \bar{0}$ and $z_j^+ - c_j \geq 0$, then nothing need be done. This possibility can occur. A second possibility is that B_+ is a basis,

$\bar{x}_{B_+} \geq \bar{0}$, but one or more $z_j^+ - c_j < 0$. In this case, further iterations of the primal simplex algorithm may be applied to obtain the optimal solution. A third possibility is that B_+ is a basis, $\bar{x}_{B_+}$ is not feasible but all the $z_j^+ - c_j \geq 0$. In this case the dual simplex algorithm can be used to obtain the optimal solution. If a situation other than the foregoing three hold, one can proceed as follows. Instead of modifying $\bar{a}_l$ to obtain $\bar{a}_l^+$, treat the problem as if one were to add a new variable, x_{n+1} with activity vector $\bar{a}_l^+$ and consider x_l as an artificial variable. Then calculate $\bar{y}_{n+1} = B^{-1}\bar{a}_l^+$ and add this to the tableau. Now we can proceed with the primal simplex algorithm. In some cases, it is best to start the whole problem all over again. It is difficult to know in advance which procedure is preferable.

7.5 THE TRANSPORTATION PROBLEM

The transportation problem is a special linear programming problem which arises in many practical applications. Its name derives from the original context in which it was formulated which was that of determining optimal shipping patterns between origins (or sources) and destinations. More generally, the name "transportation problem" refers to any linear programming problem with a certain fixed structure which is described in the following paragraphs. Many problems which have nothing to do with transportation have this structure. We shall describe the problem in its original and still most widely used context.

Suppose that m origins are to supply n destinations with a certain product. Let a_i be the amount of the product available at origin i, and b_j be the amount of product required at destination j. Further, we assume that the cost of shipping a unit amount of the product from origin i to destination j is c_{ij}. We then let x_{ij} represent the amount shipped from origin i to destination j. If shipping costs are assumed to be proportional to the amount shipped, then the formulation of the question of how much should be shipped from each origin to each destination so as to minimize total shipping costs turns out to be a linear programming problem. It is as follows:

$$\text{Min } z = \sum_{i=1}^{m} \sum_{j=1}^{n} c_{ij}x_{ij}$$

$$\sum_{j=1}^{n} x_{ij} = a_i \qquad i = 1, 2, \ldots, m \qquad a_i > 0 \tag{7-68}$$

$$\sum_{i=1}^{m} x_{ij} = b_j \qquad j = 1, 2, \ldots, n \qquad b_j > 0$$

$$x_{ij} \geq 0 \qquad \text{all } i \text{ and } j$$

The first m constraints of the problem given in (7-68) state that the sum of the amounts shipped from any origin to each of the destinations must equal the amount available at the origin. The remaining n constraints state that the sum of the amounts shipped from each of the origins to any single destination must equal the amount required at that destination. It is both allowable and reasonable to write the first m constraints as inequalities, i.e.,

$$\sum_{j=1}^{n} x_{ij} \leq a_i \qquad i = 1, 2, \ldots, m \tag{7-69}$$

However, as we shall see in a later section, there is no significant difference between these formulations when we attempt to solve the problem.

An obvious necessary and sufficient condition for the linear programming problem given in (7-68) to have a solution is that

$$\sum_{i=1}^{m} a_i = \sum_{j=1}^{n} b_j \tag{7-70}$$

This is seen as follows:

$$\sum_{i=1}^{m}\sum_{j=1}^{n} x_{ij} = \sum_{i=1}^{m} a_i \qquad i = 1, 2, \ldots, m$$

and

$$\sum_{i=1}^{m}\sum_{j=1}^{n} x_{ij} = \sum_{j=1}^{n} b_j \qquad j = 1, 2, \ldots, n$$

Therefore, (7-70) must hold. We shall assume that it is true. If it is not, a fictitious source or destination can be added, as will be explained later.

It is clear that the transportation problem, unlike linear programming problems in general, has a fixed structure as regards its constraints. This suggests, as indeed is the case, that it is probably possible to simplify the usual simplex algorithm. This follows because the matrix A is always the same for any transportation problem, viz.,

$$A = \underbrace{\begin{bmatrix} \bar{1}_n & \bar{0} & \bar{0} & \cdots & \bar{0} \\ \bar{0} & \bar{1}_n & \bar{0} & \cdots & \bar{0} \\ \bar{0} & \bar{0} & \bar{1}_n & \cdots & \bar{0} \\ \cdot & \cdot & \cdot & \cdots & \cdot \\ \cdot & \cdot & \cdot & \cdots & \cdot \\ \cdot & \cdot & \cdot & \cdots & \cdot \\ \bar{0} & \bar{0} & \bar{0} & \cdots & \bar{1}_n \\ I_n & I_n & I_n & \cdots & I_n \end{bmatrix}}_{mn \text{ columns}} \begin{matrix} \left.\vphantom{\begin{matrix}a\\a\\a\\a\\a\\a\\a\end{matrix}}\right\} m \text{ rows} \\ \} \; n \text{ rows} \end{matrix} \tag{7-71}$$

where $\bar{1}_n = \underbrace{(1, 1, \ldots, 1)}_{n \text{ components}}$

$$I_n = \begin{bmatrix} 1 & 0 & 0 & \ldots & 0 \\ 0 & 1 & 0 & \ldots & 0 \\ 0 & 0 & 1 & \ldots & 0 \\ \cdot & \cdot & \cdot & & \cdot \\ \cdot & \cdot & \cdot & & \cdot \\ \cdot & \cdot & \cdot & & \cdot \\ 0 & 0 & 0 & \ldots & 1 \end{bmatrix}$$

$$n \times n$$

If we order the components of the $\bar{x}$ vector as

$$\bar{x} = [x_{11}, x_{12}, \ldots, x_{1n}, x_{21}, x_{22}, \ldots, x_{2n}, \ldots, x_{m1}, x_{m2}, \ldots, x_{mn}]$$

and $\bar{b} = [a_1, a_2, \ldots, a_m, b_1, b_2, \ldots, b_n]$ and $\bar{c}$ is ordered as $\bar{x}$ is, then (7-68) is simply

$$\text{Max } z = \bar{c}'\bar{x}$$
$$A\bar{x} = \bar{b} \qquad \textbf{(7-72)}$$
$$\bar{x} \geq \bar{0}$$

in the usual linear programming format.

Some properties of transportation problems are easily deduced from the fixed structure of the A matrix.†

1. The matrix A has rank $= m + n - 1$. This is easily seen because of the obvious linear dependence implied by the fact that the sum of the first m rows equals the sum of the last n rows. This means the rank of $A < m + n$. That it is equal to $m + n - 1$ is seen by exhibiting a square submatrix from A of order $m + n - 1$ that is not singular. This can be done in a number of ways. One such way is to eliminate the $(m + 1)^{\text{st}}$ row and retain only columns $1, n + 1, 2n + 1, \ldots, mn - n + 1, 2, 3, \ldots, n$. For example, if $m = 3, n = 5$, we would strike out the fourth row of A and retain columns 1, 6, 11, 2, 3, 4, and 5. It is easily verified that this gives rise to a nonsingular matrix.

2. The matrix A is a *unimodular* matrix. By unimodular we mean that every square submatrix‡ of A has a determinant that is equal to $0, \pm 1$. The proof of this is easily given as follows. Suppose that T_k is a kth order submatrix formed from k rows and k columns of A. From the structure of A it is clear that each column of T_k

---

† The reader who is less interested in details of proof may ignore the following proofs and proceed to the paragraph containing (7-79).

‡ A *submatrix* of A is a matrix obtained from A by deleting row(s) or column(s) of A. The determinant of a square submatrix of A is called a *minor* of A.

must contain all zeros, a single 1 or two 1's. If T_k contains at least one column of zeros, then $|T_k| = 0$, since it could be expanded in terms of that column and hence $|T_k| = 0$. If every column of T_k contained two 1's, then as is clear from the structure of A, one of the 1's is in an origin row and the other must be in a destination row. If we sum the origin rows and subtract the destination rows we obtain the null vector. This indicates that the rows of T_k are not linearly independent and therefore T_k is singular or $|T_k| = 0$. The last case would occur if there was at least one column which had a single 1. If we expand $|T_k|$ by one of the columns† which had this 1 we would obtain

$$|T_k| = \pm |T_{k-1}|$$

where T_{k-1} is a submatrix of T_k of order $k - 1$. We now reapply the preceding arguments to T_{k-1}. Finally we reach T_1. However, it is clear that $|T_1| = 0, 1$, since every element of A is 0 or 1. Therefore, we have proven that $|T_k| = 0, \pm 1$.

Let us now consider applying the simplex method to the transportation problem. Our most formidable problem is one of notation! Let us designate the columns of A as $\bar{a}_{ij}$. The ordering is such that column $(i - 1)n + j$ of A is then $\bar{a}_{ij}$. It should be noted that because of the simple structure of A

$$\bar{a}_{ij} = \bar{e}_i + \bar{e}_{m+j} \tag{7-73}$$

Suppose we tried to solve the transportation problem by the simplex method. If $\sum_{i=1}^{m} a_i = \sum_{j=1}^{n} b_j$ holds then every transportation problem has a feasible solution. For example,

$$x_{ij} = \frac{a_i b_j}{\sum_{i=1}^{m} a_i} = \frac{a_i b_j}{\sum_{j=1}^{n} b_j} \tag{7-74}$$

is clearly a feasible solution. Hence, every transportation problem has an optimal feasible solution. We could therefore begin with the preceding or, as we shall see, many other possible initial feasible solutions for a transportation problem. They are relatively easy to generate. Hence we never need to add artificial vectors, even if we used the simplex method as previously described. There is one noteworthy feature of basic feasible solutions to transportation problems that should be mentioned. Since there are $m + n$ rows but only $m + n - 1$ linearly independent columns in A, a basis matrix need only contain $m + n - 1$ vectors. We shall henceforth assume this. Let us designate the basis matrix as

$$B_T = (\bar{d}_1, \bar{d}_2, \ldots, \bar{d}_{m+n-1})$$

---

† See Property 9 of Section 2.7.

where the $\bar{d}_l$, $l = 1, 2, \ldots, m + n - 1$ are a linearly independent set of vectors $\bar{a}_{ij}$ from A. Since B_T is a basis any vector $\bar{a}_{ij}$ in A can be expressed as a linear combination of the vectors, $\bar{d}_l$ in B_T, i.e.,

$$\bar{a}_{ij} = \sum_{l=1}^{m+n-1} y_{ij}^l \bar{d}_l \tag{7-75}$$

It is not difficult to see that all the y_{ij}^l in a transportation problem are 0, ± 1. This is seen as follows. Equations (7-75) are a set of $m + n$ linear equations in $m + n - 1$ variables, y_{ij}^l. In matrix form, they would be written

$$\bar{a}_{ij} = B_T \bar{y}_{ij} \tag{7-76}$$

where $\bar{y}_{ij} = (y_{ij}^1, y_{ij}^2, \ldots, y_{ij}^{m+n-1})$. Let us now recall that

$$\bar{a}_{ij} = \bar{e}_i + \bar{e}_{m+j}$$

If we should drop equation i from (7-76), we would now have a nonsingular matrix and we would now write the resulting set of equations

$$\bar{e}_{m+j-1} = S_T \bar{y}_{ij} \tag{7-77}$$

where $\bar{e}_{m+j-1}$ is a unit vector with $m + n - 1$ components and S_T is B_T with the ith row deleted. Since S_T is nonsingular, we can write:

$$\bar{y}_{ij} = S_T^{-1} \bar{e}_{m+j-1} \tag{7-78}$$

If we recall the definition of the inverse of a matrix and examine equation (7-78), we see that every element of $\bar{y}_{ij}$ is a minor of order $m + n - 2$ divided by $|S_T|$. However, all of these are minors of A. We proved that all minors of A are 0, ± 1. Therefore, y_{ij}^l, $l = 1, 2, \ldots, m + n - 1$ are 0, ± 1.

The preceding result enables us to greatly simplify the usual simplex calculations. First, let us consider the usual simplex transformation formulas. For the transportation problem we shall designate the variables in the basis as x_{Bl}, $l = 1, 2, \ldots, m + n - 1$ so that the transformation formulas become

$$\hat{x}_{Bl} = x_{Bl} - \frac{y_{pq}^l x_{Br}}{y_{pq}^r} \qquad l \neq r$$

$$\hat{x}_{Br} = \frac{x_{Br}}{y_{pq}^r} \tag{7-79}$$

where $\bar{a}_{pq}$ is the vector entering the basis and $\bar{d}_r$ is the vector leaving the basis. However, since all $y_{ij}^l = 0, \pm 1$, the equations in (7-79) become

$$\hat{x}_{Bl} = x_{Bl} \quad \text{or} \quad \hat{x}_{Bl} = x_{Bl} \pm x_{Br} \tag{7-80}$$

From this result it follows that only addition and subtraction are used to transform basic solutions. It follows further that *if the a_i and b_j are all integers, all subsequent solutions will be integers.*

More important, the preceding result gives us the basic clue as to how to avoid an enormous basis of order $m + n - 1$ for an m origin and n destination transportation problem. According to (7-76), any vector $\bar{a}_{ij}$ can be written in terms of the basis vectors as

$$\bar{a}_{ij} = \sum_{l=1}^{n+m-1} y_{ij}^l \bar{d}_l \tag{7-75}$$

However, we see that if we omit the y_{ij}^l that are zero this becomes

$$\bar{a}_{ij} = \sum_{l \in L} (\pm)\bar{d}_l \tag{7-81}$$

where $L = \{l \mid y_{ij}^l \neq 0\}$. Let us now recall that $\bar{d}_l$, $l = 1, 2, \ldots, m + n - 1$ are the vectors in the basis and therefore are merely some subset of the $\bar{a}_{ij}$. It is more convenient to relabel these $\bar{d}_l$ as $\bar{a}_{Sl}^B$, the $m + n - 1$ particular $\bar{a}_{ij}$ that are in the basis. Therefore

$$\bar{a}_{ij} = \sum_{l \in L} (\pm)\bar{a}_{Sl}^B \tag{7-82}$$

Each of the $\bar{a}_{ij}$, including the $\bar{a}_{St}^B$, can be represented in the form

$$\bar{a}_{ij} = \bar{e}_i + \bar{e}_{m+j} \tag{7-73}$$

This means that since each coefficient of $\bar{a}_{Sl}^B$ is ± 1, it is necessary that there is one $\bar{a}_{Sl}^B$ of the form $\bar{a}_{i\alpha}^B = \bar{e}_i + \bar{e}_{m+\alpha}$ whose coefficient is $+1$ in (7-82) so that there will indeed be a 1 in the ith component of $\bar{a}_{ij}$. If $\alpha \neq j$, there must be an $\bar{a}_{Sl}^B$ of the form $\bar{a}_{\beta\alpha}^B = \bar{e}_\beta + \bar{e}_{m+\alpha}$, $\beta \leq m$, whose coefficient is -1 in order to cancel out the $+1$ in the $(m + \alpha)^{\text{th}}$ component of $\bar{a}_{i\alpha}^B$. We can obviously continue this argument, and since $m + n - 1$ is a finite integer, we will eventually reach a vector $\bar{a}_{\delta j}^B = \bar{e}_\delta + \bar{e}_{m+j}$ whose coefficient is $+1$ so as to obtain $\bar{e}_{m+j}$. The $+1$ of the $\bar{e}_\delta$ will be cancelled out by the previous vector. There can be only one such representation of this since the representation of any vector as a linear combination of basis vectors is unique.

What we have shown is that we can represent any vector $\bar{a}_{ij}$ as follows:

$$\bar{a}_{ij} = \bar{a}_{i\alpha}^B - \bar{a}_{\beta\alpha}^B + \bar{a}_{\beta\gamma}^B - \ldots - \bar{a}_{\delta\phi}^B + \bar{a}_{\delta j}^B \tag{7-83}$$

If we now consider the calculation of the $z_{ij} - c_{ij}$ ($z_j - c_j$ for a transportation problem) and if we designate the prices in the basis as $\bar{c}_B$ then we have

$$z_{ij} - c_{ij} = \bar{c}_B' \bar{y}_{ij} - c_{ij} \tag{7-84}$$

If we designate the components of $\bar{c}_B$ as c^B_{Sl} to correspond to $\bar{a}^B_{Sl}$, then, since all $y^l_{ij} = 0, \pm 1$, (7-84) becomes

$$z_{ij} - c_{ij} = \sum (\pm) c^B_{Sl} - c_{ij} \tag{7-85}$$

By the same argument used to represent $\bar{a}_{ij}$ in terms of the vectors in the basis, $\bar{a}^B_{St}$, we can represent $z_{ij} - c_{ij}$ as

$$z_{ij} - c_{ij} = c^B_{i\alpha} - c^B_{\beta\alpha} + c^B_{\beta\gamma} - \ldots - c^B_{\delta\phi} + c^B_{\delta j} - c_{ij} \tag{7-86}$$

Because of this simple representation of the $z_{ij} - c_{ij}$, we will never have to actually deal with the y^l_{ij} or with an extended basis of size $m + n - 1$. We must now show a simple way to find the c^B_{St} without a great deal of calculation. The method given here stems from Dantzig's original work on the transportation problem. Suppose $c^B_{i\alpha}, c^B_{\beta\alpha}, c^B_{\beta\gamma}, \ldots, c^B_{\delta\phi}, c^B_{\delta j}$ are the $m + n - 1$ prices in the basis, i.e., prices corresponding to any basic feasible solution to the m origin, n destination transportation problem. Suppose we then write the following equations:

$$\begin{aligned} u_i + v_\alpha &= c^B_{i\alpha} \\ u_\beta + v_\alpha &= c^B_{\beta\alpha} \\ u_\beta + v_\gamma &= c^B_{\beta\gamma} \\ \cdot \quad \cdot \quad &\cdot \\ \cdot \quad \cdot \quad &\cdot \\ \cdot \quad \cdot \quad &\cdot \\ u_\delta + v_\phi &= c^B_{\delta\phi} \\ u_\delta + v_j &= c^B_{\delta j} \end{aligned} \tag{7-87}$$

Clearly, (7-87) is a set of $m + n - 1$ linear equations in $m + n$ variables, u_i, $i = 1, 2, \ldots, m$ and v_j, $j = 1, 2, \ldots, n$. Since the rows correspond to the vectors in the basis, $\bar{a}^B_{Sl}$ the rank of the matrix of these equations is $m + n - 1$. Hence we have an underdetermined set of equations, and one variable may be assigned arbitrarily. Because of the obviously simple nature of these equations, it is an easy matter to solve them in the order in which they occur. For example, if we arbitrarily set $u_i = 0$, then $v_\alpha = c^B_{i\alpha}$, $u_\beta = c^B_{\beta\alpha} - c^B_{i\alpha}$, $v_\gamma = c^B_{\beta\gamma} - c^B_{\beta\alpha} + c^B_{i\alpha}$, and so forth. The reason for wanting to find these u_i and v_j is as follows. Since $z_{ij} - c_{ij}$ is given by

$$z_{ij} - c_{ij} = c^B_{i\alpha} - c^B_{\beta\alpha} + c^B_{\beta\gamma} - \ldots - c^B_{\delta\phi} + c^B_{\delta j} - c_{ij} \tag{7-86}$$

if we now substitute the u_i and v_j which we have determined into (7-86) we have

$$\begin{aligned} z_{ij} - c_{ij} &= u_i + v_\alpha - u_\beta - v_\alpha + u_\beta + v_\gamma \\ &\quad - \ldots - u_\delta - v_\phi + u_\delta + v_j - c_{ij} \\ &= u_i + v_j - c_{ij} \end{aligned} \tag{7-88}$$

The simplicity of (7-88) can hardly be overemphasized. It indicates how easily the $z_{ij} - c_{ij}$ can be calculated for every $\bar{a}_{ij}$ once the u_i and v_j are all known.

We shall now put all of the preceding simplifications together into a computational algorithm for the transportation problem. It should be borne in mind, however, that it is in reality the primal simplex method, and the fact that the matrix, A, is fixed and unimodular has led to these simplifications. We shall use a greatly simplified tableau to lay out our basic data and carry out the computations. It is shown in Figure 7-2. We begin by entering the known c_{ij} and some initial basic feasible solution, a set of x_{ij}. We also enter the a_i and b_j as indicated. Now, as is true for any simplex algorithm, we must determine whether or not we have an optimal basic feasible solution. In order to do this, we must calculate the

Destinations

Origins	1	2		j		n		
1	c_{11} x_{11}	c_{12} x_{12}	⋯	c_{1j} x_{1j}	⋯	c_{1n} x_{1n}	a_1	u_1
2	c_{21} x_{21}	c_{22} x_{22}	⋯	c_{2j} x_{2j}	⋯	c_{2n} x_{2n}	a_2	u_2
	⋮	⋮	⋯	⋮	⋯	⋮	⋮	⋮
i	c_{i1} x_{i1}	c_{i2} x_{i2}	⋯	c_{ij} x_{1j}	⋯	c_{in} x_{in}	a_i	u_i
	⋮	⋮	⋯	⋮	⋯	⋮	⋮	⋮
m	c_{m1} x_{m1}	c_{m2} x_{m2}	⋯	c_{mj} x_{mj}	⋯	c_{mn} x_{mn}	a_m	u_m
	b_1	b_2	⋯	b_j	⋯	b_n	$\sum_i a_i = \sum_j b_j$	
	v_1	v_2	⋯	v_j	⋯	v_n		

Figure 7-2 Transportation problem tableau.

$z_{ij} - c_{ij}$. Our first step then is to find the u_i and v_j as we indicated. Then we use (7-88) to calculate the $z_{ij} - c_{ij}$. We recall that our statement of the transportation problem in (7-68) was a *minimization* problem. Hence, at optimality all the $z_{ij} - c_{ij} \leq 0$. Therefore, if one or more of the $z_{ij} - c_{ij} > 0$, we know that we can reduce the value of the objective function, z.† As usual, we compute the vector to enter the basis as

$$z_{pq} - c_{pq} = \text{Max}\,\{(z_{ij} - c_{ij}),\ z_{ij} - c_{ij} > 0\} \qquad \textbf{(7-89)}$$

This means that $\bar{a}_{pq}$ enters the basis, although we need never look at the $\bar{a}_{ij}$ vectors. Therefore, we know that x_{pq} will be positive in the next tableau. The vector to be removed from the basis will be determined from

$$\text{Min}\left\{\frac{x_{Bl}}{y^l_{pq}},\ y^l_{pq} > 0\right\} \qquad \textbf{(7-90)}$$

However, if $y^l_{pq} > 0$, then $y^l_{pq} = 1$. Therefore, the variable to be removed from the basis is particularly easy to determine. We simply scan the current basis variables for those which have a $y^l_{ij} = 1$ when

$$z_{pq} - c_{pq} = c^B_{p\alpha} - c^B_{\beta\alpha} + c^B_{\beta\gamma} - \cdots - c^B_{\delta\phi} + c^B_{\delta q} - c_{pq} \qquad \textbf{(7-91)}$$

The smallest of these x_{Bl} with $y^l_{ij} = 1$ in (7-91) is the vector to be removed from the basis. This is quite easy to do in practice because there is a unique "loop" involving the basis variables and the variable to enter. The finding of this loop is trivial and is best shown by an example. Suppose we had the upper tableau shown in Figure 7-3. We have omitted the u_i and v_j columns here. Suppose we have decided that $\bar{a}_{13}$ is to enter the basis and therefore, x_{13} is to become positive. We enter a quantity θ for x_{13}. However, in order to remain feasible we must subtract θ from either $x_{11} = 50$ or $x_{12} = 10$. We do not choose x_{11} because there is no other entry in column 1 to which we could add θ to balance the requirement b_1. However, if we subtract θ from $x_{12} = 10$, then we can add θ to $x_{22} = 55$. Next we can subtract θ from $x_{23} = 5$ and we have a balanced loop in which all requirements are met. We may choose any value of θ and obtain a new basic feasible solution, providing we do not choose a value of θ so large that some of the x_{ij} become negative. This is easily avoided:

$$\theta = \text{Min}\,\{x_{12}, x_{23}\} = \text{Min}\,(10, 5) = 5$$

Therefore if $\theta = 5$, $x_{13} = 5$, $x_{12} = 5$, $x_{22} = 60$, and $x_{23} = 0$. The variable corresponding to θ, in this case x_{23}, leaves the basis. All other basic variables are unchanged. We have a new basic feasible

† This, of course, assumes the absence of degeneracy. Degeneracy is easily handled theoretically in the transportation problem. See Hadley[2]. Practically, it is not a problem in computation.

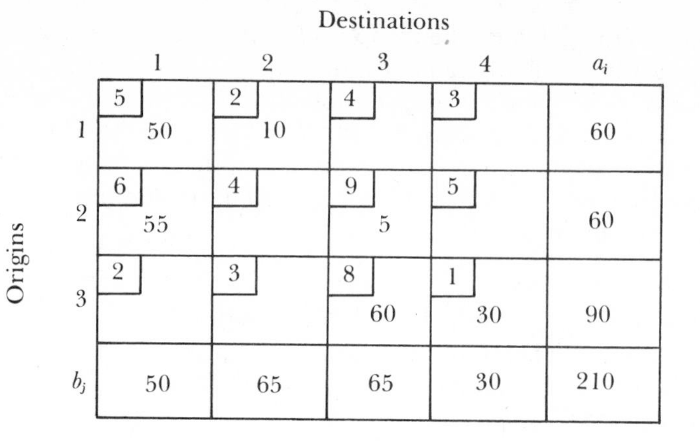

Destinations (Origins on rows; cost c_{ij} shown in the box, followed by the allocation)

Origins	1	2	3	4	a_i
1	5 · 50	2 · 10	4	3	60
2	6 · 55	4	9 · 5	5	60
3	2	3	8 · 60	1 · 30	90
b_j	50	65	65	30	210

Destinations

Origins					a_i
	5 · 50	2 · $10-\theta$	4 · θ	3	60
	6	4 · $55+\theta$	$5-\theta$		60
	2		60	30	90
b_j	50	65	65	30	210

Figure 7-3 Determination of loop.

solution and if x_{13} was chosen to become positive in accord with the criterion of (7-89), we have obtained an improved value of z. The general rule here is clear. We must determine a loop between the variable to become positive and the other basic variables so as to remain feasible. This is easily done by inspection or it can be programmed on a digital computer. The only rule necessary to keep in mind in determining the loop is that successive links of the loop be perpendicular to each other. It is perfectly possible to skip over some basic variables in a row or column. However, successive links must be perpendicular. The example will clarify this.

We have not discussed yet how to find an initial basic feasible solution to a transportation problem. Many methods have been proposed and are easy to devise. We shall discuss one of the most commonly used and refer the reader to Hadley[2] for a discussion of some others. Suppose we have the tableau given in Figure 7-4 and we seek an initial basic feasible solution. We have shown the c_{ij}, although they are not used at all in the method we describe, which

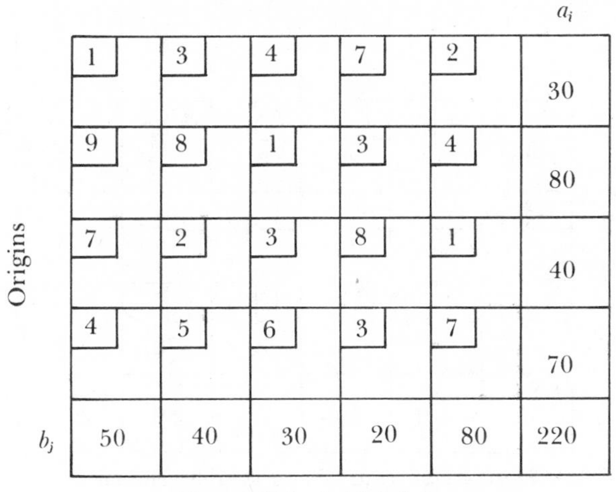

Figure 7-4 Initial data.

is known as the "Northwest Corner Rule." We start as the name implies and choose a value for x_{11}. This is determined from

$$x_{11} = \text{Min}\ (a_1, b_1) = \text{Min}\ (30, 50) = 30$$

Hence, we cannot make any other $x_{1j} > 0$ because all of a_1 has been allocated. However, destination 1 is still short 20 units so we move down column 1 and determine

$$x_{21} = \text{Min}\ (a_2, b_1 - a_1) = \text{Min}\ (80, 20) = 20$$

Now column 1 is completed, since x_{31} and x_{41} must be zero. However, there are $80 - 20 = 60$ units still available from a_2, so we now consider

$$x_{22} = \text{Min}\ (a_2 - (b_1 - a_1), b_2) = \text{Min}\ (60, 40) = 40$$

Next we find $x_{23} = \text{Min}\ (20, 30) = 20$, $x_{33} = \text{Min}\ (40, 10)$, $x_{34} = \text{min}\ (30, 20) = 20$, $x_{35} = \text{min}\ (10, 80) = 10$, $x_{45} = 70$. The last value is found by difference because of the fact that $\sum_i a_i = \sum_j b_j$. Figure 7-5 shows this initial basic feasible solution. The general rule is quite clearly to move either across a row until that origin availability is used up or down a column until the destination requirement is met, and then drop down a row or across to the next column, and so forth. The Northwest Corner Rule has great simplicity to recommend it. However, since it makes no use of the c_{ij}, there is no reason to suspect that it is a particularly

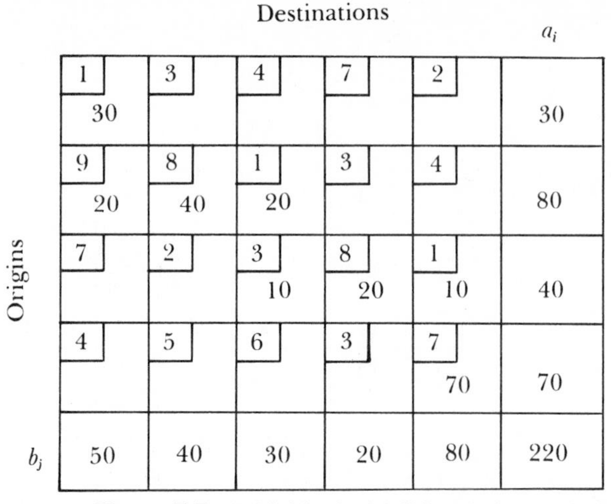

Destinations

Origins						a_i
	[1] 30	[3]	[4]	[7]	[2]	30
	[9] 20	[8] 40	[1] 20	[3]	[4]	80
	[7]	[2]	[3] 10	[8] 20	[1] 10	40
	[4]	[5]	[6]	[3]	[7] 70	70
b_j	50	40	30	20	80	220

Figure 7-5 Initial basic feasible solution.

good solution in the sense of being close to the optimum solution. For this reason, other methods have been devised for starting with a better initial basic feasible solution. However, as might be expected, they require more time to compute a solution. There is a trade-off here which has never been definitively explored.

Let us now summarize the transportation problem algorithm and then state a practical example and give the detailed solution.

Transportation Problem Algorithm

1. Determine an initial basic feasible solution by any method.

2. Set $u_1 = 0$ and then use the equations (7-87) sequentially to find all the u_i and v_j.

3. Compute the $z_{ij} - c_{ij}$ for all vectors not in the basis. If all $z_{ij} - c_{ij} \leq 0$, we have the optimal solution.

4. If one or more $z_{ij} - c_{ij} > 0$, use the criterion of Equation (7-89) to determine which vector is to enter the basis, or correspondingly, which x_{ij} presently equal to zero is to become positive.

5. Determine the loop between the x_{ij} to become positive and the other basis elements in the tableau.

6. Determine the maximum amount of change in the basis elements of the loop which we designated θ.

7. Recompute the x_{ij} for the basis elements in the loop. We now have a new basic feasible solution.

8. Return to 2.

EXAMPLE

Suppose a beer manufacturer has three breweries with weekly production capacities of 50,000, 70,000, and 90,000 barrels, respectively. Suppose he ships into 5 warehouses with weekly demands of 20,000, 60,000, 80,000, 40,000, and 10,000 barrels, respectively. If the costs of shipping (in suitable units) are given by the following table, what are the number of barrels (in units of thousands) to be shipped from each brewery to each warehouse so as to minimize total shipping costs?

Warehouses

Breweries	1	2	3	4	5
1	7	3	2	4	2
2	6	5	8	3	4
3	3	2	5	7	1

First, we construct the initial tableau, using the Northwest Corner Rule. Figure 7-6 shows this first tableau. Capacities are given in thousands of barrels. Let us see how it was constructed. We start with $x_{11} = \text{Min}(20, 50) = 20$. This leaves requirement 1 satisfied and 30 units remaining at brewery 1. Next we determine $x_{12} = \text{Min}(30, 60) = 30$. Now brewery 1's capacity is used up and 30 units remain to be satisfied for warehouse 2. We then determine $x_{22} = \text{Min}(30, 70) = 30$. Now warehouse 2's requirement is satisfied and brewery 1 has 40 units still available. We next determine $x_{23} = \text{Min}(40, 80) = 40$. Now brewery 2's capacity is used up and warehouse 3 still requires 40 units. We therefore next determine $x_{33} = \text{Min}(40, 90) = 40$. Obviously then, $x_{34} = \text{Min}(40, 50) = 40$ and $x_{35} = 10$. We have circled the basic feasible solution. It will be seen that $m + n - 1 = 3 + 5 - 1 = 7$ $x_{ij} > 0$.

Our next step is to calculate the $z_{ij} - c_{ij}$. First we calculate the u_i and v_j. We have

$$u_1 + v_1 = c_{11} = 7 \qquad u_1 + v_2 = c_{12} = 3 \qquad u_2 + v_2 = c_{22} = 5$$

$$u_2 + v_3 = c_{23} = 8 \qquad u_3 + v_3 = c_{33} = 5 \qquad u_3 + v_4 = c_{34} = 7$$

$$u_3 + v_5 = c_{35} = 1$$

Setting $u_1 = 0$, we calculate sequentially, $v_1 = 7$, $v_2 = 3$, $u_2 = 2$, $v_3 = 6$, $u_2 = -1$, $v_4 = 8$, and $v_5 = 2$.* These are then placed in the u_i and v_j columns of Figure 7-6. We are now ready to compute the

Warehouses

Breweries	1	2	3	4	5	a_i	u_i
1	[7] (20)	[3] (30)	[2] 4	[4] 4	[2] 0	50	0
2	[6] 3	[5] (30)	[8] (40)	[3] 7	[4] 0	70	2
3	[3] 3	[2] 0	[5] (40)	[7] (40)	[1] (10)	90	−1
b_j	20	60	80	40	10	210	
v_j	7	3	6	8	2		

$z = 1190$

Figure 7-6 Transportation Tableau I.

$z_{ij} - c_{ij} = u_i + v_j - c_{ij}$ for those vectors which are not in the basis. For example,

$$z_{13} - c_{13} = 0 + 6 - 2 = 4$$

$$z_{21} - c_{21} = 2 + 7 - 6 = 3$$

and so forth. The uncircled numbers represent the $z_{ij} - c_{ij}$ in Tableau I.

We see that there are $z_{ij} - c_{ij} > 0$. We determine the vector to enter from

$$z_{pq} - c_{pq} = \text{Max}\,(z_{ij} - c_{ij}) \qquad z_{ij} - c_{ij} > 0$$

The maximum of these is $z_{24} - c_{24} = 7$. Therefore, x_{24} is to become positive. The loop involving the x_{24} cell in the tableau and the basis elements is as follows:

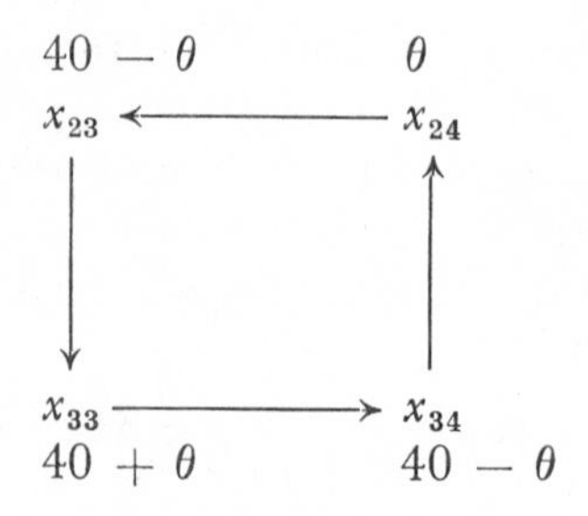

$\theta = \text{Min}\,(40, 40) = 40$. Therefore, the new values of these variables are $x_{24} = 40$, $x_{23} = 0$, $x_{33} = 80$, and $x_{34} = 0$. Because of the tie in determining θ, not one but two basic variables are forced to zero. We therefore have chosen one of them, x_{34}, to remain in the basis. We have a degenerate basic solution.

It is worth comparing the two solutions. Our initial solution had a value of z given by

$$\begin{aligned} z &= 7(20) + 3(30) + 5(30) + 8(40) + 5(40) + 7(40) + 1(10) \\ &= 140 + 90 + 150 + 320 + 200 + 280 + 10 = 1190 \end{aligned}$$

The solution of Tableau II is

$$\hat{z} = 7(20) + 3(30) + 5(30) + 3(40) + 5(80) + 1(10) = 910$$

We see that $\hat{z} < z$.

We next compute the u_i and v_j of Tableau II (Figure 7-7) as before, and they are shown, as well as the new values of $z_{ij} - c_{ij}$. We see that there are still $z_{ij} - c_{ij} > 0$. The largest of these is $z_{31} - c_{31} = 10$. Therefore, x_{31} becomes positive. The "loop" in this case is more complicated. It is as follows:

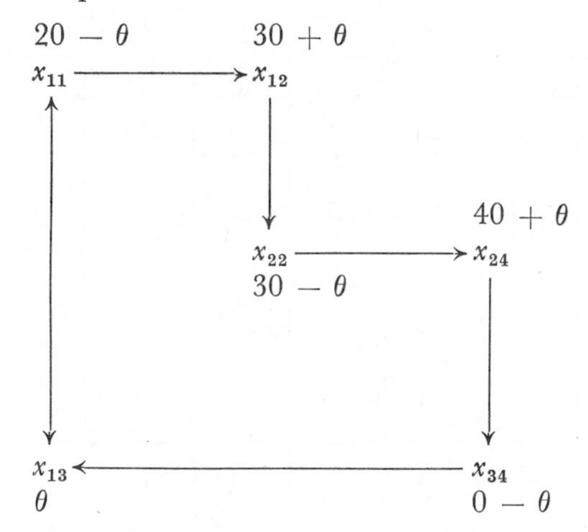

(Each cell shows the cost c_{ij} in the box, followed by the circled allocation (in parentheses) or the value of $z_{ij} - c_{ij}$.)

	1	2	3	4	5	a_i	u_i	
1	7 (20)	3 (30)	2 −3	4 −3	2 −7	50	0	
2	6 3	5 (30)	8 −7	3 (40)	4 −7	70	2	
3	3 10	2 7	5 (80)	7 (0)	1 (10)	90	6	$z = 910$
b_j	20	60	80	40	10	210		
v_j	7	3	−1	1	−5			

Figure 7-7 Transportation Tableau II.

	1	2	3	4	5	a_i	u_i
1	[7] (20)	[3] (30)	[2] 7	[4] −3	[2] 3	50	0
2	[6] 3	[5] (30)	[8] 3	[3] (40)	[4] 3	70	2
3	[3] (0)	[2] 3	[5] (80)	[7] −10	[1] (10)	90	−4
b_j	20	60	80	40	10	210	
v_j	7	3	9	1	5		

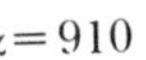
$z = 910$

Figure 7-8 Transportation Tableau III.

	1	2	3	4	5	a_i	u_i
1	[7] −7	[3] (30)	[2] (20)	[4] −3	[2] −4	50	0
2	[6] −4	[5] (30)	[8] −4	[3] (40)	[4] −4	70	2
3	[3] (20)	[2] 4	[5] (60)	[7] −3	[1] (10)	90	3
b_j	20	60	80	40	10	210	
v_j	0	3	2	1	−2		

$z = 770$

Figure 7-9 Transportation Tableau IV.

	1	2	3	4	5	a_i	u_i
1	[7] −7	[3] −4	[2] (50)	[4] −7	[2] −4	50	0
2	[6] 0	[5] (30)	[8] 0	[3] (40)	[4] 0	70	6
3	[3] (20)	[2] (30)	[5] (30)	[7] −7	[1] (10)	90	3
b_j	20	60	80	40	10	210	
v_j	0	−1	2	−3	−2		

$z = 650$

Figure 7-10 Transportation Tableau V.

We see that the min (20, 30, 0) $= 0 = \theta$. Therefore, x_{31} enters the basis at a zero level and $x_{34} = 0$ but is no longer in the basis. The objective function remains unchanged. The new tableau is shown in Figure 7-8. The remaining tableaux are given in Figures 7-9 and 7-10. Since no $z_{ij} - c_{ij} > 0$ in Tableau V, we have the optimal solution, which is

$$\begin{aligned} x_{13} &= 50 \\ x_{22} &= 30 \\ x_{24} &= 40 \\ x_{31} &= 20 \qquad z = 650 \\ x_{32} &= 30 \\ x_{33} &= 30 \\ x_{35} &= 10 \end{aligned}$$

We can see that our initial basic solution was far from optimal.

One last point should be mentioned with respect to transportation problems. Frequently they may be phrased as follows:

$$\text{Min } z = \sum_{i=1}^{m} \sum_{j=1}^{n} c_{ij} x_{ij}$$

$$\sum_{j=1}^{n} x_{ij} \leq a_i \qquad i = 1, 2, \ldots, m \tag{7-92}$$

$$\sum_{i=1}^{m} x_{ij} = b_j \qquad j = 1, 2, \ldots, n$$

$$x_{ij} \geq 0 \qquad \text{all } i, j$$

A formulation such as given in (7-92) implies that $\sum_{i=1}^{m} a_i > \sum_{j=1}^{n} b_j$, i.e., there is more available at the origins than is required at the destinations. However, we know that in order for a solution to exist to a transportation problem, $\sum_i a_i = \sum_j b_j$. The way to handle a problem such as (7-92) is quite simple. We define

$$b_{n+1} = \sum_{i=1}^{m} a_i - \sum_{j=1}^{n} b_j \tag{7-93}$$

and further, convert the inequality constraints to equalities by adding slack variables:

$$\sum_{j=1}^{n} x_{ij} + x_{i,n+1} = a_i, \qquad i = 1, 2, \ldots, m \tag{7-94}$$

We now have

$$\sum_{i=1}^{m} a_i = \sum_{i=1}^{m} \sum_{j=1}^{n} x_{ij} + \sum_{i=1}^{m} x_{i,n+1}$$
$$\sum_{j=1}^{n} b_j = \sum_{i=1}^{m} \sum_{j=1}^{n} x_{ij} \tag{7-95}$$

Therefore, since by (7-93), $b_{n+1} = \sum_{i=1}^{m} a_i - \sum_{j=1}^{n} b_j$, by (7-95) we have

$$\sum_{i=1}^{m} x_{i,n+1} = b_{n+1} \tag{7-96}$$

Now, $\sum_{i=1}^{m} a_i = \sum_{j=1}^{n+1} b_j$ and we can solve the problem. Hence, we add the additional constraint (7-96) which adds one more destination to which all the slack is to be sent. We simply add one additional column to the tableau with zero costs, i.e., $c_{i,n+1} = 0$, since if we do not ship these amounts we incur no cost, and solve the problem as before.

7.6 NETWORK FLOW PROBLEMS

In this section we shall introduce the ideas behind the analysis of flows in networks. A "network" is an idealized mathematical abstraction from the real world in which certain "points" or "nodes" are connected by "lines" or "branches" or "arcs" and through which the flow of some "material" occurs. A collection of nodes and branches is called a "graph" in the branch of mathematics known as the theory of graphs. If a "flow" occurs in the graph, it is generally referred to as a network. Figure 7-11 gives an example of a graph.

As examples of networks, if the nodes are compressor stations and the branches are pipes and the flow is natural gas, we have a network that is a natural gas pipeline system. If the nodes are telephone switching points and the branches are wires and the flow is "information" we have a telephone communication system. Similarly, we can regard a connecting system of roads as a network or a system of work stations on manufacturing assembly lines as a network. There are many real world situations that may, more or less, be characterized as networks.

It will be noted in Figure 7-11 that there is no "orientation" of the branches. If there is a particular sense of direction imparted to each branch of the graph, then we speak of an "oriented graph." Figure 7-12 shows an oriented graph. In this graph, if we had considered each of the oriented branches as a direction along which

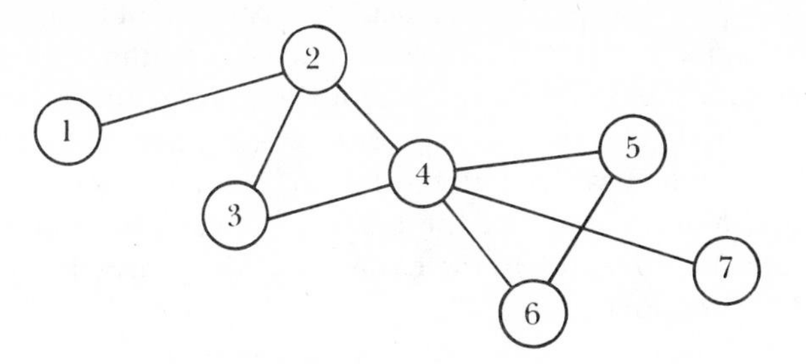

Figure 7-11 An example of a graph.

some kind of flow could occur, then we would call it a network. It is possible to have an unoriented network in the sense that flow can be permitted in both directions.

We shall now define some terms commonly used in connection with networks and network flow problems. The "capacity" of a branch in one (or both) directions is the maximum allowable flow that the branch can accommodate. A flow capacity is a nonnegative number. If there is no finite upper bound on an allowable flow, we consider the flow capacity to be infinite. It is usually convenient to indicate flow capacities in both directions between nodes. For example, if x_{ij} is the flow from node i to node j, then we might say $x_{ij} = 10$ and $x_{ji} = 10$ for equal flow capacities in both directions. If $x_{ij} = 10$ and $x_{ji} = 0$, we have a completely oriented flow branch, in that there is no allowable flow from node j to node i, but only from node i to node j.

There are usually in a network one or more nodes which are distinguished as "sources." A source is a node such that all branches connected to the node are oriented so that the flows are away from the node. Similarly, a "sink" is a node such that the flows in each of the branches joined to the node are into the node. In short, flows originate in sources and terminate at sinks.

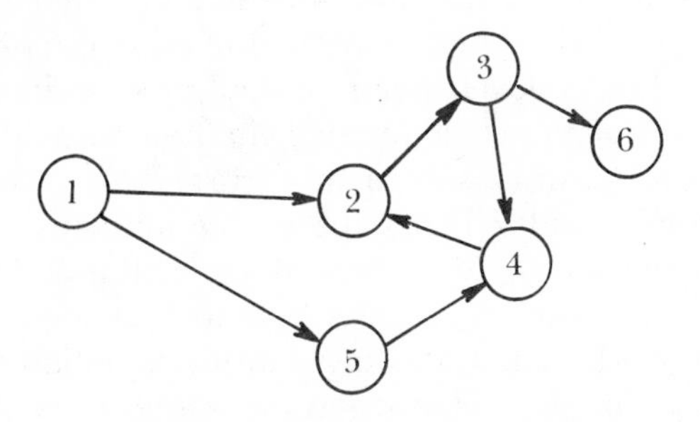

Figure 7-12 An oriented graph.

Let us now consider the *maximal flow problem*. In this problem, we have a single source and a single sink. We assume that conservation of flow holds for each node (except the source and the sink), i.e., flow into the node equals flow out of the node. Suppose there are N nodes, where node 1 is the source and node N is the sink. For each branch, designated (i, j), i.e., the branch between nodes i and j, we let x_{ij} denote the flow and c_{ij} the capacity. There are obviously flow constraints of the form

$$0 \leq x_{ij} \leq c_{ij} \qquad \text{all } i, j \tag{7-97}$$

In addition there are constraints which relate to the conservation of flow, i.e., the flow into a node equals the flow out of the node:

$$\sum_i x_{ik} = \sum_j x_{kj} \qquad k = 2, 3, \ldots, N-1$$

or

$$\sum_i x_{ik} - \sum_j x_{kj} = 0 \qquad k = 2, 3, \ldots, N-1 \tag{7-98}$$

The total flow which is to be maximized is given by

$$z = \sum_j x_{1j} \tag{7-99}$$

The objective function is the sum of flows over each branch connecting the source to the nodes connected to it. It could equally well be represented by

$$z = \sum_i x_{iN} \tag{7-100}$$

which is the corresponding flow into the sink. In summary then, the maximal flow problem can be written

$$\begin{gathered}\text{Max } z = \sum_j x_{1j} \\ \sum_i x_{ik} - \sum_j x_{kj} = 0 \qquad k = 2, 3, \ldots, N-1 \\ 0 \leq x_{ij} \leq c_{ij} \qquad \text{all } i, j\end{gathered} \tag{7-101}$$

It will be recognized that (7-101) is a linear programming problem. However, for a network with many branches and nodes, it is inefficient to use the simplex method to solve the problem, even though this could be done. Instead we shall present simpler methods for solving the problem of maximizing the flow through a network.

Consider the network shown in Figure 7-13. We have seven nodes with nodes 1 and 7 being the source and sink, respectively. The branch capacities, c_{ij}, are shown along each branch connecting the nodes. The problem we face is how to find the maximal flow through this network. An appealingly intuitive notion of how to do this is to reason as follows. Starting at the source we could seek out a path each of whose branches had positive flow capacities to the sink.

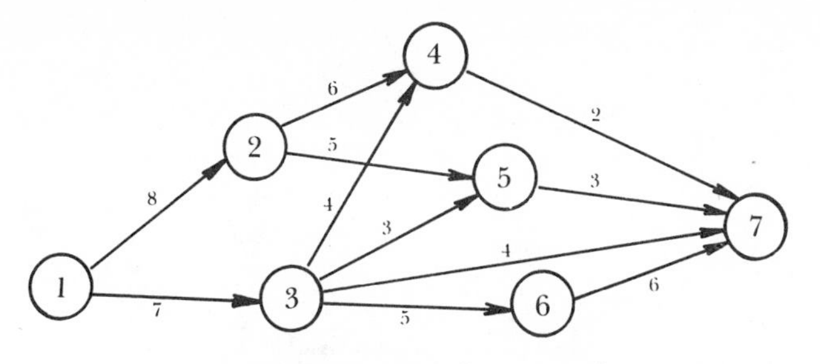

Figure 7-13 Network flow example.

The branch with the smallest capacity in the path to the sink would limit the flow. Call this limiting capacity, F_1. If we now subtract F_1 from each of the capacities in the path selected, one of the capacities is reduced to zero and the remaining capacities are reduced by F_1. We also have, thus far, an allowable flow of F_1. For example, for the network of Figure 7-13, if we trace out a path consisting of (1, 3), (3, 6), and (6, 7) we have a minimum capacity of $F_1 = 5$ and we have accordingly shown in Figure 7-14 the network with the modified capacities and our flow thus far.

We now repeat the preceding process, seeking out another path with branches of positive capacity between the source and the sink. Such a path is (1, 2), (2, 4) and (4, 7). This gives an additional flow $F_2 = 2$. We again reduce the capacities and now have the network shown in Figure 7-15. We now choose the path (1, 3), (3, 5), and (5, 7) and achieve an additional flow of $F_3 = 2$. This is shown in Figure 7-16. We now choose path (1, 2), (2, 5), and (5, 7) which gives $F_4 = 1$ and the network in Figure 7-17.

It can now be seen that there are no more paths whose branches have positive capacity between the source and sink. Does the maximal flow then equal 10? The answer is, unfortunately, no. This is easily seen by carrying out the same process but in the sequence shown in Figure 7-18. We have now achieved a flow equal to 12, so it clearly matters in what order this process is carried out. This is not tolerable if we are to have an algorithm, so that this naive

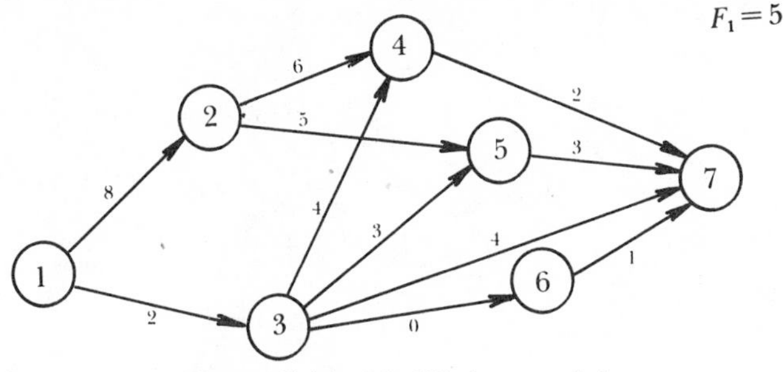

Figure 7-14 Modified network I.

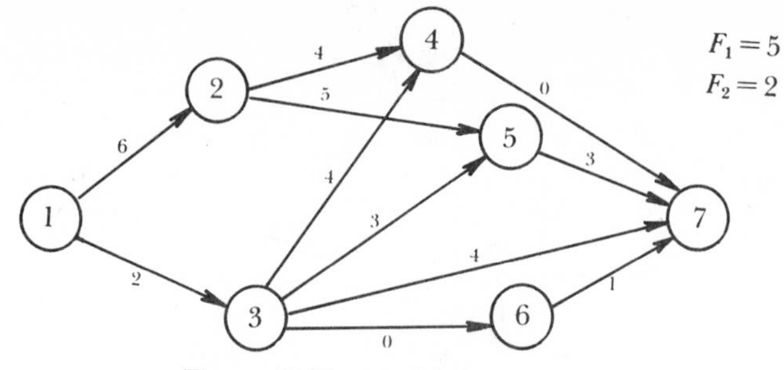

Figure 7-15 Modified network II.

method fails. It can be saved, however, by a modification which allows fictional flows in the wrong direction. This is done so as to reduce part or all of a flow previously assigned in the allowable direction. However, this is a cumbersome procedure, at best, and it will not be described here. Instead we shall describe a more systematic method for solving the problem. It is sometimes referred to as a labeling process and is due to Ford and Fulkerson.[11]

The labeling process is different from what we previously described in that we will "fan out" from the source in order to reach the sink, rather than sequentially search out particular paths from the source to the sink. First, we need some nomenclature. Suppose we have a network with capacities c_{ij} and c_{ji}. We do not preclude the possibility of flow in either direction. If flow is allowed only from some particular node i to node j and not the reverse then $c_{ij} \neq 0$ and $c_{ji} = 0$. We shall now define *excess capacities*, d_{ij}, for each branch as follows:

$$\begin{aligned} d_{ij} &= c_{ij} - x_{ij} + x_{ji} \\ d_{ji} &= c_{ji} - x_{ji} + x_{ij} \end{aligned} \qquad \text{all } i, j \qquad \textbf{(7-102)}$$

for any set of flows x_{ij} and x_{ji}. Initially, all flows are zero. Assume all branches are designated with their excess capacities. We wish to increase the flow. We begin at the source and consider all nodes which are connected to the source by branches of positive *excess capacity*. We designate the source as node 1 just as we did previously

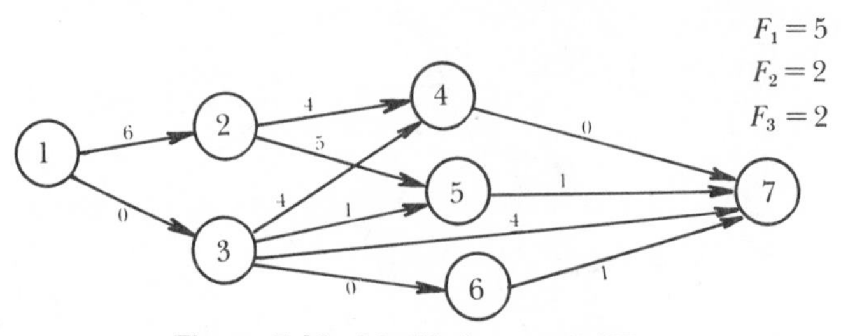

Figure 7-16 Modified network III.

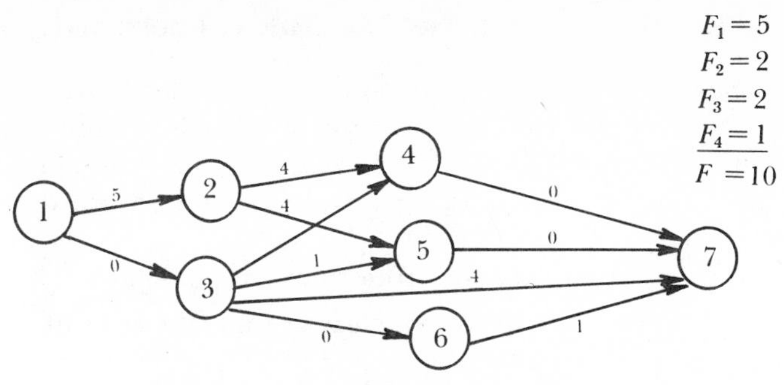

Figure 7-17 Modified network IV.

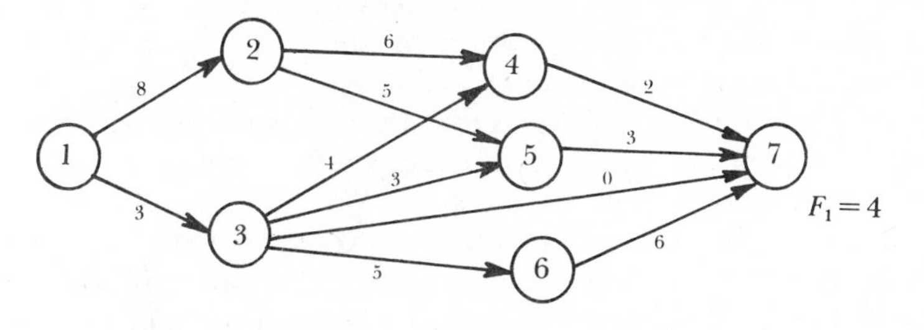

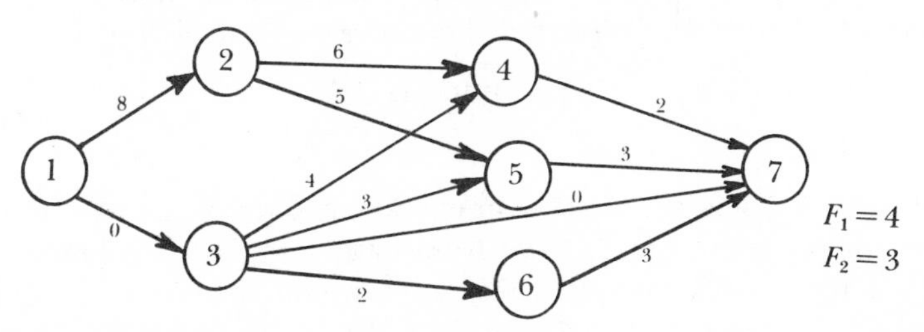

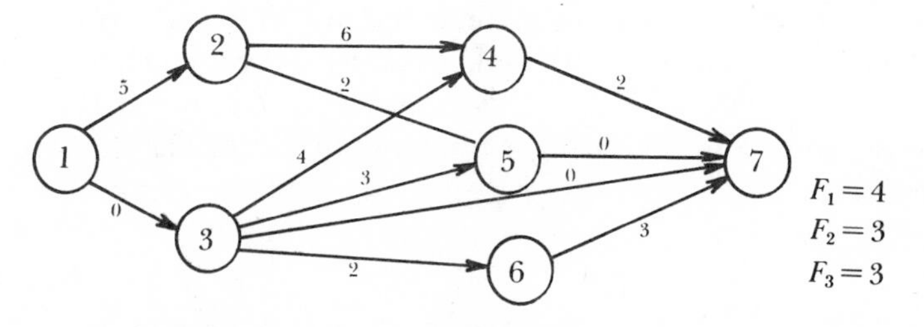

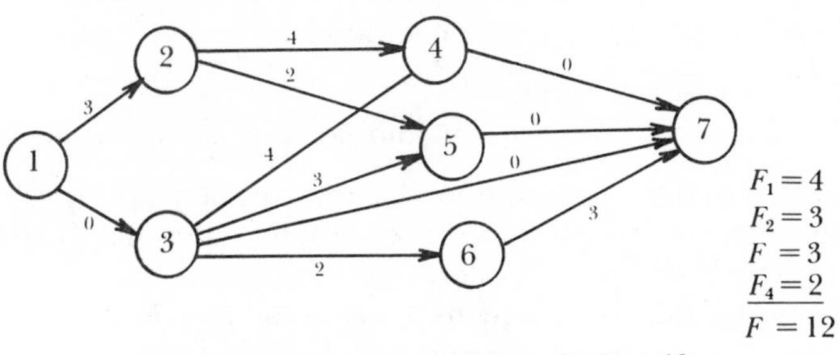

Figure 7-18 Sequence of flows for $F = 12$.

and the sink is designated as node N. Let us designate the nodes which are connected to the source this way with the index p. Call this set of nodes P. We then label each node $p \in P$ on our network with two numbers or labels, (α_p, β_p). The labels are defined as follows:

$$\alpha_p = \text{excess capacity from source to node } p$$

$$\beta_p = \text{the node from which we came (the source)}$$

Therefore,

$$(\alpha_p, \beta_p) = (d_{1p}, 1) \qquad p \in P \tag{7-103}$$

If we labeled the sink in the first step, which is unlikely but possible, we move to the final part of this algorithm. In this part, we have a method for increasing the flow.

What is usually the case is that the sink is not reached (labeled) in the first step. From the set of nodes $p \in P$, we choose $p = 1$. Then we look for nodes not yet labeled which are joined to the node $p = 1$, $p \in P$ by branches of positive excess capacity. If there are none, we go to $p = 2$, $p \in P$ and repeat. If unlabeled nodes can be reached let us designate this set of nodes with the index q and this set will be called Q. We label the nodes $q \in Q$ as follows:

$$\begin{aligned} \alpha_q &= \min\,(d_{pq}, \alpha_p) \\ \beta_q &= p \end{aligned} \tag{7-104}$$

The label α_q gives the minimum excess capacity of the two branches from the source to node p, and from node p to node q. The label β_q tells us the node from which we came to node q. We carry out this labeling process for all nodes $p \in P$. Again, if we have labeled the sink, we go to the last part of this algorithm.

The general step is the same as the previous step. If we have a set of labeled nodes $r \in R$, we look for nodes not yet labeled which are connected to nodes r. We do this for each $r \in R$, in turn. If we find a set of unlabeled nodes $t \in T$, we label these nodes t as follows:

$$\begin{aligned} \alpha_t &= \min\,(d_{rt}, \alpha_r) \\ \beta_t &= r \end{aligned} \tag{7-105}$$

We repeat this general step. We must, in a finite number of steps, reach one of the following two conditions:

1. The sink is labeled.
2. We cannot label the sink and no other nodes can be labeled.

As before, if we reach condition (1), we can increase the flow. We will indicate how to do this. If we reach condition (2), then the existing flow is the maximal flow and we are done.

Let us now consider how to increase the flow if condition (1) is reached. The label on the sink, α_N, indicates how much positive excess capacity exists from source to sink over the path traversed and hence how much the flow can be increased. It is a simple matter to traverse the path, since the second label on the nodes, β_j, indicates the preceding node leading to node j. Hence we can trace the path backwards.

If we designate by d_{uv} the excess capacities of the branches in this particular path from source to sink which enabled us to label the sink we can increase the flow by α_N and calculate the new excess capacities as

$$\hat{d}_{uv} = d_{uv} - \alpha_N$$

$$\hat{d}_{vu} = d_{vu} + \alpha_N \qquad \textbf{(7-106)}$$

$$\hat{d}_{ij} = d_{ij} \qquad \text{for branches not in the sink labelling path}$$

We now repeat the entire labeling process for the network with the foregoing excess capacities, starting again at the source. It is clear that in a finite number of steps we must reach condition (2), if we are dealing with finite maximal flows.

The net flows in each branch can be computed from (7-102) as follows. The net flow in the branch from node i to node j is

$$x_{ij} = c_{ij} - d_{ij} \quad \text{if} \quad c_{ji} = 0$$

If both c_{ij} and $c_{ji} \neq 0$ then either

$$x_{ij} = c_{ij} - d_{ij} \quad \text{and} \quad x_{ji} = 0$$

or

$$x_{ji} = c_{ji} - d_{ji} \quad \text{and} \quad x_{ij} = 0$$

depending upon which of $c_{ij} - d_{ij}$ and $c_{ji} - d_{ji}$ is positive.

Let us illustrate this labeling process for the network given in Figure 7-13. We give that network in Figure 7-19 with the excess capacities shown. Initially, all flows are zero. We start at node 1 and seek all nodes connected to 1 with positive excess capacity. These are nodes 2 and 3. Therefore $\alpha_2 = d_{12} = 8$ and $\beta_2 = 1$. Similarly, $\alpha_3 = d_{13} = 7$ and $\beta_3 = 1$. These labels are indicated as (α_2, β_2) and (α_3, β_3) at these nodes. We now choose node 2 and look for nodes of positive excess capacity connected to it. These are nodes 4 and 5. We calculate the labels for nodes 4 and 5 as follows:

$$\alpha_4 = \min(\alpha_2, d_{24}) = \min(8, 6) = 6$$

$$\beta_4 = 2$$

$$\alpha_5 = \min(\alpha_2, d_{25}) = \min(8, 5) = 5$$

$$\beta_5 = 2$$

These labels are shown at nodes 4 and 5 as (6, 2) and (5, 2). Similarly the labels for unlabeled nodes reached from node 3, i.e., 6 and 7

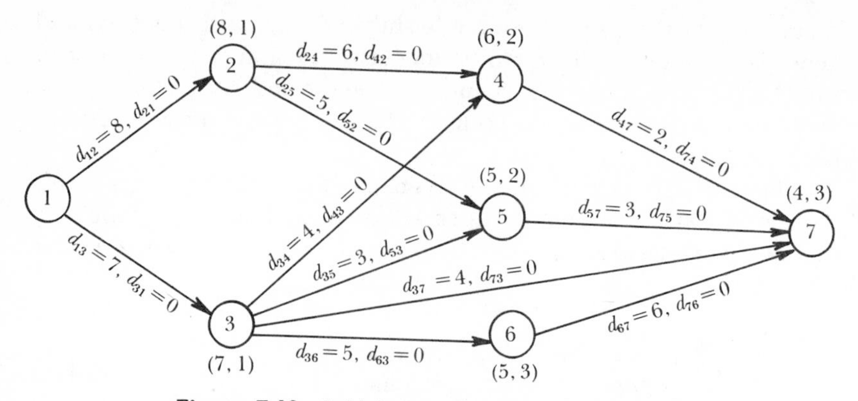

Figure 7-19 Initial network with excess capacities.

are shown as (5, 3) and (4, 3). We see that we have labeled the sink, node 7. Therefore we can increase the flow by $\alpha_4 = 4$ units. We now recompute the excess capacities in the path that is traced, first backward, from node 7. Since $\beta_7 = 3$, we look at node 3, $\beta_3 = 1$. Therefore the path is 1 – 3 –7.

$$\hat{d}_{13} = 7 - 4 = 3 \qquad \hat{d}_{31} = 0 + 4 = 4$$
$$\hat{d}_{37} = 4 - 4 = 0 \qquad \hat{d}_{73} = 0 + 4 = 4$$

The network with the new excess capacities is shown in Figure 7-20. If we now repeat the labeling process, we arrive at the set of labels shown on the nodes of Figure 7-20. Again, we have labeled the sink and we modify the flows, and so forth. The subsequent networks are shown in Figures 7-21 to 7-23. In the final network, Figure 7-23, we see that we cannot label the sink. Hence, the flow is maximal and is equal to the sum of the α_N for each network, i.e.,

$$\text{Maximal flow} = 4 + 2 + 3 + 3 = 12$$

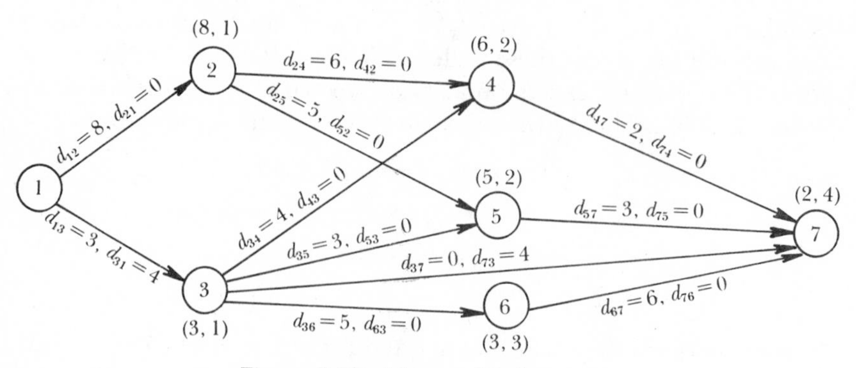

Figure 7-20 First modified network.

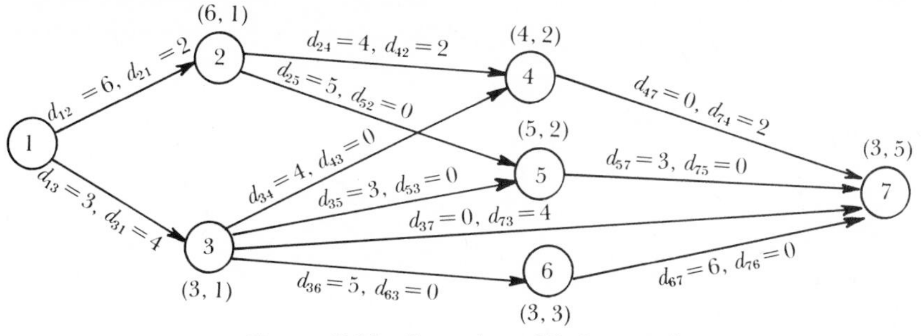

Figure 7-21 Second modified network.

This procedure does not require a set of network diagrams. It is a simple matter to represent this same information in a matrix and perform the calculations with matrix tableaux. See Hadley[2] for a discussion of this. Since it is not different in principle from what we have done, we will not discuss it further.

While we have given what appears to be an algorithm for the maximal flow problem on intuitive grounds, we have not yet proven that this is so. We will do that here. First, let us define a *cut* of a network as *a set of oriented branches of the network such that every path of oriented branches from the source to the sink contains at least one branch in the set.* There are many cuts for any network, but the number is obviously finite. The importance of this notion of cuts is that it is obvious that the maximal flow could not be greater than the sum of the branch capacities in any cut. We shall use this idea in this proof.

If we examine the *last modified network* resulting from the application of the algorithm, we see that we have a disjoint set of nodes with respect to nodes we have been able to label and those we cannot. Let us call the set of nodes that we have been able to reach in the labeling process, R^+, and those we have not been able to reach, R^-. If we now consider the set of oriented branches that connect members of R^+ to members of R^-, this set of branches, say J, is clearly a cut of

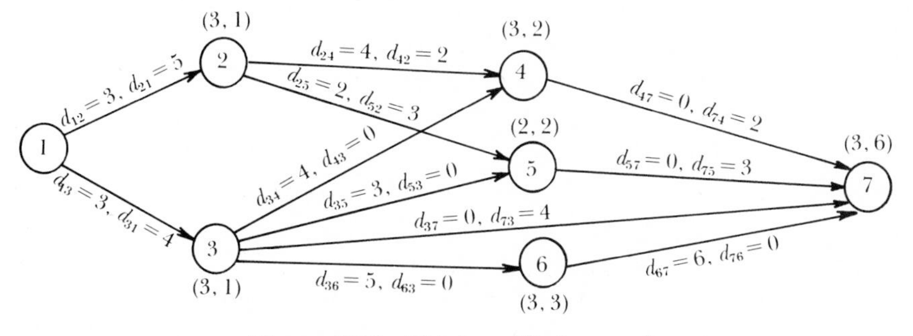

Figure 7-22 Third modified network.

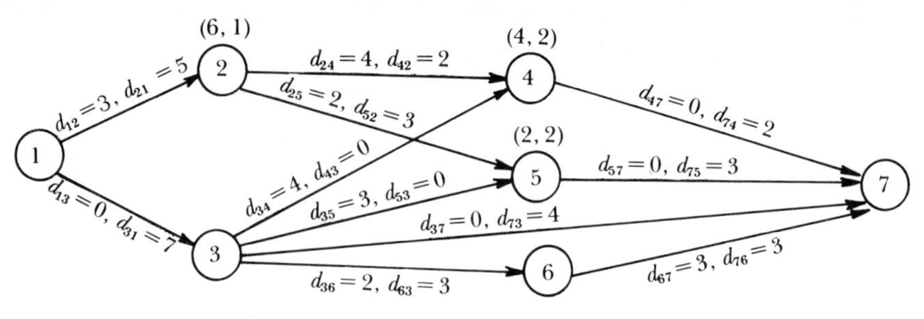

Figure 7-23 Fourth modified network.

the network. What we must now show is that the sum of the capacities of the branches in J is equal to the flow in the network. Because of the way in which excess capacities are defined, and referring to the equations for conservation of flow at each node (7-98), we can write

$$\sum_{\substack{j=1\\ j\neq i}}^{N} (c_{ij} - d_{ij}) = 0 \qquad i = 2, \ldots, N-1 \tag{7-107}$$

For the source, $i = 1$, we have

$$\sum_{\substack{j=1\\ j\neq i}}^{N} (c_{ij} - d_{ij}) = \sum_{j=2}^{N} x_{ij} = \text{total flow} \tag{7-108}$$

If we now sum the left hand side of (7-107) over nodes in R^+ (labeled nodes) we have, recalling that the source is, by definition, a labeled node:

$$\sum_{i\in R^+} \sum_{\substack{j=1\\ j\neq i}}^{N} (c_{ij} - d_{ij}) = \sum_{j=2}^{N} x_{ij} = \text{total flow} \tag{7-109}$$

If both $j \in R^+$ and $i \in R^+$, then both $c_{ij} - d_{ij}$ and $c_{ji} - d_{ji}$ are present and cancel each other in (7-109). Therefore, all terms cancel for $j \in R^+$ and only terms for $j \in R^-$ are present. However, if $i \in R^+$ and $j \in R^-$ then by the fact that the sets are disjoint by definition, $d_{ij} = 0$. Hence, (7-109) reduces to

$$\sum_{i\in R^+} \sum_{j\in R^-} c_{ij} = \text{total flow} \tag{7-110}$$

As we previously noted, a set of branches such as shown in (7-110) is a cut and this cut *equals* the total flow. We know that the total flow cannot exceed the sum of the capacities in a cut and here we have found a cut in which the total flow *equals* the capacities of the branches in a cut. Hence, for this particular cut the flow is maximal. This proves that our algorithm will indeed obtain the maximal flow.

Table 7.1 *List of Some Cuts for Network in Figure 7-13*

Cut	Sum of Capacities
(1, 2), (1, 3)	15
(3, 7), (4, 7), (5, 7), (6, 7)	15
(1, 2), (3, 4), (3, 5), (3, 6), (3, 7)	24
(1, 2), (3, 4), (3, 5), (3, 7), (6, 7)	25
(1, 2), (3, 6), (3, 7), (4, 7), (5, 7)	22
(1, 3), (2, 4), (2, 5)	18
(1, 3), (4, 7), (5, 7)	12
(1, 3), (2, 4), (5, 7)	16
(1, 3), (2, 5), (4, 7)	14

What we have proved is often stated as the *max flow–min cut theorem.*

Max flow–min cut theorem. If we calculate the sum of the capacities of the branches in every cut of a finite network, the smallest sum is equal to the maximal possible flow in the network.

We can illustrate the meaning of this theorem by means of the network in our sample problem. It should be borne in mind however that this is not a practical computational procedure for large networks. If we consider the network in Figure 7-13, let us list some of the sets of branches that are cuts for this network. We also list the sum of the capacities for the branches in each cut (Table 7.1). We should, of course, list all possible cuts. We have not done this but we have listed all the "irredundant" cuts. If we now choose the minimum cut: (1, 3), (4, 7), and (5, 7), we see that the sum of the branch capacities is 12. This was indeed the maximum flow that we obtained previously.

There are many other kinds of network flow problems and techniques but we shall not discuss them in this introductory work. The interested reader may consult Ford and Fulkerson.[12]

EXERCISES

Duality Theory

1. Formulate the dual problem of the following linear programming problem so that the dual variables are nonnegative:

$$\text{Max } z = 2x_1 + 3x_2 + 5x_3$$
$$3x_1 + 4x_2 + 5x_3 \leq 50$$
$$2x_1 + x_2 + 2x_3 \geq 12$$
$$5x_1 + 3x_2 + x_3 \leq 30$$
$$x_1, x_2, x_3 \geq 0$$

2. Formulate the dual problem of the following linear programming problem so that the dual variables are unrestricted:

$$\begin{aligned} \text{Min } z &= 3x_1 + 5x_2 + 2x_3 \\ 2x_1 + 3x_2 + x_3 &= 21 \\ 3x_1 + x_2 + 4x_3 &= 38 \\ x_1, x_2, x_3 &\geq 0 \end{aligned}$$

3. Formulate the dual problem of the following linear programming problem. Which dual variables, if any, are unrestricted?

$$\begin{aligned} \text{Min } z &= 2x_1 + 3x_2 - 4x_3 + 5x_4 \\ 3x_1 + 2x_2 + x_3 - 2x_4 &\leq 19 \\ 2x_1 + 3x_2 - x_3 + 3x_4 &\geq 22 \\ x_1 - x_2 + 2x_3 - 3x_4 &= 38 \\ x_1, x_2, x_3, x_4 &\geq 0 \end{aligned}$$

4. Solve the following problem (1) graphically and (2) formulate the dual problem and solve by the simplex method.

$$\begin{aligned} \text{Min } z &= 20x_1 + 30x_2 \\ x_1 + 2x_2 &\leq 22 \\ 2x_1 - x_2 &\leq 3 \\ x_1 + 3x_2 &\geq 4 \\ -2x_1 + x_2 &\leq 8 \\ 5x_1 + 4x_2 &\geq -1 \\ x_1, x_2 &\geq 0 \end{aligned}$$

Other Simplex Algorithms

5. Solve the following problem by the revised simplex method:

$$\begin{aligned} \text{Max } z &= 3x_1 + 2x_2 \\ 3x_1 + 5x_2 &\leq 12 \\ 2x_1 + 3x_2 &\leq 10 \\ x_1, x_2 &\geq 0 \end{aligned}$$

6. Solve the following problem by the dual simplex algorithm:

$$\begin{aligned} \text{Min } z &= 2x_1 + 3x_2 + 4x_3 + x_4 \\ x_1 + 2x_2 + 6x_3 + 2x_4 &\geq 25 \\ 4x_1 + 3x_2 - x_3 + 3x_4 &\geq 10 \\ 2x_1 + 5x_2 + x_3 + 6x_4 &\geq 15 \\ x_1, x_2, x_3, x_4 &\geq 0 \end{aligned}$$

7. Solve the following problem by the decomposition principle:

$$\begin{aligned}
\text{Max } z &= 5x_1 + 6x_2 + 7x_3 + 4x_4 + 8x_5 \\
3x_1 + 5x_2 + 3x_3 + 6x_4 + 4x_5 &\leq 30 \\
x_1 + 6x_2 &\leq 10 \\
3x_1 + 2x_2 &\leq 7 \\
2x_3 + 8x_4 + 2x_5 &\geq 3 \\
3x_3 + 2x_4 + x_5 &\leq 18 \\
x_1, x_2, x_3, x_4, x_5 &\geq 0
\end{aligned}$$

8. Consider the following linear programming problem:

$$\begin{aligned}
\text{Max } z &= \bar{c}'\bar{x} \\
A\bar{x} &\leq \bar{b} \\
\bar{0} \leq \bar{x} &\leq \bar{d}
\end{aligned}$$

Explain in detail how you could use the decomposition principle to solve this problem.

Postoptimal Analysis

9. Consider Table 6-10, which is the final tableau for the problem given in equations (6-94). Suppose c_1 is increased from 3 to 6. Find the optimal solution. Suppose c_1 is changed from 3 to 7. Find the optimal solution. Suppose c_2 is changed from 8 to 5. Is the solution still optimal? Suppose it is dropped to 3. Is the solution still optimal? If not, find the optimal solution.

10. If in the final tableau given in Table 6-10, the requirements vector is changed from $\bar{b} = \begin{bmatrix} 20 \\ 6 \end{bmatrix}$ to $\bar{b} = \begin{bmatrix} 10 \\ 6 \end{bmatrix}$ is the solution still optimal?

11. If for the problem given in equations (6-94), the final tableau for which is given in Table 6-10, we change the coefficient $a_{12} = 4$ to $a_{12} = 5$, will this affect the optimal solution given in the final tableau?

The Transportation Problem

12. A company has four warehouses containing 12,000, 5,000, 8,000 and 14,000 units of its products. In the next week it must ship to each of its eight retail stores 3,000; 2,000; 1,000; 10,000; 8,000; 4,000; 3,000, and 2,000 units, respectively.

The unit costs of shipment are shown in the following chart:

	Retail Stores							
Warehouses 1	10	5	8	20	18	10	5	10
2	20	10	12	5	16	12	6	12
3	30	15	20	15	14	20	7	12
4	25	20	26	10	12	15	8	20

Find the amounts to be shipped from each warehouse to each retail store so as to minimize total shipping cost.

13. If a constant α is added to each cost in one row, say row k, of a transportation tableau, how is the optimal solution to the transportation problem affected? If the same is done to the costs in one column, column l, how is the optimal solution affected?

14. Suppose we have a transportation problem of the type

$$\text{Min } z = \sum_{i=1}^{m} \sum_{j=1}^{n} c_{ij} x_{ij}$$

$$\sum_{j=1}^{n} x_{ij} \leq a_i \qquad i = 1, \ldots, m$$

$$\sum_{i=1}^{m} x_{ij} \geq b_j \qquad j = 1, \ldots, n$$

$$x_{ij} \geq 0 \qquad \text{all } i, j$$

Convert the constraints to equalities and add two equations so that $\sum_{i=1}^{m} a_i = \sum_{j=1}^{n} b_j$. Show that the matrix for this problem is also unimodular.

Network Flow Problems

15. Find the maximal flow in the following network:

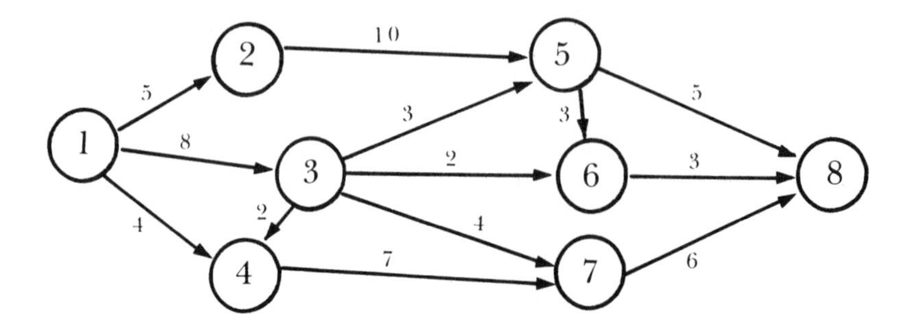

16. Find the maximal flow in the following network by the maximal flow algorithm and also by applying the max flow–min cut theorem.

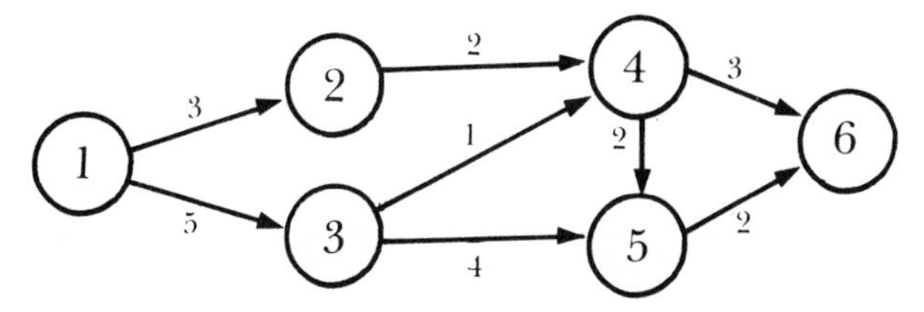

17. Consider the directed "flow" networks of the type discussed in this chapter. Now suppose that the c_{ij} are not branch capacities but rather the time that is required for some commodity to move through the branch. Develop a labeling procedure for finding the path through the network which requires the least amount of time.

18. Devise a matrix scheme to carry out, in a more compact form, the calculations associated with the maximal flow algorithm presented in this chapter.

REFERENCES

1. Orchard-Hays, W.: *Advanced Linear Programming Computing Techniques*. McGraw-Hill, New York (1968).
2. Hadley, G.: *Linear Programming*. Addison-Wesley, Reading, Massachusetts (1962).
3. Simmonard, M.: *Linear Programming*. Prentice-Hall, Englewood Cliffs, N.J. (1966).
4. Chung, An-min: *Linear Programming*. Charles E. Merrill, Columbus, Ohio (1963).
5. Gass, S. I.: *Linear Programming Methods and Applications*, Third Edition, McGraw-Hill, New York (1969).
6. Lemke, C. E.: The dual method of solving the linear programming problem. Naval Res. Log. Quart. *1*, 48–54 (1954).
7. Dantzig, G. B., Ford, L. R., and Fulkerson, D. R.: A Primal-Dual Algorithm for Linear Programming. *In* Kuhn and Tucker (eds.): *Linear Inequalities and Related Systems*. Princeton University Press, Princeton (1956).
8. Mueller, R. K., and Cooper, L.: A comparison of the primal simplex and primal-dual algorithms for linear programming. Comm. of the ACM, *8*, 682–686 (1965).
9. Marble, George F.: A comparison of the primal-dual and composite simplex algorithms. M.S. thesis, Washington University, St. Louis (1967).
10. Dantzig, G. B., and Wolfe, P.: A decomposition principle for linear programs. Oper. Res., *8*, 101–111 (1960).
11. Ford, L. R., and Fulkerson, D. R.: A simple algorithm for finding maximal network flows and an application to the Hitchcock problem. Canadian J. of Math., *9*, 210–218 (1957).
12. Ford, L. R., and Fulkerson, D. R.: *Flows in Networks*. Princeton University Press, Princeton (1962).

Chapter 8

NONLINEAR PROGRAMMING

8.1 INTRODUCTION

In Chapter 4 we investigated certain classical optimization procedures for maximizing (or minimizing) a function of n variables subject to constraint equations on the variables. Unfortunately, as we have noted previously, these procedures cannot be used to solve problems involving a large number of variables. In this chapter we shall present several computationally feasible methods for solving certain types of *nonlinear programming problems*. A nonlinear programming problem is a problem in which we seek to maximize (or minimize) a function $f(\bar{x})$ subject to a set of constraint equations or inequalities, in which either $f(\bar{x})$ or at least one of the functions appearing in the constraint set (or both) is a nonlinear function. The general nonlinear programming problem may be stated as

$$\text{Maximize} \quad z = f(\bar{x}) \tag{8-1}$$

$$\text{Subject to} \quad g_i(\bar{x})\{\leq, =, \geq\}b_i \qquad i = 1, 2, \ldots, m \tag{8-2}$$

where one of the three relations $\{\leq, =, \geq\}$ is assigned to each of the m constraints (8-2). The function $f(\bar{x})$ in (8-1) is called the *objective function*. In addition, nonnegativity restrictions ($x_j \geq 0$) on some or all of the variables may be stated separately, or may be assumed to be included in the constraints (8-2). There is no known method of determining the global maximum to the general nonlinear programming problem. However, if the objective function and the constraints satisfy certain properties, the global maximum can sometimes be found. For example, we proved in Chapter 4 (Theorem 4.9) that the global maximum of a convex function over a convex set bounded from below occurs at an extreme point of the convex set.

In Section 8.3, we shall investigate the "Kuhn-Tucker conditions," a set of necessary conditions for a local maximum to a nonlinear programming problem. These conditions in certain cases also yield the global maximum. The Kuhn-Tucker conditions are particularly useful in the derivation of methods for solving some kinds of nonlinear programming problems. In Section 8.4 we shall see how the use of the Kuhn-Tucker conditions leads to the development of an algorithm, using the simplex method, for solving quadratic programming problems. In a quadratic programming problem the objective function is quadratic and the constraints are linear.

8.2 CONVEX CONSTRAINT SETS

Before beginning our study of the Kuhn-Tucker conditions, let us first determine the properties which the functions $g_i(\bar{x})$ should have in order that the constraints set (8-2) be a convex set.

Theorem 8.1. If $g(\bar{x})$ is a convex function, then the set $C_1 = \{\bar{x} \mid g(\bar{x}) \le b\}$ is a convex set; if $h(\bar{x})$ is a concave function, then the set $C_2 = \{\bar{x} \mid h(\bar{x}) \ge b\}$ is a convex set.

Proof: We need to show that given any two points $\bar{x}_1$, $\bar{x}_2 \in C_1$, the point $\hat{\bar{x}} = \lambda\bar{x}_1 + (1 - \lambda)\bar{x}_2$ is in C_1. Since $g(\bar{x})$ is convex, it follows that

$$g(\hat{\bar{x}}) \le \lambda g(\bar{x}_1) + (1 - \lambda)g(\bar{x}_2) \tag{8-3}$$

since $\bar{x}_1$, $\bar{x}_2 \in C_1$, $g(\bar{x}_1) \le b$ and $g(\bar{x}_2) \le b$. Substituting these inequalities into (8-3) yields

$$g(\hat{\bar{x}}) \le \lambda b + (1 - \lambda)b = b \tag{8-4}$$

Thus, $g(\hat{\bar{x}}) \le b$ and so $\hat{\bar{x}} \in C_1$.

The proof that C_2 is a convex set is completely analogous.

Two examples of C_1 and C_2 are shown in Figure 8-1. The shaded region C_1 in Figure 8-1a represents the set of points satisfying $g(x_1, x_2) = x_1^2 + x_2 \le 4$. The reader may verify that $g(x_1, x_2)$ is a convex function; clearly, C_1 is a convex set. In Figure 8-1b, the shaded region C_2 represents the set of points satisfying

$$h(x_1, x_2) = x_2 - 2x_1^2 \ge 1$$

$h(x_1, x_2)$ is a concave function and C_2 is a convex set. The intersection of C_1 and C_2, the convex set C, is shown in Figure 8-2.

Combining the results of Theorem 8.1 with the fact that the intersection of convex sets is a convex set, we see that the constraint

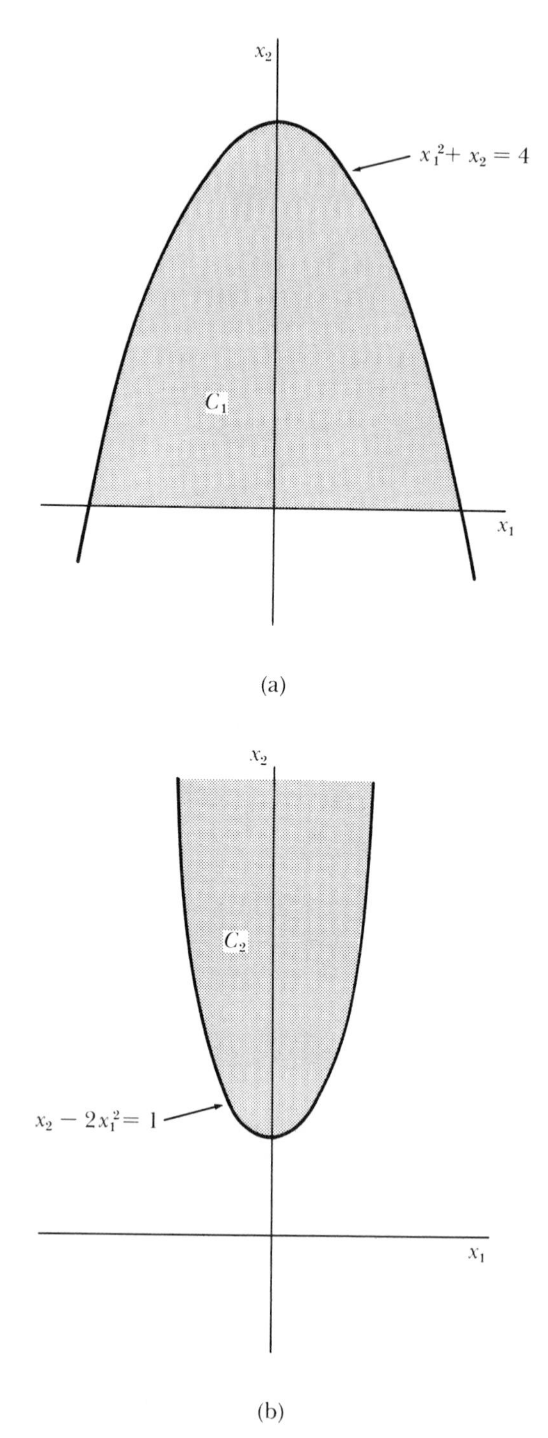

Figure 8-1

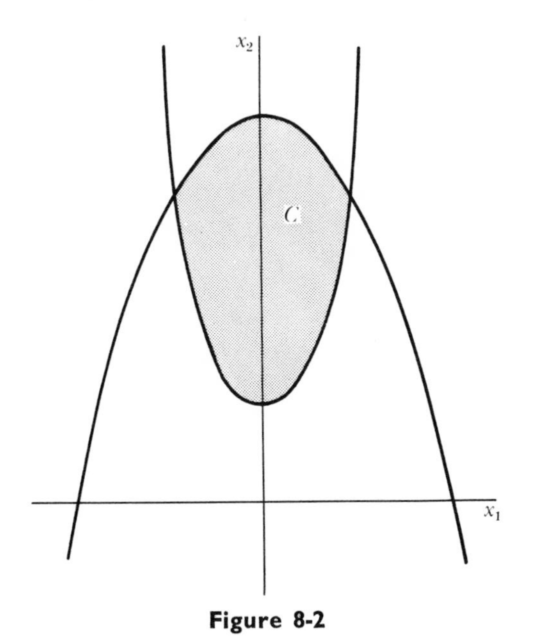

Figure 8-2

set (8-2) will be a convex set if the "$\leq$" sign applies whenever $g_i(\bar{x})$ is a convex function and the "$\geq$" sign applies whenever $g_i(\bar{x})$ is a concave function. We have said nothing yet about the case where the equality sign holds for some constraint (8-2). Looking at Figure 8-1, we see that if the constraints were $x_1^2 + x_2 = 4$ and $x_2 - 2x_1^2 = 1$, then the resulting sets would be the boundaries of C_1 and C_2, respectively. Neither of these sets is convex. In general, the only time an equality constraint, $g(\bar{x}) = b$, will produce a convex set is if $g(\bar{x})$ is linear; that is, $g(\bar{x}) = \bar{a}'\bar{x} = b$ (a hyperplane).

The preceding paragraph, then, describes a set of sufficient conditions for the constraint set (8-2) to be convex. The addition of nonnegativity restrictions, of course, does not affect the convexity. The foregoing set of conditions for a convex constraint set are not necessary, however, since, for example, it is possible for the intersection of several nonconvex sets to produce a convex set. Consider the two constraints

$$x_2 - x_1^3 + 2x_1^2 + x_1 \leq 2 \tag{8-5}$$

$$x_2 - x_1 \geq 1 \tag{8-6}$$

The shaded region of Figure 8-3a indicates the set of points satisfying (8-5). Clearly, this is not a convex set. However, the set of points lying in the intersection of constraints (8-5) and (8-6), shown in Figure 8-3b, is a convex set.

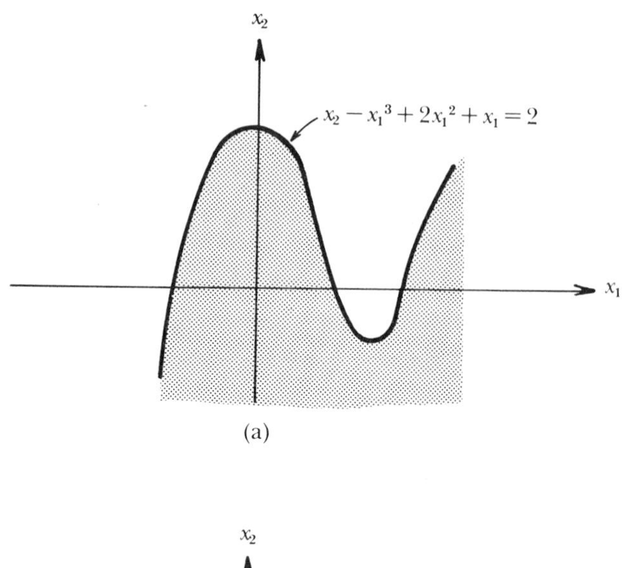

(a)

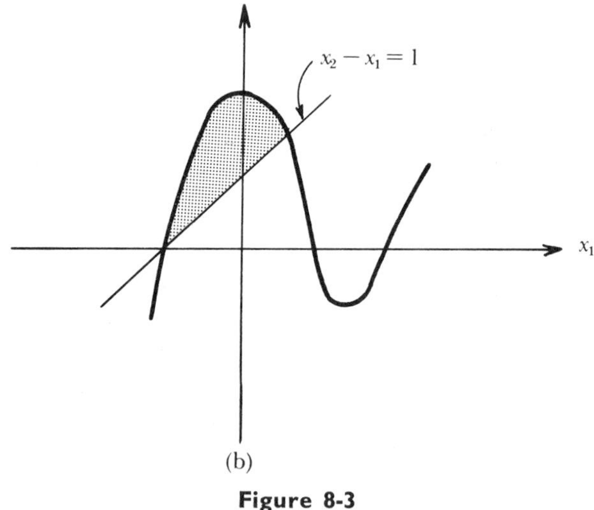

(b)

Figure 8-3

8.3 THE KUHN-TUCKER CONDITIONS

In attempting to develop algorithms for solving nonlinear programming problems, it is useful to have some information concerning the characteristics of an optimal solution. In Chapter 4 a set of relations were obtained which an optimal solution must satisfy for a constrained optimization problem in which the constraints were all equalities. However, it is desirable to consider the case of inequality constraints as well. In particular, we wish to consider the effect on the optimality conditions of nonnegativity restrictions on the variables.

Let us consider the following nonlinear programming problem:

$$\text{Maximize} \quad z = f(\bar{x}) \equiv f(x_1, x_2, \ldots, x_n) \tag{8-7}$$

$$\text{Subject to} \quad g_i(\bar{x}) \le b_i, \qquad i = 1, 2, \ldots, p \tag{8-8}$$

$$g_i(\bar{x}) \ge b_i, \qquad i = p + 1, \ldots, m \tag{8-9}$$

$$x_j \ge 0, \qquad j = 1, 2, \ldots, n \tag{8-10}$$

Furthermore, let us assume for convenience that the functions in constraints (8-8) are convex and the functions in constraints (8-9) are concave, so that the set of feasible solutions is a convex set. We shall also assume that all the first partial derivatives of $f(\bar{x})$ and $g_i(\bar{x})$, $i = 1, 2, \ldots, m$, are defined.

Now, let us consider the following Lagrangian function:

$$F(\bar{x}, \bar{\lambda}, \bar{\mu}) = f(\bar{x}) + \sum_{i=1}^{p} \lambda_i[b_i - g_i(\bar{x})] + \sum_{i=p+1}^{m} \mu_i[b_i - g_i(\bar{x})] \tag{8-11}$$

Suppose $\bar{x}_0$ is a relative maximum to the nonlinear programming problem described by (8-7) through (8-10). A constraint of the form (8-8) or (8-9) is said to be *active* at $\bar{x}_0$ if strict equality holds at $\bar{x}_0$ (i.e., if $g_i(\bar{x}_0) = b_i$) and *inactive* at $\bar{x}_0$ if strict inequality holds. Now, suppose that we maximize $F(\bar{x}, \bar{\lambda}, \bar{\mu})$ subject to certain upper or lower bound constraints on the variables $\bar{x}$, $\bar{\lambda}$, $\bar{\mu}$; we wish to determine what upper or lower bounds to impose in order that the maximum $[\bar{x}_0, \bar{\lambda}_0, \bar{\mu}_0]$ obtained will also yield a maximum, $\bar{x}_0$, to the nonlinear programming problem. Obviously, we must require that

$$\bar{x} \ge \bar{0} \tag{8-12}$$

Also, for the constraints (8-8)

$$b_i - g_i(\bar{x}) \ge 0 \tag{8-13}$$

Thus, in maximizing $F(\bar{x}, \bar{\lambda}, \bar{\mu})$, we would like to require that each term in the sum

$$\sum_{i=1}^{p} \lambda_i[b_i - g_i(\bar{x})]$$

be nonnegative; this will occur if

$$\lambda_i \ge 0 \qquad i = 1, 2, \ldots, p \tag{8-14}$$

Similarly, we would like each term in the sum

$$\sum_{i=p+1}^{m} \mu_i[b_i - g_i(\bar{x})]$$

to be nonnegative; however, for $i = p + 1, \ldots, m$,

$$b_i - g_i(\bar{x}) \leq 0 \tag{8-15}$$

Therefore, if we require that

$$\mu_i \leq 0 \qquad i = p + 1, \ldots, m, \tag{8-16}$$

then each $\mu_i[b_i - g_i(\bar{x})]$ will be nonnegative. Thus, in maximizing $F(\bar{x}, \bar{\lambda}, \bar{\mu})$ subject to the constraints (8-12), (8-14), and (8-16) while ignoring the constraints (8-8) and (8-9) to the nonlinear programming problem, we see that if any of these latter constraints is violated, the value of $F(\bar{x}, \bar{\lambda}, \bar{\mu})$ will be smaller than it would be if the constraint was satisfied. Thus, a necessary condition for the maximum of $F(\bar{x}, \bar{\lambda}, \bar{\mu})$ subject to $\bar{x} \geq \bar{0}$, $\bar{\lambda} \geq \bar{0}$, $\bar{\mu} \leq \bar{0}$ is that each of the constraints (8-8) and (8-9) be satisfied.

Now, recall that a set of necessary conditions for the existence of a maximum to the *unconstrained* function $F(\bar{x}, \bar{\lambda}, \bar{\mu})$ is that

$$\frac{\partial F}{\partial x_j} = 0 \qquad j = 1, 2, \ldots, n \tag{8-17}$$

$$\frac{\partial F}{\partial \lambda_i} = 0 \qquad i = 1, 2, \ldots, p \tag{8-18}$$

$$\frac{\partial F}{\partial \mu_i} = 0 \qquad i = p + 1, \ldots, m \tag{8-19}$$

However, these conditions must be modified when $F(\bar{x}, \bar{\lambda}, \bar{\mu})$ is maximized† subject to $\bar{x} \geq \bar{0}$, $\bar{\lambda} \geq \bar{0}$, $\bar{\mu} \leq \bar{0}$. The modifications are as follows: Let $(\bar{x}_0, \bar{\lambda}_0, \bar{\mu}_0)$ be a maximum point of $F(\bar{x}, \bar{\lambda}, \bar{\mu})$ subject to $\bar{x} \geq \bar{0}$, $\bar{\lambda} \geq \bar{0}$, $\bar{\mu} \leq \bar{0}$, where $\bar{x}_0 = [x_{01}, x_{02}, \ldots, x_{0n}]$, $\bar{\lambda}_0 = [\lambda_{01}, \lambda_{02}, \ldots, \lambda_{0p}]$, and $\bar{\mu}_0 = [\mu_{0,p+1}, \ldots, \mu_{0_m}]$. Then, the following conditions must be satisfied:

If $x_{0_j} > 0$,

$$\frac{\partial F}{\partial x_j} = 0 \tag{8-20a}$$

If $x_{0_j} = 0$,

$$\frac{\partial F}{\partial x_j} \leq 0 \tag{8-20b}$$

If $\lambda_{0_i} > 0$,

$$\frac{\partial F}{\partial \lambda_i} = 0 \tag{8-21a}$$

† In many texts e.g., Hadley,[2] the function $F(\bar{x}, \bar{\lambda}, \bar{\mu})$ is maximized with respect to $\bar{x}$ and minimized with respect to the Lagrange multipliers $\bar{\lambda}$ and $\bar{\mu}$, yielding what is called a *saddle point*. However, in such instances the λ_i's and μ_i's are the negatives of those used here. The authors feel that the foregoing presentation, while perhaps somewhat limiting theoretically, will enable the reader to grasp more readily the fundamental ideas behind the Kuhn-Tucker conditions.

If $\lambda_{0_i} = 0$,

$$\frac{\partial F}{\partial \lambda_i} \leq 0 \tag{8-21b}$$

If $\mu_{0_i} < 0$,

$$\frac{\partial F}{\partial \mu_i} = 0 \tag{8-22a}$$

If $\mu_{0_i} = 0$,

$$\frac{\partial F}{\partial \mu_i} \geq 0 \tag{8-22b}$$

Equations (8-20a), (8-21a), and (8-22a) follow since if x_{0j}, λ_{0i}, and μ_{0i} are not zero, then the constraints ($x_{0j} \geq 0, \lambda_{0i} \geq 0$, or $\mu_{0i} \leq 0$) have no effect on the maximization of $F(\bar{x}, \bar{\lambda}, \bar{\mu})$. However, if any of these variables is zero, then the optimal solution occurs at a boundary. If, for example, $x_{0_j} = 0$, then $\dfrac{\partial F}{\partial x_j} \leq 0$ must hold at $[\bar{x}_0, \bar{\lambda}_0, \bar{\mu}_0]$. This is illustrated in Figure 8-4a. If, on the other hand, $\mu_{0i} = 0$, then $\dfrac{\partial F}{\partial \mu_i} \geq 0$ must hold at $[\bar{x}_0, \bar{\lambda}_0, \bar{\mu}_0]$. See Figure 8-4b.

Now, let us express equations (8-20) to (8-22) in terms of $F(\bar{x})$ and $g_i(\bar{x})$, $i = 1, 2, \ldots, m$. From equation (8-11) and the preceding discussion, we see that a point $\bar{x}_0$ is a maximum for the nonlinear programming problem of (8-7) through (8-10) only if there exist vectors $\bar{\lambda}_0$, $\bar{\mu}_0$ such that the conditions in Figure 8-4 exist. If $x_{0j} > 0$,

$$\frac{\partial F}{\partial x_j} \equiv \frac{\partial f}{\partial x_j} - \sum_{i=1}^{p} \lambda_{0i} \frac{\partial g}{\partial x_j} = 0 \qquad \text{at } \bar{x} = \bar{x}_0 \tag{8-23a}$$

If $x_{0_j} = 0$,

$$\frac{\partial F}{\partial x_j} \equiv \frac{\partial f}{\partial x_j} - \sum_{i=1}^{p} \lambda_{0_j} \frac{\partial g}{\partial x_j} \leq 0 \qquad \text{at } \bar{x} = \bar{x}_0 \tag{8-23b}$$

If $\lambda_{0_i} > 0$,

$$\frac{\partial F}{\partial \lambda_i} \equiv b_i - g_i(\bar{x}) = 0 \qquad \text{at } \bar{x} = \bar{x}_0 \tag{8-24a}$$

If $\lambda_{0_i} = 0$,

$$\frac{\partial F}{\partial \lambda_i} \equiv b_i - g_i(\bar{x}) \leq 0 \qquad \text{at } \bar{x} = \bar{x}_0 \tag{8-24b}$$

If $\mu_{0_i} < 0$,

$$\frac{\partial F}{\partial \mu_i} \equiv b_i - g_i(\bar{x}) = 0 \qquad \text{at } \bar{x} = \bar{x}_0 \tag{8-25a}$$

If $\mu_{0_i} = 0$,

$$\frac{\partial F}{\partial \mu_i} \equiv b_i - g_i(\bar{x}) \geq 0 \qquad \text{at } \bar{x} = \bar{x}_0 \tag{8-25b}$$

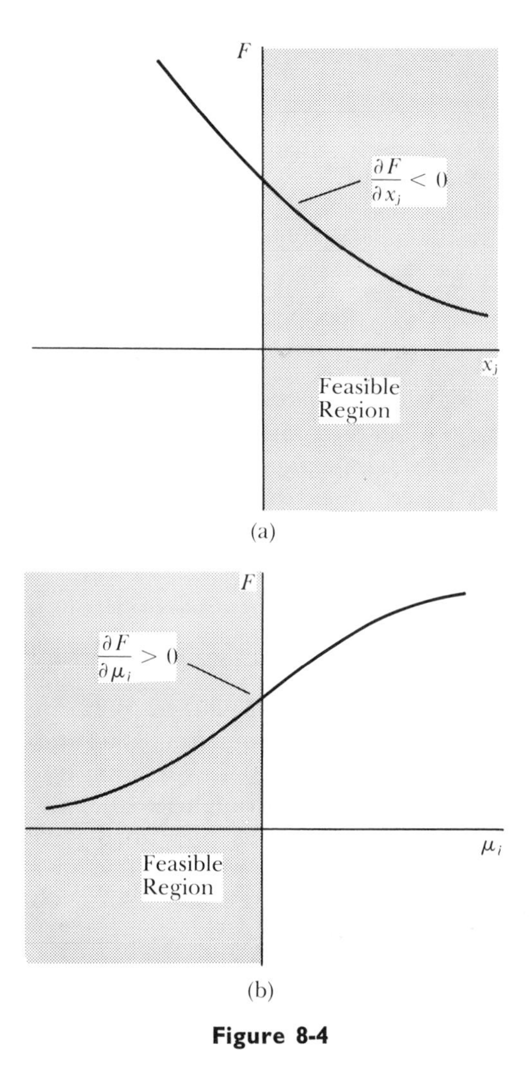

Figure 8-4

These conditions are called the Kuhn-Tucker conditions.[4] One of relations (8-23a) and (8-23b) must hold for each j, $j = 1, 2, \ldots, n$; one of relations (8-24a) and (8-24b) must hold for each i, $i = 1, 2, \ldots, p$; one of relations (8-25a) and (8-25b) must hold for each i, $i = p + 1, \ldots, m$. These latter four relations ensure that $\bar{x}_0$ will satisfy the constraints (8-8) and (8-9).

Although we have restricted out attention to the case where the constraints (8-8) and (8-9) form a convex set, Kuhn and Tucker actually derived the foregoing conditions for a more general case with less stringent restrictions on the $g_i(\bar{x})$. The reader is referred to Hadley[2] for a thorough discussion of the Kuhn-Tucker conditions.

It should be emphasized that, in general, the Kuhn-Tucker conditions provide only a set of necessary conditions for a maximum. A point satisfying all the Kuhn-Tucker conditions may be either a global maximum, a local maximum, or neither.

However, if $f(\bar{x})$ is a concave function and the constraints (8-8) and (8-9) form a convex set, then Kuhn and Tucker have proved:

Theorem 8.2. If $f(\bar{x})$ is a concave function and the constraints (8-8) and (8-9) form a convex set, then any point $\bar{x}_0$ satisfying the conditions (8-23) through (8-25) is the global maximum solution.

8.4 QUADRATIC PROGRAMMING

A *quadratic programming problem* is a nonlinear programming problem of the following form:

$$\text{Maximize} \quad f(\bar{x}) = \bar{c}'\bar{x} + \bar{x}'P\bar{x} \tag{8-26}$$

$$\text{Subject to} \quad \begin{Bmatrix} A\bar{x} \leq \bar{b} \\ \bar{x} \geq \bar{0} \end{Bmatrix} \tag{8-27}$$

As before, we shall assume that there are m constraints and n variables; thus A is an $m \times n$ matrix, $\bar{b}$ an m-component vector, $\bar{c}$ an n-component vector, and P an $n \times n$ matrix, all with known elements, and $\bar{x}' = [x_1, x_2, \ldots, x_n]$ is the vector of unknowns. The only nonlinear part of this problem is the objective function, which is the sum of a linear term, $\bar{c}'\bar{x}$, and a quadratic form, $\bar{x}'P\bar{x}$. Recall from Section 2.12 that in a quadratic form it can be assumed without loss in generality that P is symmetric; hence, we shall assume here that P is symmetric.

The first step in devising a scheme for solving a problem is to obtain as much information as possible about the location of any maxima. Since the constraint set (8-27) is a convex set, we know that if $f(\bar{x})$ is a convex function, then the global maximum will occur at an extreme point of the convex set of feasible solutions. (See Section 4.11, Theorem 4.9). On the other hand, if $f(\bar{x})$ is a concave function, then any point $\bar{x}_0$ satisfying the Kuhn-Tucker conditions is a global maximum. We shall now determine under what conditions $f(\bar{x})$ is convex or concave.

Theorem 8.3. The function $f(\bar{x}) = \bar{c}'\bar{x} + \bar{x}'P\bar{x}$ is concave if P is negative semidefinite or negative definite.

Proof: We wish to show that, given any two points $\bar{x}_1$ and $\bar{x}_2$, then if $\bar{x}_3 = \lambda\bar{x}_1 + (1 - \lambda)\bar{x}_2$, $0 \leq \lambda \leq 1$,

$$f(\bar{x}_3) \geq \lambda f(\bar{x}_1) + (1 - \lambda)f(\bar{x}_3) \tag{8-28}$$

But,

$$
\begin{aligned}
f(\bar{x}_3) &= \bar{c}'[\lambda\bar{x}_1 + (1-\lambda)\bar{x}_2] + [\lambda\bar{x}_1 + (1-\lambda)\bar{x}_2]'P[\lambda\bar{x}_1 + (1-\lambda)\bar{x}_2] \\
&= \lambda\bar{c}'\bar{x}_1 + (1-\lambda)\bar{c}'\bar{x}_2 + [\bar{x}_2 + \lambda(\bar{x}_1 - \bar{x}_2)]'P[\bar{x}_2 + \lambda(\bar{x}_1 - \bar{x}_2)] \\
&= \lambda\bar{c}'\bar{x}_1 + (1-\lambda)\bar{c}'\bar{x}_2 + \bar{x}_2'P\bar{x}_2 + \lambda\bar{x}_2'P(\bar{x}_2 - \bar{x}_2) \\
&\quad + \lambda(\bar{x}_1 - \bar{x}_2)'P\bar{x}_2 + \lambda^2(\bar{x}_1 - \bar{x}_2)'P(\bar{x}_1 - \bar{x}_2)
\end{aligned} \tag{8-29}
$$

Since P is symmetric, it follows that

$$
\begin{aligned}
\bar{x}_2'P(\bar{x}_1 - \bar{x}_2) &= x_2'P'(\bar{x}_1 - \bar{x}_2) \\
&= (\bar{x}_1 - \bar{x}_2)'P\bar{x}_2
\end{aligned} \tag{8-30}
$$

Substituting (8-30) into (8-29) yields

$$
\begin{aligned}
f(\bar{x}_3) = \lambda\bar{c}'\bar{x}_1 + (1-\lambda)\bar{c}'\bar{x}_2 + \bar{x}_2'P\bar{x}_2 \\
+ 2\lambda(\bar{x}_1 - \bar{x}_2)'P\bar{x}_2 + \lambda^2(\bar{x}_1 - \bar{x}_2)'P(\bar{x}_1 - \bar{x}_2)
\end{aligned} \tag{8-31}
$$

Now, since P is negative semidefinite or negative definite,

$$(\bar{x}_1 - \bar{x}_2)'P(\bar{x}_1 - \bar{x}_2) \leq 0$$

and since $0 \leq \lambda \leq 1$, it follows that $\lambda \geq \lambda^2$, and so we have

$$\lambda(\bar{x}_1 - \bar{x}_2)'P(\bar{x}_1 - \bar{x}_2) \leq \lambda^2(\bar{x}_1 - \bar{x}_2)'P(\bar{x}_1 - \bar{x}_2) \tag{8-32}$$

Substituting the result of (8-32) into (8-31) yields

$$
\begin{aligned}
f(\bar{x}_3) &\geq \lambda\bar{c}'\bar{x}_1 + (1-\lambda)\bar{c}'\bar{x}_2 + \bar{x}_2'P\bar{x}_2 \\
&\quad + 2\lambda(\bar{x}_1 - \bar{x}_2)'P\bar{x}_2 + \lambda(\bar{x}_1 - \bar{x}_2)'P(\bar{x}_1 - \bar{x}_2) \\
&= \lambda\bar{c}'\bar{x}_1 + (1-\lambda)\bar{c}'\bar{x}_2 + \bar{x}_2'P\bar{x}_2 \\
&\quad + 2\lambda(\bar{x}_1 - \bar{x}_2)'P\bar{x}_2 + \lambda\bar{x}_1P\bar{x}_1 - \lambda(\bar{x}_1 - \bar{x}_2)'P\bar{x}_2 - \lambda\bar{x}_2'P\bar{x}_1 \\
&= \lambda\bar{c}'\bar{x}_1 + (1-\lambda)\bar{c}'\bar{x}_2 + \lambda\bar{x}_1P\bar{x}_1 \\
&\quad + \bar{x}_2'P\bar{x}_2 + \lambda(\bar{x}_1 - \bar{x}_2)'P\bar{x}_2 - \lambda\bar{x}_2'P\bar{x}_1 \\
&= \lambda\bar{c}'\bar{x}_1 + (1-\lambda)\bar{c}'\bar{x}_2 + \lambda\bar{x}_1'P\bar{x}_1 + \bar{x}_2'P\bar{x}_2 - \lambda\bar{x}_2'P\bar{x}_2 \\
&= (\lambda\bar{c}'\bar{x}_1 + \lambda\bar{x}_1'P\bar{x}_1) + (1-\lambda)(\bar{c}'\bar{x}_2 + \bar{x}_2'P\bar{x}_2) \\
&= \lambda f(\bar{x}_1) + (1-\lambda)f(\bar{x}_2)
\end{aligned}
$$

Thus, $f(\bar{x})$ is concave.

We leave it to the reader to prove:

Theorem 8.4. The function $f(\bar{x}) = \bar{c}'\bar{x} + \bar{x}'P\bar{x}$ is convex if P is positive semidefinite or positive definite.

In general, there are no known computationally feasible methods for obtaining the global maximum to a quadratic programming problem unless $f(\bar{x})$ is concave. The fact that the optimal solution lies at an extreme point if $f(\bar{x})$ is convex turns out to be of little practical use because of the large number of possible extreme points and the fact that the extreme points are relative maxima, not global maxima.

We shall restrict our attention, therefore, to the case where $f(\bar{x})$ is concave. In this case, we know that the Kuhn-Tucker conditions will be satisfied by a global maximum. Let us then apply the Kuhn-Tucker conditions to the quadratic programming problem (8-26) and (8-27), assuming that $f(\bar{x})$ is concave.

From (8-26) and (8-11) we see that

$$F(\bar{x}, \bar{\lambda}) = \bar{c}'\bar{x} + \bar{x}'P\bar{x} + \sum_{i=1}^{m} \lambda_i \left[b_i - \sum_{j=1}^{n} a_{ij}x_j \right] \tag{8-33}$$

Thus, the global maximum to the quadratic programming problem is any point $\bar{x}_0$ which satisfies the following:

If $x_{0_j} > 0$,

$$\frac{\partial F}{\partial x_j} = c_j + 2\sum_{k=1}^{n} p_{jk}x_k - \sum_{i=1}^{m} \lambda_i a_{ij} = 0 \qquad \text{at } \bar{x} = \bar{x}_0 \tag{8-34a}$$

If $x_{0_j} = 0$,

$$\frac{\partial F}{\partial x_j} = c_j + 2\sum_{k=1}^{n} p_{jk}x_k - \sum_{i=1}^{m} \lambda_i a_{ij} \le 0 \qquad \text{at } \bar{x} = \bar{x}_0 \tag{8-34b}$$

If $\lambda_{0_i} > 0$,

$$\frac{\partial F}{\partial \lambda_i} = b_i - \sum_{j=1}^{n} a_{ij}x_j = 0 \qquad \text{at } \bar{x} = \bar{x}_0 \tag{8-35a}$$

If $\lambda_{0_i} = 0$,

$$\frac{\partial F}{\partial \lambda_i} = b_i - \sum_{j=1}^{n} a_{ij}x_j \le 0 \qquad \text{at } \bar{x} = \bar{x}_0 \tag{8-35b}$$

For convenience, let us rewrite these relations by adding slack variables y_j, $j = 1, 2, \ldots, n$, to the inequalities (8-34b) and slack variables x_{si}, $i = 1, 2, \ldots, m$, to the inequalities (8-35b). The resulting equations are

$$2\sum_{k=1}^{n} p_{jk}x_k - \sum_{i=1}^{m} a_{ij}\lambda_i + y_j = -c_j \qquad j = 1, 2, \ldots, n \tag{8-36}$$

$$\sum_{j=1}^{n} a_{ij}x_j + x_{si} = b_i \qquad i = 1, 2, \ldots, m \tag{8-37}$$

Notice that equations (8-34a) and (8-36) are equivalent, provided that $y_j = 0$ if $x_j > 0$; similarly, equations (8-35a) and (8-37) are equivalent, provided that $x_{si} = 0$ if $\lambda_i > 0$. These two requirements may be expressed as

$$x_j y_j = 0 \qquad j = 1, 2, \ldots, n \tag{8-38a}$$

$$\lambda_i x_{si} = 0 \qquad i = 1, 2, \ldots, m \tag{8-38b}$$

To summarize the foregoing discussion: a point $\bar{x}_0$ will be a global maximum solution if and only if there exist nonnegative numbers $\lambda_1, \lambda_2, \ldots, \lambda_m, x_{s1}, x_{s2}, \ldots, x_{sm}, y_1, y_2, \ldots, y_n$ such that equations (8-36), (8-37), and (8-38) are satisfied at $x = x_0$. Except for the constraints (8-38), these conditions are linear. This fact suggests that the simplex method might be applicable. This is in fact the case. We need only to modify the simplex method so that the (nonlinear) constraints (8-38) are also satisfied. Thus, we wish to find an initial feasible solution to the following linear constraints:

$$\begin{bmatrix} 2P & 0_{(n\times m)} & -A' & I_n \\ & & & \\ A & I_m & 0_{(m\times m)} & 0_{(m\times n)} \end{bmatrix} \begin{bmatrix} \bar{x} \\ \bar{x}_s \\ \bar{\lambda} \\ \bar{y} \end{bmatrix} = \begin{bmatrix} -\bar{c} \\ \\ \bar{b} \end{bmatrix} \tag{8-39}$$

$$\bar{x} \geq \bar{0}, \bar{x}_s \geq \bar{0}, \bar{\lambda} \geq \bar{0}, \bar{y} \geq \bar{0} \tag{8-40}$$

where $\bar{x}' = (x_1, x_2, \ldots, x_n)$, $\bar{x}_s' = (x_{s1}, x_{s2}, \ldots, x_{sm})$,

$$\bar{\lambda}' = (\lambda_1, \lambda_2, \ldots, \lambda_m), \quad \text{and} \quad \bar{y}' = (y_1, y_2, \ldots, y_n)$$

The reader should verify that equation (8-39) is the matrix form of the $(m + n)$ equations (8-36) and (8-37). In addition, of course, we require that equations (8-38) are satisfied.

A basic solution to (8-39) will have no more than $(m + n)$ positive variables, and the remaining variables will be zero. Let us now note that *any* feasible solution to (8-39) which also satisfies (8-38) will also be a *basic* feasible solution. Consider, first, equation (8-38a). Of the $2n$ variables $x_j, y_j, j = 1, 2, \ldots, n$, for each $x_j > 0$, the corresponding $y_j = 0$. Thus, if k x_j's are positive and $(n - k)$ x_j's are zero, then k y_j's must be zero and at most $(n - k)$ y_j's can be positive. Thus, at most n of these $2n$ variables can be positive. Similarly, at most m of the $2m$ variables $x_{si}, \lambda_i, i = 1, 2, \ldots, m$, can be positive. Therefore, in attempting to find a solution to (8-38) and (8-39), we need only consider basic feasible solutions.

We are now ready to describe the algorithm for solving a quadratic programming problem with a concave objective function. Using a Phase I* of the simplex method, obtain an initial basic feasible solution to (8-39) and (8-40) with the following modifications in the selection of vectors to enter the basis:

1. If a variable x_j is currently in the basis at a positive level, do not consider y_j as a candidate for entry into the basis; if x_j is currently in the basis at a zero level, y_j may enter the basis only if x_j remains at a zero level.

2. If a variable λ_i is currently in the basis at a positive level, do not consider x_{si} as a candidate for entry into the basis; if λ_i is

*Phase I of the simplex method consists of adding artificial variables with costs of -1 in the objective function; all other variables have costs of 0. Thus, at optimality the Phase I objective function $Z = -\Sigma x_{a_i}$ will have value $Z = 0$, and all artificial variables will be zero.

currently in the basis at a zero level, λ_i may enter the basis only if x_{si} remains at a zero level.

These modifications will ensure that equations (8-38) will be satisfied. When the Phase I procedure is finished, the initial basic feasible solution $[\bar{x}_0, \bar{x}_{0s}, \bar{\lambda}_0, \bar{y}_0]$ to (8-39) contains the optimal solution $\bar{x}_0$ to the quadratic programming problem. The foregoing algorithm was developed by Wolfe.[7] In Hadley[2] a detailed proof is given which shows that the algorithm will always yield the optimal solution, provided that P is negative definite; that is, the modifications we have introduced into the Phase I procedure do not prevent us from obtaining an initial basic feasible solution. This fact is not at all obvious.

If P is only negative semidefinite, however, it is possible that the solution is unbounded. This case causes no difficulties in practice, however, since small perturbations in the diagonal elements of a semidefinite matrix will yield a definite matrix (See Problem 11, page 319).

EXAMPLE

$$\begin{array}{ll} \text{Maximize} & f(\bar{x}) = x_1 + x_2 - \tfrac{1}{2}x_1^2 + x_1x_2 - x_2^2 \\ \text{Subject to} & x_1 + x_2 \leq 3 \\ & 2x_1 + 3x_2 \geq 6 \\ & x_1, x_2 \geq 0 \end{array}$$

We can rewrite $f(\bar{x})$ as

$$\begin{aligned} f(\bar{x}) &= (1, 1)\begin{bmatrix} x_1 \\ x_2 \end{bmatrix} + [x_1x_2]\begin{bmatrix} -\tfrac{1}{2} & \tfrac{1}{2} \\ \tfrac{1}{2} & -1 \end{bmatrix}\begin{bmatrix} x_1 \\ x_2 \end{bmatrix} \\ &= \bar{c}'\bar{x} + \bar{x}'P\bar{x} \end{aligned}$$

Note that P is negative definite, since

$$\bar{x}'P\bar{x} = \tfrac{1}{2}(-x_1^2 + 2x_1x_2 - 2x_2^2) = \tfrac{1}{2}[-(x_1 - x_2)^2 - x_2^2] < 0$$

unless $x_1 = x_2 = 0$. Equation (8-39) yields

$$\left[\begin{array}{cc:cc:cc:cc} -1 & 1 & 0 & 0 & -1 & -2 & 1 & 0 \\ 1 & -2 & 0 & 0 & -1 & 3 & 0 & 1 \\ \hdashline 1 & 1 & 1 & 0 & 0 & 0 & 0 & 0 \\ -2 & -3 & 0 & 1 & 0 & 0 & 0 & 0 \end{array}\right]\begin{bmatrix} x_1 \\ x_2 \\ x_{s1} \\ x_{s2} \\ \lambda_1 \\ \lambda_2 \\ y_1 \\ y_2 \end{bmatrix} = \left[\begin{array}{c} -1 \\ -1 \\ \hdashline 3 \\ -6 \end{array}\right]$$

Table 8.1

$\bar{c}_B$	Variables in Basis	$\bar{x}_B$	x_1	x_2	x_{s1}	x_{s2}	λ_1	λ_2	y_1	y_2	w_1	w_2	w_3	w_4
-1	w_1	1	(1)	-1	0	0	1	2	-1	0	1	0	0	0
-1	w_2	1	-1	2	0	0	1	-3	0	-1	0	1	0	0
-1	w_3	3	1	1	1	0	0	0	0	0	0	0	1	0
-1	w_4	6	2	3	0	-1	0	0	0	0	0	0	0	1
	$z_j - c_j$	-11	-3	-5	-1	1	-2	1	1	1	0	0	0	0

Table 8.2

$\bar{c}_B$	Variables in Basis	$\bar{x}_B$	x_1	x_2	x_{s1}	x_{s2}	λ_1	λ_2	y_1	y_2	w_1	w_2	w_3	w_4
0	x_1	1	1	-1	0	0	1	2	-1	0	1	0	0	0
-1	w_2	2	0	1	0	0	2	-1	-1	-1	1	1	0	0
-1	w_3	2	0	2	1	0	-1	-2	1	0	-1	0	1	0
-1	w_4	4	0	(5)	0	-1	-2	-4	2	0	-2	0	0	1
	$z_j - c_j$	-8	0	-8	-1	1	1	7	-2	1	4	0	0	0

Table 8.3

$\bar{c}_B$	Variables in Basis	$\bar{x}_B$	x_1	x_2	x_{s1}	x_{s2}	λ_1	λ_2	y_1	y_2	w_2	w_3
0	x_1	$\frac{9}{5}$	1	0	0	$-\frac{1}{5}$	$\frac{3}{5}$	$\frac{6}{5}$	$-\frac{3}{5}$	0	0	0
-1	w_2	$\frac{6}{5}$	0	0	0	$\frac{1}{5}$	$\frac{12}{5}$	$-\frac{1}{5}$	$-\frac{7}{5}$	-1	1	0
-1	w_3	$\frac{2}{5}$	0	0	(1)	$\frac{2}{5}$	$-\frac{1}{5}$	$-\frac{2}{5}$	$\frac{1}{5}$	0	0	1
0	x_2	$\frac{4}{5}$	0	1	0	$-\frac{1}{5}$	$-\frac{2}{5}$	$-\frac{4}{5}$	$\frac{2}{5}$	0	0	0
	$z_j - c_j$	$-\frac{8}{5}$	0	0	-1	$-\frac{3}{5}$	$-\frac{11}{5}$	$\frac{3}{5}$	$\frac{6}{5}$	1	0	0

Table 8.4

$\bar{c}_B$	Variables in Basis	$\bar{x}_B$	x_1	x_2	x_{s1}	x_{s2}	λ_1	λ_2	y_1	y_2	w_2
0	x_1	$\frac{9}{5}$	1	0	0	$-\frac{1}{5}$	$\frac{3}{5}$	$\frac{6}{5}$	$-\frac{3}{5}$	0	0
-1	w_2	$\frac{6}{5}$	0	0	0	$\frac{1}{5}$	$\frac{12}{5}$	$-\frac{1}{5}$	$-\frac{7}{5}$	-1	1
0	x_{s1}	$\frac{2}{5}$	0	0	1	($\frac{2}{5}$)	$-\frac{1}{5}$	$-\frac{2}{5}$	$\frac{1}{5}$	0	0
0	x_2	$\frac{4}{5}$	0	1	0	$-\frac{1}{5}$	$-\frac{2}{5}$	$-\frac{4}{5}$	$\frac{2}{5}$	0	0
	$z_j - c_j$	$-\frac{6}{5}$	0	0	0	$-\frac{1}{5}$	$-\frac{12}{5}$	$\frac{1}{5}$	$\frac{7}{5}$	1	0

Table 8.5

$\bar{c}_B$	Variables in Basis	$\bar{x}_B$	x_1	x_2	x_{s1}	x_{s2}	λ_1	λ_2	y_1	y_2	w_2
0	x_1	2	1	0	$\frac{1}{2}$	0	$\frac{1}{2}$	1	$-\frac{1}{2}$	0	0
-1	w_2	1	0	0	$-\frac{1}{2}$	0	($\frac{5}{2}$)	0	$-\frac{3}{2}$	-1	1
0	x_{s2}	1	0	0	$\frac{5}{2}$	1	$-\frac{1}{2}$	-1	$\frac{1}{2}$	0	0
0	x_2	1	0	1	$\frac{1}{2}$	0	$-\frac{1}{2}$	-1	$\frac{1}{2}$	0	0
	$z_j - c_j$	-1	0	0	$\frac{1}{2}$	0	$-\frac{5}{2}$	0	$\frac{3}{2}$	1	0

Table 8.6

$\bar{c}_B$	Variables in Basis	$\bar{x}_B$	x_1	x_2	x_{s1}	x_{s2}	λ_1	λ_2	y_1	y_2
0	x_1	$\frac{9}{5}$	1	0	$\frac{3}{5}$	0	0	1	$-\frac{1}{5}$	$\frac{1}{5}$
0	λ_1	$\frac{2}{5}$	0	0	$-\frac{1}{5}$	0	1	0	$-\frac{3}{5}$	$-\frac{2}{5}$
0	x_{s2}	$\frac{6}{5}$	0	0	$\frac{12}{5}$	1	0	-1	$\frac{1}{5}$	$-\frac{1}{5}$
0	x_2	$\frac{6}{5}$	0	1	$\frac{2}{5}$	0	0	-1	$\frac{4}{5}$	$-\frac{1}{5}$
	$z_j - c_j$	0	0	0	0	0	0	0	0	0

We now add four artificial variables w_1, w_2, w_3, and w_4, and set up the initial Phase I simplex tableau, given in Table 8.1. We can choose any variable whose corresponding $z_j - c_j$ is negative to enter the basis. Choosing x_1 to enter, we see that w_1 then leaves the basis. The resulting tableau is given by Table 8.2 (with the column corresponding to w_1 deleted, since w_1 will never be considered as a candidate for re-entry into the basis). Next, we choose x_2 to enter the basis replacing w_4. The resulting tableau is given by Table 8.3. Now, the variables eligible for entry into the basis are x_{s1}, x_{s2}, and λ_1. We choose x_{s1} to enter, replacing w_3; the new tableau is given by Table 8.4. Now, the only remaining variables with negative $z_j - c_j$ are x_{s2} and λ_1. However, λ_1 may not enter the basis, since $x_{s1} > 0$; therefore, x_{s2} enters the basis, replacing x_{s1}. The new tableau is given by Table 8.5. Now, we bring λ_1 into the basis, replacing w_4, and we have found the desired feasible solution. From Table 8.6, the final tableau, we see that the optimal solution is $x_{01} = \frac{9}{5}$, $x_{02} = \frac{6}{5}$, and the optimal value of the objective function is $f(\bar{x}_0) = -\frac{21}{10}$.

8.5 SEPARABLE PROGRAMMING

In the first three sections of this chapter we investigated certain properties relating to nonlinear programming problems and their solutions. However, we have not yet considered any methods of solution for such problems (except for quadratic programming problems with concave objective functions). In this section we wish to present a technique which may be used to obtain, at least approximately, optimal solutions for a relatively large class of nonlinear programming problems. Moreover, this technique, as was the case in the previous section, employs a modification of the simplex method and hence is computationally reasonably efficient. However, as we shall see, the size of the linear programming problems involved may be rather large, and, in general, the method will obtain only a local maximum.

Let us again consider the nonlinear programming problem

$$\begin{aligned} &\text{Maximize} \quad z = f(\bar{x}) \equiv f(x_1, x_2, \ldots, x_n) \\ &\text{Subject to} \quad g_i(\bar{x})\{\leq, =, \geq\}b_i, \qquad i = 1, 2, \ldots, m \\ &\qquad\qquad\quad \bar{x} \geq \bar{0} \end{aligned}$$

It will also be convenient to assume that there are known upper bounds on each variable x_j:

$$x_j \leq d_j \qquad j = 1, 2, \ldots, n \tag{8-41}$$

(If this is not the case, some arbitrarily large value for d_j may be chosen.) Suppose that $f(\bar{x})$ may be expressed as a sum of n functions $f_j(x_j), j = 1, 2, \ldots, n$, each of which is a function of a single variable. Thus,

$$f(\bar{x}) = \sum_{j=1}^{n} f_j(x_j) \tag{8-42}$$

Such a function is said to be *separable*. Let us also assume that each $g_i(\bar{x})$, $i = 1, 2, \ldots, m$, is separable. Let

$$g_i(\bar{x}) = \sum_{j=1}^{n} g_{ij}(x_j) \qquad i = 1, 2, \ldots, m \tag{8-43}$$

We can now formulate the nonlinear programming problem in terms of a set of functions $\{f_j(x_j)\}$ and $\{g_{ij}(x_j)\}$ each of which is a function of a single variable. In order to be able to make use of the simplex method, we will need to make some additional changes in the formulation of the foregoing separable nonlinear programming problem. In particular, we shall approximate each function $f_j(x_j)$ and each function $g_{ij}(x_j)$ by a *piecewise linear function.* A piecewise linear function is described by a set of connected line segments. This is also referred to as a "polygonal line," and the resulting approximation as a "polygonal approximation."

Before continuing with the development of the method, let us consider the function $f(x)$ shown in Figure 8-5. An example of a polygonal approximation of $f(x)$ is given by $\hat{f}(x)$. $\hat{f}(x)$ is composed of five line segments; at the end points of each line segment, $\hat{f}(x) = f(x)$. The equation of the line segment which represents $\hat{f}(x)$ in the interval $[a_1, a_2]$ is given by

$$\hat{f}(x) = f(a_1) + \frac{f(a_2) - f(a_1)}{a_2 - a_1}(x - a_1) \qquad a_1 \leq x \leq a_2 \tag{8-44}$$

Similarly, the equation of the line segment which represents $\hat{f}(x)$ in the interval $[a_2, a_3]$ is given by

$$\hat{f}(x) = f(a_2) + \frac{f(a_3) - f(a_2)}{a_3 - a_2}(x - a_2) \tag{8-45}$$

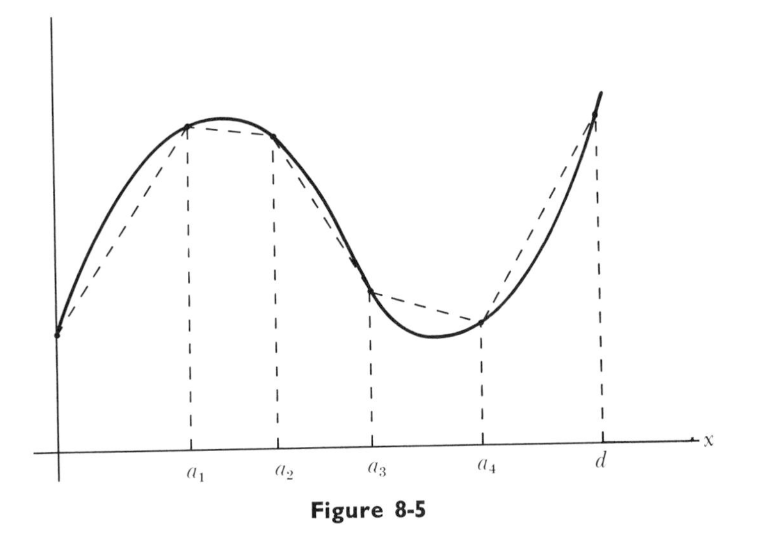

Figure 8-5

Now, let us see how we can combine equations (8-44) and (8-45) so that we can represent $\hat{f}(x)$ over the interval $a_1 \leq x \leq a_3$ by one equation. First, note that for any x in the interval $[a_1, a_2]$ we can determine a value of λ, $0 \leq \lambda \leq 1$, such that

$$x = \lambda a_1 + (1 - \lambda)a_2 \tag{8-46}$$

Let us denote this value of λ by λ_1 and let us define $\lambda_2 = 1 - \lambda_1$. Thus, equation (8-46) becomes

$$x = \lambda_1 a_1 + \lambda_2 a_2 \tag{8-47}$$

where

$$\lambda_1 + \lambda_2 = 1, \qquad \lambda_1 \geq 0, \qquad \lambda_2 \geq 0 \tag{8-48}$$

Upon subtracting a_1 from both sides of equation (8-47), we see that

$$\begin{aligned} x - a_1 &= \lambda_2 a_2 - (1 - \lambda_1)a_1 \\ &= \lambda_2 a_2 - \lambda_2 a_1 \\ &= \lambda_2(a_2 - a_1) \end{aligned} \tag{8-49}$$

Substitution of the right-hand side of equation (8-49) into equation (8-44) yields

$$\begin{aligned} \hat{f}(x) &= f(a_1) + \frac{f(a_2) - f(a_1)}{a_2 - a_1} \lambda_2(a_2 - a_1) \\ &= \lambda_2 f(a_2) + (1 - \lambda_2) f(a_1) \\ &= \lambda_2 f(a_2) + \lambda_1 f(a_1) \end{aligned} \tag{8-50}$$

Similarly, any x in the interval $[a_2, a_3]$ can be expressed as

$$x = \lambda_2 a_2 + \lambda_3 a_3 \tag{8-51}$$

and, as before, we have that, for $a_2 \leq x \leq a_3$

$$\hat{f}(x) = \lambda_3 f(a_3) + \lambda_2 f(a_2) \quad \textbf{(8-52)}$$

Let us consider one additional interval. The representation of $\hat{f}(x)$ for any x in the interval $[a_3, a_4]$ is given by

$$\hat{f}(x) = \lambda_4 f(a_4) + \lambda_5 f(a_3) \quad \textbf{(8-53)}$$

Now, in order that the right-hand side of equations (8-50), (8-52), and (8-53) be valid expressions for $\hat{f}(x)$, it must be true that

$$\lambda_1 + \lambda_2 = 1 \quad \textbf{(8-54a)}$$

$$\lambda_2 + \lambda_3 = 1 \quad \textbf{(8-54b)}$$

$$\lambda_3 + \lambda_4 = 1 \quad \textbf{(8-54c)}$$

$$\lambda_1, \lambda_2, \lambda_3, \lambda_4 \geq 0 \quad \textbf{(8-55)}$$

(One of equations (8-54) must hold, depending upon which interval x is in.) Let us now see how we can combine these representations of x and of $\hat{f}(x)$ into a single equation for each. Consider the following equations:

$$x = \lambda_1 a_1 + \lambda_2 a_2 + \lambda_3 a_3 + \lambda_4 a_4 \quad \textbf{(8-56)}$$

$$\hat{f}(x) = \lambda_1 f(a_1) + \lambda_2 f(a_2) + \lambda_3 f(a_3) + \lambda_4 f(a_4) \quad \textbf{(8-57)}$$

$$\lambda_1 + \lambda_2 + \lambda_3 + \lambda_4 = 1 \quad \textbf{(8-58)}$$

$$\lambda_1, \lambda_2, \lambda_3, \lambda_4 \geq 0 \quad \textbf{(8-59)}$$

Suppose that in each of the equations (8-56), (8-57), and (8-58) we require that

1. At most two λ_i can be positive, with the rest zero, and
2. Only *adjacent* λ_i are allowed to be positive.

Requirement (1) is self-explanatory; requirement (2) means that if, for example, λ_2 is positive, then either λ_1 or λ_3 can be positive, but not both and not λ_4. When these requirements are satisfied, equations (8-56) through (8-59) are equivalent to equations (8-46) through (8-55). However, in order to use equations (8-56) through (8-59), we do not have to know, in advance, in which interval x lies; this will be determined by the values we assign to the λ_i's. Note also that equations (8-57) and (8-58) are linear.

Let us now return to the separable nonlinear programming problem. Suppose that each x_j is subdivided into p_j intervals, and let x_{jk} denote the value of x_j at the kth subdivision. Thus,

$$0 = x_{j_0} < x_{j_1} < x_{j_2} < \ldots < x_{jk} < \ldots x_{jp_j} = d_j \quad \textbf{(8-60)}$$

Also, suppose that each function $f_j(x_j)$ and $g_{ij}(x_j)$ is approximated by a piecewise linear function, denoted by $\hat{f}_j(x_j)$ and $\hat{g}_{ij}(x_j)$, respectively.

Then, let

$$x_j = \sum_{k=0}^{p_j} \lambda_{jk} x_{jk} \quad \textbf{(8-61)}$$

where

$$\sum_{k=0}^{p_j} \lambda_{jk} = 1 \quad \textbf{(8-62)}$$

and

$$\lambda_{jk} \geq 0, \qquad k = 0, 1, \ldots, p_j \quad \textbf{(8-63)}$$
$$j = 1, 2, \ldots, n$$

Also,

$$\hat{f}_j(x_j) = \sum_{k=0}^{p_j} f_j(x_{jk}) \lambda_{jk}$$

and

$$\hat{g}_{ij}(x_j) = \sum_{k=0}^{p_j} g_{ij}(x_{jk}) \lambda_{jk}$$

Now, letting $f_j(x_{jk}) = f_{jk}$ and $g_{ij}(x_{jk}) = g_{ijk}$, we obtain a new problem:

Maximize

$$z = \sum_{j=1}^{n} \sum_{k=0}^{p_j} f_{jk} \lambda_{jk} \quad \textbf{(8-64)}$$

Subject to

$$\sum_{j=1}^{n} \sum_{k=0}^{p_j} g_{ijk} \lambda_{jk} \{\leq, =, \geq\} b_i, \qquad i = 1, 2, \ldots, m \quad \textbf{(8-65)}$$

$$\sum_{k=0}^{p_j} \lambda_{jk} = 1 \qquad j = 1, 2, \ldots, n \quad \textbf{(8-66)}$$

$$\lambda_{jk} \geq 0 \qquad k = 0, 1, \ldots, p_j \text{ and } j = 1, 2, \ldots, n \quad \textbf{(8-67)}$$

The requirements (1) and (2) on the λ_{jk} also must be satisfied, for each j. This problem is called the *approximating problem* to the separable nonlinear programming problem. Except for the requirements (1) and (2), it is an ordinary linear programming problem, with $(m + n)$ constraints and $\left(\sum_{j=1}^{n} p_j + n\right)$ variables. As was the case with Wolfe's quadratic programming algorithm, the approximating problem can be solved by the simplex method, with one modification made to the rules for selecting vectors to enter the basis. The modification needed is merely that no vector may enter the basis if either of requirements (1) and (2) cannot be satisfied.

It is important to note that the optimal solution obtained to the approximating problem may not even be a feasible solution to the original problem, since the polygonal approximations may yield a constraint set that is somewhat larger than the original constraint set. However, if the original constraint set was a convex set, then any straight line approximation to a curved boundary will lie inside the convex set, and so the constraint set for the approximating problem will lie wholly within the original convex set.

Another important observation concerning the approximating problem is that it might be necessary to subdivide the intervals $0 \leq x_j \leq d_j$ into many subintervals in order to obtain reasonably good approximations for the functions $f_j(x_j)$ and $g_{ij}(x_j)$. Thus, the numbers p_j might become quite large, and hence there may be a larger number of variables in the approximating problem. However, the number of constraints remains the same, no matter how small or large the p_j's are. Since currently available computer codes for the simplex method can handle problems with a great many variables, one encounters no practical difficulties in implementing these procedures.

In general, when the simplex method is used to solve the approximating problem (with the restricted basis entry rules) only a local maximum is obtained. It is quite possible that better local maxima exist; that is, even though a point is a local maximum, it might indeed be a very poor local maximum. For example, consider the function $f(x)$ shown in Figure 8-6. There are three local maxima, at $x = a_1$, $x = a_2$, and $x = a_3$. There is no way of determining which local maxima will be obtained. Hadley[2] suggests that a useful technique for avoiding poor local maxima is as follows: First, solve the problem using a small number of subdivisions for each x_j. Then, in the neighborhood of the local maximum obtained, increase the number of subdivisions and resolve the problem. This procedure, of starting with a poor approximation, will in many cases eliminate many of the poorer local optima. For example, if the function of

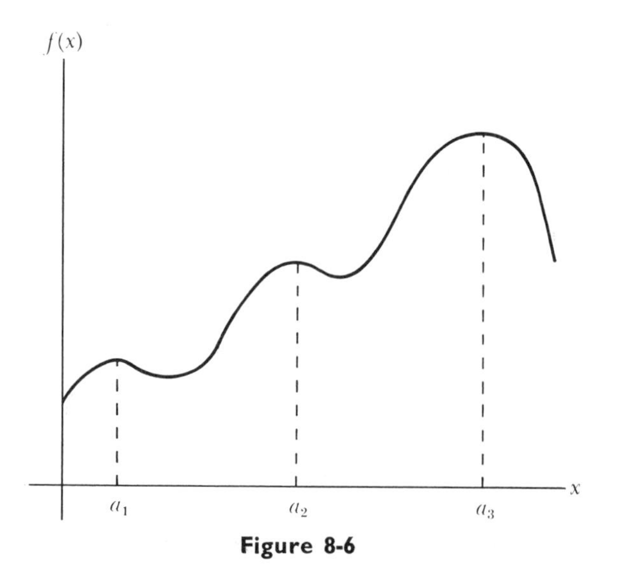

Figure 8-6

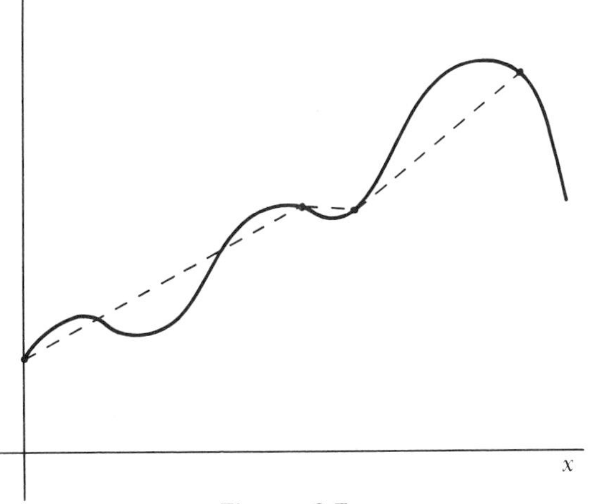

Figure 8-7

Figure 8-6 were approximated with only those subintervals in Figure 8-7 being used, then the local maximum at $x = a_1$ is eliminated.

In closing, we wish to indicate, without proof, a very useful result:

Theorem 8.5. If $f(\bar{x})$ is a concave function, and the constraints (8-43) form a convex set, then the simplex method can be applied to the approximating problem, ignoring the requirements (1) and (2), and the resulting solution will be the global maximum for the approximating problem.

We shall indicate intuitively why this is so. Consider the concave function $f(x)$ shown in Figure 8-8, along with its polygonal approximation $\hat{f}(x)$. Thus,

$$x = \lambda_1 a_1 + \lambda_2 a_2 + \lambda_3 a_3 + \lambda_4 a_4$$

$$\hat{f}(x) = \lambda_1 f(a_1) + \lambda_2 f(a_2) + \lambda_3 f(a_3) + \lambda_4 f(a_4)$$

Suppose, for example, that $\lambda_1 = \lambda_3 = 0$, $\lambda_2 = \lambda_4 = \frac{1}{2}$. Then,

$$x = \tfrac{1}{2}(a_2 + a_4) \equiv b$$

$$\hat{f}(x) = \tfrac{1}{2}[f(a_2) + f(a_4)]$$

However, this value of $\hat{f}(x)$ is clearly smaller than the true value of $\hat{f}(b)$, which lies on the polygonal approximation. Note that, for this example, $x = b$ lies in the interval $[a_2, a_3]$; thus, for some λ, $0 \leq \lambda \leq 1$,

$$b = \lambda a_2 + (1 - \lambda)a_3$$

letting $\lambda_2 = \lambda$, $\lambda_3 = 1 - \lambda$, and $\lambda_1 = \lambda_4 = 0$, we obtain the true value of $\hat{f}(b)$. In other words, even though

$$\tfrac{1}{2}(a_2 + a_4) = b = \lambda_2 a_2 + \lambda_3 a_3$$

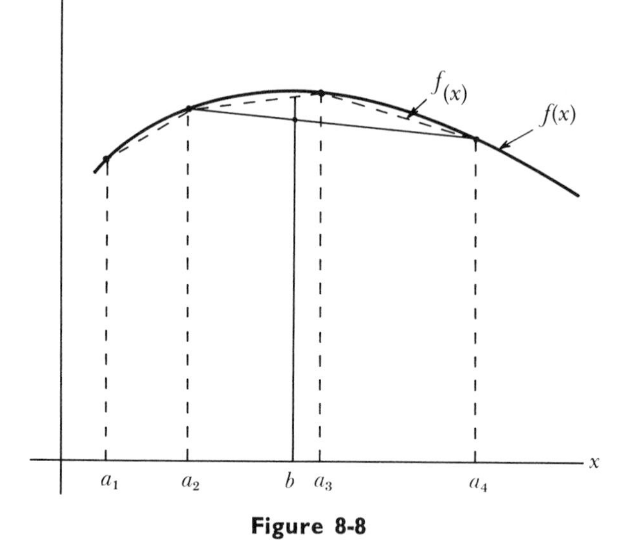

Figure 8-8

we have seen that

$$\hat{f}(\tfrac{1}{2}[a_2 + a_4]) < \hat{f}(\lambda_2 a_2 + \lambda_3 a_3)$$

Therefore, in any *optimal* solution to the approximating problem, the requirements (1) and (2) will be satisfied automatically, provided that $f(\bar{x})$ is concave and the constraint set is convex.

8.6 GRADIENT METHODS

In Section 5.14 we considered a search algorithm for finding the unconstrained optimum of a function $f(\bar{x})$, which made use of the gradient vector, $\overline{\nabla f}(\bar{x})$. This algorithm, the method of steepest ascent,† can be modified so that each of the successive points calculated is feasible and will yield an improved value of the objective function.

Let us assume that we are solving the problem

$$\begin{aligned} \text{Maximize} \quad & z = f(\bar{x}) \\ \text{Subject to} \quad & g_i(\bar{x}) \le b_i, \qquad i = 1, 2, \ldots, m \\ & \bar{x} \ge \bar{0} \end{aligned}$$

Recall that the basic step in the method of steepest ascent, starting from some point $\bar{x}_s$, is to calculate $\overline{\nabla f}(\bar{x}_s)$, the gradient vector

† In a minimization problem, the method is called the method of steepest descent; in a maximization problem, the method of steepest ascent.

evaluated at $\bar{x}_s$, and then to find a value for λ, denoted by λ_s, such that

$$f(\bar{x}_s + \lambda_s \overline{\nabla} f(\bar{x}_s)) = \underset{\lambda}{\text{Maximum}}\, f(\bar{x}_s + \lambda \overline{\nabla} f(\bar{x}_s))$$

The next point, $\bar{x}_{s+1}$, is then calculated by

$$\bar{x}_{s+1} = \bar{x}_s + \lambda_s \overline{\nabla} f(\bar{x}_s)$$

As long as $\bar{x}_s$ is an interior point of the feasible set of solutions, we will always be able to find a λ_s such that $\bar{x}_{s+1}$ is also feasible, no matter in what direction the gradient vector points. However, if $\bar{x}_s$ is a boundary point, it is possible that the gradient vector will point away from the feasible region. This does not necessarily indicate that $\bar{x}_s$ is a local maximum: The gradient vector merely points in the direction of the greatest rate of increase of $f(\bar{x})$; there might be other directions lying within the feasible region along which $f(\bar{x})$ increases in value. Thus, if $\bar{x}_s$ is a boundary point, we would like to find the direction of greatest rate of increase of $f(\bar{x})$ within the feasible region. This direction can be found by "projecting" the gradient vector onto the surface which forms that part of the boundary upon which $\bar{x}_s$ lies and choosing this projection as the desired direction.

However, the calculation of this direction may be rather difficult or time consuming. For this reason, when we cannot move in the direction of the gradient, we can choose to move in *any* direction in which $f(\bar{x})$ increases. There are a wide variety of methods for choosing such a direction; see, for example, Rosen[5,6] and Zoutendijk.[8] An excellent discussion of the computational details of several gradient methods appears in Hadley.[2]

Let us consider the following example, illustrated graphically in Figure 8-9.

$$\text{Maximize} \quad z = 3x_1^2 + 2x_2^2 \tag{8-68}$$

$$\text{Subject to} \quad g_1(x_1, x_2) = x_1^2 + x_2^2 \leq 25 \tag{8-69}$$

$$g_2(x_1, x_2) = 9x_1 - x_2^2 \leq 27 \tag{8-70}$$

$$x_1, x_2 \geq 0 \tag{8-71}$$

From the level curves of the objective function (8-68) which are shown in Figure 8-9, we see that the optimal solution (x_1^*, x_2^*) occurs at the point (4, 3), with $z^* = f(4, 3) = 66$.

Suppose we wish to solve this problem by a gradient method. Suppose also that we have found an initial feasible solution, $\bar{x}_0 = (1.9578, 0.5)$. Let us calculate the gradient of $f(x_1, x_2)$ at $\bar{x}_0$:

$$\overline{\nabla} f(x_1, x_2) = (6x_1, 4x_2)$$

$$\overline{\nabla} f(1.9578, 0.5) = (11.7468, 2.0)$$

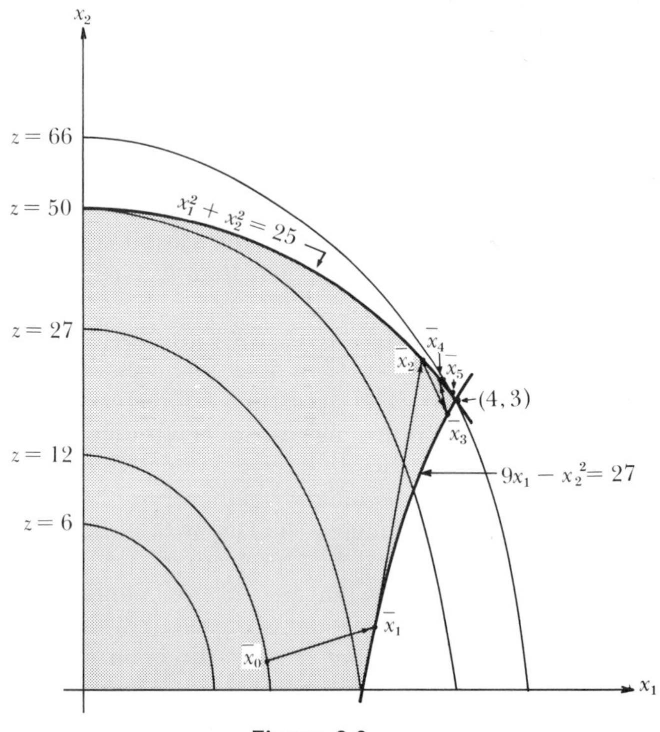

Figure 8-9

Thus, the maximum rate of increase of $f(x_1, x_2)$ occurs along the line

$$\begin{aligned}\bar{x} &= \bar{x}_0 + \lambda\overline{\nabla}f(x_0)\\ &= (1.9578, 0.5) + \lambda(11.7468, 2.0)\end{aligned}$$

The value of λ for which $f[\bar{x}_0 + \lambda\overline{\nabla}f(\bar{x}_0)]$ is a maximum, *within the constraint set*, is $\lambda_0 = 0.0936$, from which we obtain

$$\begin{aligned}\bar{x}_1 &= \bar{x}_0 + \lambda_0\overline{\nabla}f(x_0)\\ &= (3.0573, 0.6872)\end{aligned}$$

Now $\bar{x}_1$ is on the boundary of the constraint set, and $\overline{\nabla}f(\bar{x}_1)$ points away from the constraint set, so we must choose another direction in which $f(x_1, x_2)$ increases. One such direction is indicated in Figure 8-9; if we denote the unit vector pointing in this direction by $\bar{r}$, then we wish to find the value of λ for which $f(\bar{x}_1 + \lambda\bar{r})$ is a maximum, such that the point $(\bar{x}_1 + \lambda\bar{r})$ is feasible. In this example, we have chosen $\bar{r}$ to be the unit vector tangent to the boundary at $\bar{x}_1$. Thus,

$$\overline{\nabla}g_2(\bar{x}_1) = (9, -2x_2) = (9, -1.3744) \qquad \textbf{(8-72)}$$

and the equation of the tangent line to $g_2(\bar{x})$ at $\bar{x}_1$ is given by

$$9x_1 - 1.3744x_2 = 26.5712 \quad \textbf{(8-73)}$$

or

$$x_2 = 2.9524 + 0.15217x_1$$

Thus,

$$\begin{aligned}\bar{r} &= (0.15271,\ 1)/\sqrt{(0.15271)^2 + 1} \\ &= (0.15096,\ 0.98854)\end{aligned}$$

and

$$\bar{x}_2 = \bar{x}_1 + \lambda_1\bar{r} \quad \textbf{(8-74)}$$

We find that $\lambda_1 = 2.919$ and hence $\bar{x}_2 = (3.4980, 3.5727)$.

The points $\bar{x}_3$, $\bar{x}_4$, and $\bar{x}_5$, shown in Figure 8-9, are then calculated, respectively. Since $f(\bar{x}_4)$ is approximately the same as $f(\bar{x}_5)$, we have terminated the process, with $\bar{x}_5$ being the final approximation to the optimal solution.

8.7 THE METHOD OF GRIFFITH AND STEWART

In this section we shall present a method developed by Griffith and Stewart[1] which solves nonlinear programming problems by formulating and solving a sequence of linear programming problems. The basic idea employed in the method of Griffith and Stewart is that, given any feasible solution $\bar{x}_0$ to a nonlinear programming problem, a linear programming problem can be constructed by expanding each nonlinear function in a Taylor Series about $\bar{x}_0$ and ignoring terms of higher order than linear.

For simplicity, let us consider the following nonlinear programming problem:

$$\text{Maximize} \quad z = f(\bar{x}) \quad \textbf{(8-75)}$$

$$\text{Subject to} \quad g_i(\bar{x}) = b_i \qquad i = 1, 2, \ldots, m \quad \textbf{(8-76)}$$

$$0 \le x_j \le d_j \qquad j = 1, 2, \ldots, n \quad \textbf{(8-77)}$$

The fact that each constraint of (8-76) has been expressed as an equality here is of no consequence; any of these may be replaced by an inequality without affecting the development which follows. We do, however, require the upper bound constraints (8-77). We shall assume only that the functions $f(\bar{x})$ and $g_i(\bar{x})$, $i = 1, 2, \ldots, m$, are differentiable; they need not be separable. Then, if $\bar{x}_s$ is a feasible solution, the corresponding linear approximations to the nonlinear functions are

$$f(\bar{x}) \cong f(\bar{x}_s) + [\overline{\nabla} f(\bar{x}_s)]'(\bar{x} - \bar{x}_s)$$

$$= f(x_{s1}, x_{s2}, \ldots, x_{sn}) + \sum_{j=1}^{n} \frac{\partial f(\bar{x}_s)}{\partial x_j}(x_j - x_{sj}) \qquad \textbf{(8-78)}$$

$$b_i = g_i(\bar{x}) \cong g_i(\bar{x}_s) + [\overline{\nabla} g_i(\bar{x}_s)]'(\bar{x} - \bar{x}_s) \qquad \textbf{(8-79)}$$

Let us define a new vector of variables $\bar{y}$:

$$\bar{y} = \bar{x} - \bar{x}_s \qquad \textbf{(8-80)}$$

Thus, $\bar{y}$ represents the change from the current solution $\bar{x}_s$ to the next solution $\bar{x}$. The components of $\bar{y}$ are unrestricted in sign. Let us, then, define a set of nonnegative variables $\bar{y}^N$, $\bar{y}^P$:

$$\bar{y} = \bar{y}^P - \bar{y}^N \qquad \textbf{(8-81)}$$

Thus, $\bar{y}^P$ contains those components of $\bar{y}$ which are positive, and $\bar{y}^N$ contains those components of $\bar{y}$ which are negative. In other words,

$$y_j^P = \begin{cases} y_j & \text{if } y_j > 0 \\ 0 & \text{if } y_j \leq 0 \end{cases}$$

$$y_j^N = \begin{cases} -y_j & \text{if } y_j < 0 \\ 0 & \text{if } y_j \geq 0 \end{cases}$$

For convenience, let us write

$$\bar{c} = \overline{\nabla} f(\bar{x}_s) \qquad \textbf{(8-82)}$$

$$\bar{a}_i = \overline{\nabla} g_i(\bar{x}_s) \qquad i = 1, 2, \ldots, m \qquad \textbf{(8-83)}$$

$$\hat{b}_i = b_i - g_i(\bar{x}_s) \qquad i = 1, 2, \ldots, m \qquad \textbf{(8-84)}$$

Substitution of equations (8-80) through (8-84) into equations (8-78) and (8-79) yields

$$f(\bar{x}) \cong f(\bar{x}_s) + \bar{c}'(\bar{y}^P - \bar{y}^N) \qquad \textbf{(8-85)}$$

$$\hat{b}_i \cong \bar{a}_i'(\bar{y}^P - \bar{y}^N) \qquad i = 1, 2, \ldots, n \qquad \textbf{(8-86)}$$

Since $f(\bar{x}_s)$ is a constant, it can be ignored, and the linear programming approximating problem becomes

$$\text{Maximize} \quad \hat{z} = \bar{c}'(\bar{y}^P - \bar{y}^N) \qquad \textbf{(8-87)}$$

$$\text{Subject to} \quad \bar{a}_i'(\bar{y}^P - \bar{y}^N) = \hat{b}_i \qquad i = 1, 2, \ldots, m \qquad \textbf{(8-88)}$$

$$\bar{y}^P \geq \bar{0}, \ \bar{y}^N \geq \bar{0} \qquad \textbf{(8-89)}$$

However, there are two factors which we have not yet taken into account in the foregoing formulation. The first factor is the upper bound constraints (8-77); the second is the fact that the linear approximations which we have employed are only likely to be reasonable approximations in a small neighborhood of $\bar{x}_s$. To ensure that the solutions to (8-87), (8-88), and (8-89) remain close to $\bar{x}_s$, we need to impose upper and lower bounds on each y_j, since $\bar{y}$ represents the change from the solution $\bar{x}_s$. Let

m_j = maximum distance x_j is allowed to move before the linear approximations are recalculated

then,

$$-m_j \leq y_j \leq m_j \qquad j = 1, 2, \ldots, n$$

$$-m_j \leq y_j^P - y_j^N \leq m_j \qquad j = 1, 2, \ldots, n \tag{8-90}$$

Because of the nonnegativity restrictions on y_j^P and y_j^N, equation (8-90) may be expressed as

$$0 \leq y_j^P \leq m_j \qquad j = 1, 2, \ldots, n \tag{8-91}$$

$$0 \leq y_j^N \leq m_j \qquad j = 1, 2, \ldots, n \tag{8-92}$$

However, from (8-77) and (8-80), we see that

$$0 \leq y_j + x_{sj} \leq d_j \tag{8-93}$$

$$-x_{sj} \leq y_j \leq d_j - x_{sj}$$

$$-x_{sj} \leq y_j^P \leq d_j - x_{sj} \tag{8-94}$$

$$x_{sj} - d_j \leq y_j^N \leq x_{sj} \tag{8-95}$$

Upon comparing inequalities (8-91) and (8-93) we see that, if both are to be satisfied simultaneously, then y_j^P must satisfy

$$0 \leq y_j^P \leq \text{minimum } \{m_j, d_j - x_{sj}\} \tag{8-96}$$

Similarly, y_j^N must satisfy

$$0 \leq y_j^N \leq \text{minimum } \{m_j, x_{sj}\} \tag{8-97}$$

The addition of the upper bound constraints (8-96) and (8-97) ensures that the optimal solution to the resulting linear programming problem (defined by equations (8-87), (8-88), (8-96) and (8-97)) will not violate the upper bound constraints (8-77) and will be reasonably close to $\bar{x}_s$.

Let us now outline the method of Griffith and Stewart:

1. Beginning with a point $\bar{x}_s$, formulate the linear programming problem approximation, defined by (8-87), (8-88), (8-96), and (8-97).

2. Using the simplex method, find the optimal solution to this problem. Denote the optimal solution by $\bar{x}_{s+1}$.

3. If $|\bar{x}_{s+1} - \bar{x}_s| < \varepsilon$ (that is, the two successive solutions agree to within the desired accuracy) terminate and call $\bar{x}_{s+1}$ the optimal solution. If $|x_{s+1} - x_s| > \varepsilon$, return to Step 1, reformulating the linear approximations about $\bar{x}_{s+1}$.

Note that the termination rule in Step 3 ignores the objective function, in contrast with the gradient search methods of the previous section. The criterion $|\bar{x}_{s+1} - \bar{x}_s| < \varepsilon$ can be easily replaced by $|f(\bar{x}_{s+1}) - f(\bar{x}_s)| < \delta$, where δ is another prescribed tolerance number for the desired accuracy. In fact, both criteria could be employed simultaneously, if desired.

There are $n + 1$ parameters which must be chosen in advance: ε and $m_j, j = 1, 2, \ldots, n$. As noted in Step 3, ε is determined by the desired accuracy. Values for the m_j, however, are somewhat more difficult to determine. If the m_j are large, the approximating linear programming problem may well lead to an infeasible solution. On the other hand, if the m_j are small, the allowable changes in $\bar{x}$ will be small, and hence a great many iterations may be required for convergence. (Each "iteration" here consists of Step 1 to Step 3.)

Although Griffith and Stewart offer no suggestions for determination of the m_j, it seems reasonable to assume that the values of m_j are related to the length of the interval $0 \leq x_j \leq d_j$. Thus, as was the case with the separable programming algorithm, we might initially choose the m_j so that the interval $0 \leq x_j \leq d_j$ is subdivided into a small number of subintervals (e.g., let $m_j = d_j/10$ or $d_j/5$). If the optimal solution to the resulting linear programming problem is infeasible, decrease the m_j's accordingly and resolve the problem.

Let us again consider the example of the previous section:

$$\text{Maximize} \quad z = 3x_1^2 + 2x_2^2 \tag{8-68}$$

$$\text{Subject to} \quad g_1(x_1, x_2) = x_1^2 + x_2^2 \leq 25 \tag{8-69}$$

$$g_2(x_1, x_2) = 9x_1 - x_2^2 \leq 27 \tag{8-70}$$

$$x_1, x_2 \geq 0 \tag{8-71}$$

From constraints (8-69) and (8-71) it is easy to see that

$$0 \leq x_1 \leq 5$$

$$0 \leq x_2 \leq 5$$

Suppose we choose $m_1 = m_2 = 2$, and suppose also that $\bar{x}_0 = (1, 1)$.

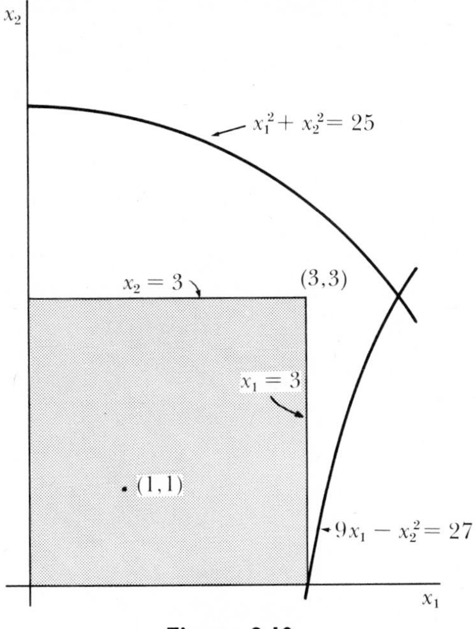

Figure 8-10

Thus, the first linear programming problem to be solved is

$$
\begin{aligned}
\text{Maximize} \quad & z = 6(y_1^P - y_1^N) + 4(y_2^P - y_2^N) \\
\text{Subject to} \quad & 2(y_1^P - y_1^N) + 2(y_2^P - y_2^N) \le 23 \\
& 9(y_1^P - y_1^N) - 2(y_2^P - y_2^N) \le 19 \\
& 0 \le y_1^P \le 2 \\
& 0 \le y_2^P \le 2 \\
& 0 \le y_1^N \le 1 \\
& 0 \le y_2^N \le 1
\end{aligned}
$$

This problem is illustrated (in terms of x_1, x_2) in Figure 8-10; the optimal solution is $y_1^P = y_2^P = 2, y_1^N = y_2^N = 0$, and so

$$\bar{x}_1 = (1, 1) + (2, 2) = (3, 3)$$

The new linear approximating problem becomes

$$
\begin{aligned}
\text{Maximize} \quad & z = 18(y_1^P - y_1^N) + 12(y_2^P - y_2^N) \\
\text{Subject to} \quad & 6(y_1^P - y_1^N) + 6(y_2^P - y_2^N) \le 7 \\
& 9(y_1^P - y_1^N) - 6(y_2^P - y_2^N) \le 9 \\
& 0 \le y_1^P \le 2 \\
& 0 \le y_2^P \le 2 \\
& 0 \le y_1^N \le 2 \\
& 0 \le y_2^N \le 2
\end{aligned}
$$

This problem is illustrated in Figure 8-11; the optimal solution is $y_1^P = 2, y_2^N = \frac{5}{6}, y_1^N = y_2^P = 0$; thus,

$$\bar{x}_2 = (3, 3) + (2, -\tfrac{5}{6}) = (5, \tfrac{13}{6})$$

However, $\bar{x}_2 = (5, \frac{13}{6})$ violates the constraints; let us now reduce the m_j's. If we let $m_1 = m_2 = 1$, the resulting problem (see Figure 8-12) has an optimal solution of $(4, \frac{19}{6})$ which is still slightly infeasible (but very close to optimal). If we let $m_1 = m_2 = \frac{1}{2}$, the optimal solution to the resulting problem is $(\frac{7}{2}, \frac{7}{2})$, which is feasible. Thus, letting $\bar{x}_2 = (\frac{7}{2}, \frac{7}{2})$, the new linear approximating problem, shown in Figure 8-13, becomes

$$\begin{aligned} \text{Maximize} \quad & z = 21(y_1^P - y_1^N) + 14(y_2^P - y_2^N) \\ \text{Subject to} \quad & 7(y_1^P - y_1^N) + 7(y_2^P - y_2^N) \le 0.5 \\ & 9(y_1^P - y_1^N) - 7(y_2^P - y_2^N) \le 8.75 \\ & 0 \le y_1^P \le 0.5 \\ & 0 \le y_2^P \le 0.5 \\ & 0 \le y_1^N \le 0.5 \\ & 0 \le y_2^N \le 0.5 \end{aligned}$$

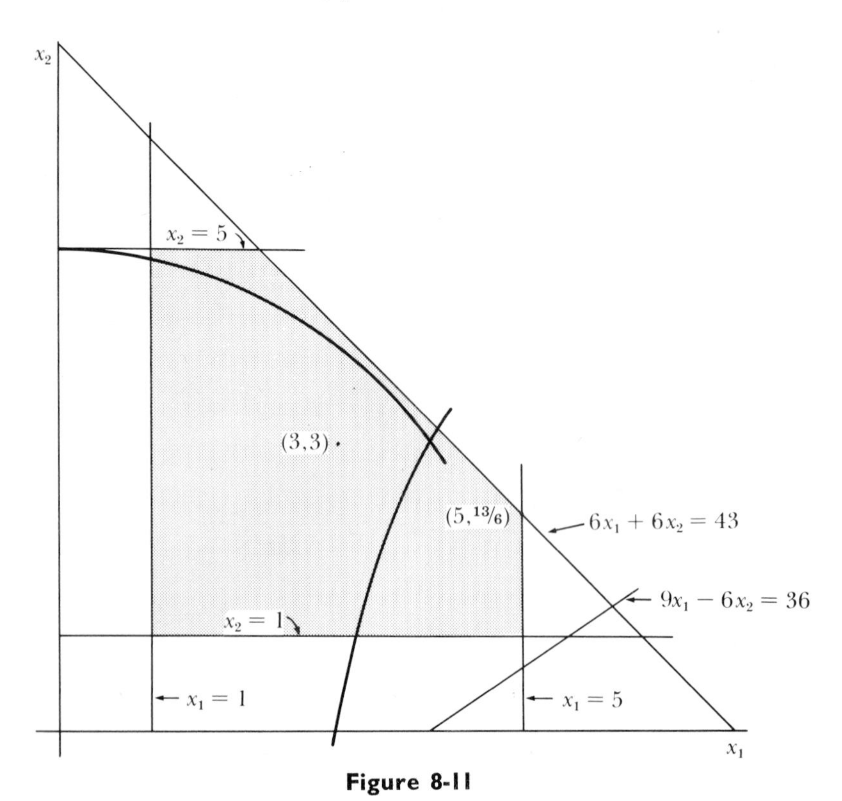

Figure 8-11

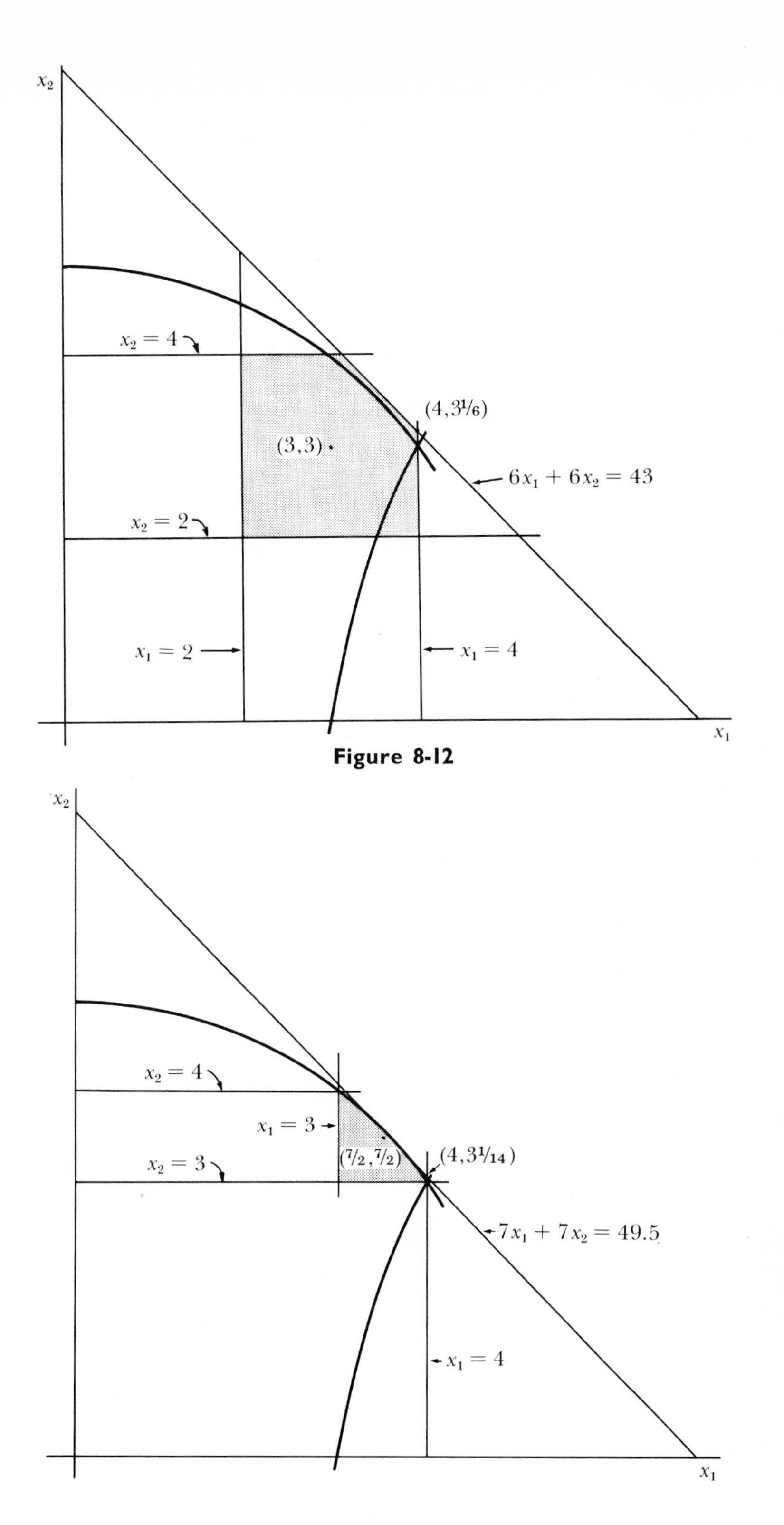

x_2
$x_2 = 4$
$(4, 3^1/_6)$
$(3,3)$
$6x_1 + 6x_2 = 43$
$x_2 = 2$
$x_1 = 2$
$x_1 = 4$
x_1

Figure 8-12

x_2
$x_2 = 4$
$x_1 = 3$
$(^7/_2, ^7/_2)$
$(4, 3^1/_{14})$
$x_2 = 3$
$7x_1 + 7x_2 = 49.5$
$x_1 = 4$
x_1

The optimal solution is $y_1^P = \frac{1}{2}$, $y_2^N = -\frac{3}{7}$, $y_2^P = y_1^N = 0$. Thus, $\bar{x}_3 = (4, 3\frac{1}{14})$. The new approximating problem is

$$\text{Maximize} \quad z = 8(y_1^P - y_1^N) + \tfrac{43}{7}(y_2^P - y_2^N)$$

$$\text{Subject to} \quad 8(y_1^P - y_1^N) + \tfrac{43}{7}(y_2^P - y_2^N) \leq 0.43369$$

$$9(y_1^P - y_1^N) - \tfrac{43}{7}(y_2^P - y_2^N) \leq 0.43369$$

and the same bounds on y_1^P, y_2^P, y_1^N, y_2^N as before. The optimal solution to this problem is $\bar{x}_4 \simeq (4, 3)$. Since $\bar{x}_4 \approx \bar{x}_3$, we terminate the method, calling $\bar{x}_4 = (4, 3)$ the optimal solution.

EXERCISES

Nonlinear Programming Problems

1. Solve graphically:

$$\text{Maximize} \quad z = (x_1 - 1)^2 + (x_2 - 2)^2$$

$$\text{Subject to} \quad \begin{Bmatrix} x_1 + x_2 \leq 6 \\ 3x_1 + 2x_2 \geq 7 \\ x_1, x_2 \geq 0 \end{Bmatrix}$$

2. Solve graphically:

$$\text{Minimize} \quad z = (x_1 - 1)^2 + (x_2 - 2)^2$$

$$\text{Subject to} \quad \begin{Bmatrix} x_1 + x_2 \leq 6 \\ 3x_1 + 2x_2 \geq 7 \\ x_1, x_2 \geq 0 \end{Bmatrix}$$

3. Solve graphically:

$$\text{Maximize} \quad z = 3x_1^2 + 2x_2^2$$

$$\text{Subject to} \quad \begin{Bmatrix} x_1^2 + x_2^2 \leq 9 \\ x_1 + x_2 \leq 3 \\ x_1, x_2 \geq 0 \end{Bmatrix}$$

4. (a) Determine all local maxima, by graphically solving:

$$\text{Maximize} \quad z = x_1 + x_2$$

$$\text{Subject to} \quad \begin{Bmatrix} x_1^3 - 2x_1^2 - x_1 - x_2 \geq -2 \\ |x_1 - 1| \leq 3 \end{Bmatrix}$$

   (b) What is the global maximum?
   (c) Are there any local minima? If so, how many?

Kuhn-Tucker Conditions

5. Verify that the optimal solution to the problem of exercise 1 satisfies the Kuhn-Tucker conditions.

6. Verify that the optimal solution to the problem of exercise 2 satisfies the Kuhn-Tucker conditions.

7. Verify that the optimal solution to the problem of exercise 3 satisfies the Kuhn-Tucker conditions.

8. Do each of the local maxima and local minima for the problem of exercise 4 satisfy the Kuhn-Tucker conditions?

9. What modifications must be made in the Kuhn-Tucker conditions if one of the constraints (8-8) or (8-9) is an equality constraint?

Quadratic Programming

10. Prove Theorem 8.4: The function $f(\bar{x}) = \bar{c}'\bar{x} + \bar{x}'P\bar{x}$ is convex if P is positive semidefinite or positive definite.

11. Show that if P is a negative semidefinite matrix, then, for any positive number ε, the matrix $(P - \varepsilon I)$ is negative definite.

12. Solve by Wolfe's algorithm:

$$\text{Maximize} \quad z = -x_1^2 - x_2^2$$

$$\text{Subject to} \quad \begin{Bmatrix} x_1 + x_2 \leq 8 \\ 3x_1 + 2x_2 \geq 6 \\ x_1, x_2 \geq 0 \end{Bmatrix}$$

13. Solve by Wolfe's algorithm:

$$\text{Maximize} \quad z = 3x_1 + 2x_2 - x_1^2 + 4x_1x_2 - x_2^2$$

$$\text{Subject to} \quad \begin{Bmatrix} x_1 + x_2 \leq 8 \\ 3x_1 + 2x_2 \geq 6 \\ x_1, x_2 \geq 0 \end{Bmatrix}$$

14. Prepare a flow chart of Wolfe's algorithm.

Separable Programming

15. Set up and solve as a separable programming problem the problem of exercise 1. Graphically determine upper bounds for x_1, x_2.

16. Set up and solve the problem of exercise 3 by separable programming methods.

17. Prepare a flow chart for a separable programming algorithm, indicating carefully the modifications in the simplex method necessary to solve the approximating problem.

Method of Griffith and Stewart

18. Solve the problem of exercise 1 by the method of Griffith and Stewart.

19. Solve the problem of exercise 3 by the method of Griffith and Stewart.

20. Solve the problem of exercise 4 by the method of Griffith and Stewart.

21. Prepare a flow chart for the method of Griffith and Stewart. Incorporate a feature allowing for the periodic recalculation of the parameters $m_1, m_2, \ldots, m_n$, when necessary.

REFERENCES

1. Griffith, R. E., and Stewart, R. A.: A nonlinear programming technique for the optimization of continuous processing systems. *Manage. Sci.*, *7*, 379–392 (1961).
2. Hadley, G.: *Nonlinear Programming*. Addison-Wesley, Reading, Massachusetts (1964).
3. Hillier, F. S., and Lieberman, G. J.: *Introduction to Operations Research*. Holden-Day, San Francisco (1967).
4. Kuhn, H. W., and Tucker, A. W.: Nonlinear programming. *In* Neyman, J. (ed.): *Proceedings of the Second Berkeley Symposium on Mathematical Statistics and Probability*, University of California Press, Berkeley, California, 481–492 (1951).
5. Rosen, J. B.: The gradient projection method for nonlinear programming, Part I. Linear constraints. *J. Soc. Ind. & Appl. Math.*, *8*, 181–217 (1960).
6. Rosen, J. B.: The gradient projection method for nonlinear programming, Part II. Nonlinear constraints. *J. Soc. Ind. & Appl. Math.*, *9*, 514–532 (1961).
7. Wolfe, P.: The simplex method for quadratic programming. *Econometrica*, *27*, 382–398 (1959).
8. Zoutendijk, G.: *Methods of Feasible Directions*. Elsevier, Amsterdam (1960).

Chapter 9

INTEGER PROGRAMMING

9.1 INTRODUCTION

Throughout the major portion of this text we have restricted our attention to the theory and the methods of solving optimization problems in which the unknown variables were allowed to take on any values (within the feasible region of the particular problem). That is, these variables were considered to be *continuous*. However, there are many situations in which it would not make sense for these variables to assume other than integer values. For example, if the variables in a given problem represent numbers of men to be hired, or numbers of machines to be purchased, then fractional solutions would be of little use.

One might at first suppose that a logical way to obtain integer solutions to a problem would be to solve it as if it were a continuous problem and then to round the optimal values of the variables to the nearest integer. However, such a procedure is not always valid, since the rounded solution may no longer be feasible, and, even if it is feasible, it may not be optimal. In fact, the rounded solution may turn out to be a rather poor solution, with respect to the true optimal integer solution.

To illustrate let us consider the following problem (Figure 9-1):

$$\begin{aligned} \text{Maximize} \quad & z = 15x_1 + 32x_2 \\ \text{Subject to} \quad & 7x_1 + 16x_2 \leq 52 \\ & 3x_1 - 2x_2 \leq 9 \\ & x_1, x_2 \geq 0 \end{aligned}$$

The optimal solution occurs at $(4, \frac{3}{2})$, with $z = 108$. However, if we

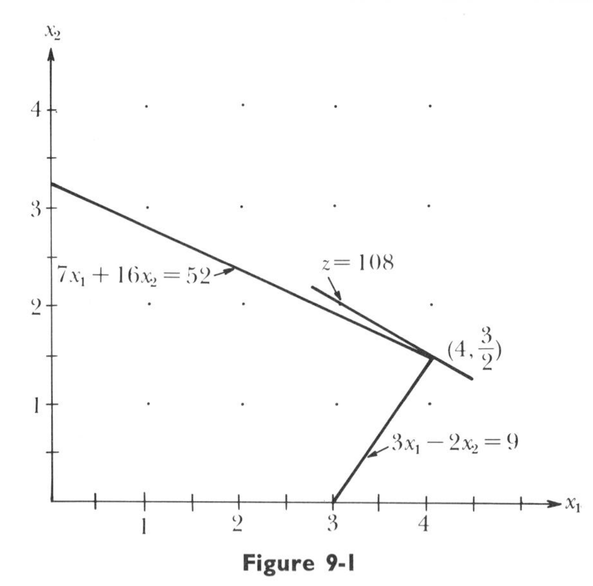

Figure 9-1

require an all-integer solution and attempt to round x_2 to either 1 or 2, the resulting solution is infeasible. The nearest all-integer feasible solution to $(4, \frac{3}{2})$ is $(3, 1)$, with corresponding $z = 77$. The true optimal integer solution, however, is seen to be $(0, 3)$ with $z^* = 96$. Note also that we cannot restrict our attention to extreme points of the set of solutions; in this example, there are only two extreme points with all-integer components: $(0, 0)$ and $(3, 0)$ with corresponding values of z of 0 and 45, respectively. The optimal solution $(0, 3)$ was a boundary point; however, it is also possible that an interior point may be optimal. If, for example, we change the objective function in the problem to

$$z = 3x_1 + 4x_2$$

then the optimal integer solution becomes $(2, 2)$.

The difficulties we encountered in this example illustrate the need for special algorithms for handling integer requirements. In this chapter we shall investigate several methods which have been developed for solving linear programming problems in which some or all of the variables are required to have integral values. Such problems are called *integer linear programming problems*, or sometimes, just *integer programming problems*. If all of the variables are required to be integers, the problem is referred to as an "all-integer" problem; if some of the variables may be nonintegral, the problem is called a "mixed integer-continuous variable problem."

Very little productive research has been done regarding methods of solution for nonlinear integer programming problems; however, the method of Griffith and Stewart can easily be applied to such problems, at least theoretically, since this method only involves solving a sequence of linear programming problems. Instead, these linear programming problems could be solved as integer linear programming problems.

Computational experience with the currently known integer programming methods indicates that problems involving more than 50 or 60 integer variables cannot be solved in a reasonable amount of time, even on the fastest available computers. Occasionally, however, larger problems have been solved. For very large problems, the only practical method of handling integer requirements is to attempt some type of rounding procedure, unsatisfactory as that is.

9.2 FORMULATION OF MODELS AS INTEGER PROGRAMMING PROBLEMS

Before proceeding to investigate some algorithms for solving integer programming problems, let us develop some optimization models which are most readily formulated in terms of integer variables.

Model 1: Capital Budgeting

A manufacturing company has a sum of money, D dollars, available for investment. The company's investment counselor has determined that there are N projects suitable for investment and at least d_j dollars must be invested in project j if it is decided that project j is worthy of investment. The counselor has also determined that the net profit which can be made by investment in project j is P_j dollars. The company's dilemma is that it cannot invest in all N projects, because $\sum_{j=1}^{N} d_j > D$. Thus, the company must decide in which of the projects it wishes to invest, in order to maximize its profit. To solve this problem, the counselor formulates the following problem: Let

$$x_j = \begin{cases} 1, & \text{if the company invests in project } j \\ 0, & \text{if the company does not invest in project } j \end{cases}$$

The total amount that will be invested is then $\sum_{j=1}^{N} d_j x_j$, and since this amount cannot exceed D dollars, we have the constraint

$$\sum_{j=1}^{N} d_j x_j \leq D \tag{9-1}$$

Also, from the definition of the x_j's, we see that we need the following constraints as well:

$$\begin{Bmatrix} 0 \leq x_j \leq 1 \\ x_j \text{ an integer} \end{Bmatrix} \qquad j = 1, 2, \ldots, N \tag{9-2}$$

Moreover, the total profit will be $\sum_{j=1}^{N} P_j x_j$. Thus, the company desires the solution to the problem

$$\text{Maximize } z = \sum_{j=1}^{N} P_j x_j \tag{9-3}$$

subject to the constraints (9-1) and (9-2).

Model 2: Assignment Problems

A certain company manufactures small metal parts, made to order. Currently, it has orders for N different parts. The company has available a total of M men who are each capable of producing any of the parts; however, each man requires a different amount of time in which to produce a given part.

In particular, let us define the following quantities:

d_j = number of parts of type j which have been ordered

x_{ij} = number of parts of type j produced by man i

t_{ij} = amount of time required for man i to produce a part of type j.

Then, if the company wishes to minimize the amount of time it takes to produce all the desired parts, the problem may be formulated as follows:

$$\text{Minimize} \quad z = \sum_{i=1}^{M} \sum_{j=1}^{N} t_{ij} x_{ij}$$

$$\text{Subject to} \quad \sum_{i=1}^{M} x_{ij} \geq d_j \qquad j = 1, 2, \ldots, N$$

$$x_{ij} \text{ a nonnegative integer, all } i \text{ and } j.$$

Suppose, on the other hand, that each man is allowed a total of r_i hours in which to complete his work (e.g., if it is desired that the work be completed in one 8-hour working day, then $r_i = 8$, $i = 1, 2, \ldots, M$) and that the cost of having man i produce a part of type j is C_{ij} dollars (this cost is not necessarily proportional to time, but may also be related to the over-all efficiency of man i). The problem would then become

$$\text{Minimize} \quad z = \sum_{i=1}^{M} \sum_{j=1}^{N} C_{ij} x_{ij}$$

$$\text{Subject to} \quad \sum_{i=1}^{M} x_{ij} \geq d_j \qquad j = 1, 2, \ldots, N$$

$$\sum_{i=1}^{N} t_{ij} x_{ij} \leq r_i \qquad i = 1, 2, \ldots, M$$

$$x_{ij} \text{ a nonnegative integer, all } i \text{ and } j$$

There are numerous additional features which can be added to the model, such as incorporating the extra cost of overtime, the time limitations of the available machinery required to produce each of the parts, and so forth.

Model 3: The Traveling Salesman Problem

A salesman for a small chemical firm has been given a list of N cities, each of which he must visit exactly once, starting from and returning to the company's corporate headquarters (hereinafter called city 1). The salesman would like to plan his route so that the traveling costs are a minimum. Clearly, there are $(N - 1)!$ possible routes. This problem, usually referred to as "The Traveling Salesman Problem," is extremely easy to describe verbally, but actually is very difficult to solve. We shall present an integer programming formulation. However, because of the large numbers of constraints and variables required by this formulation, more efficient methods of solution have been devised. Several of these are discussed in Section 9.5.

Let us define the following quantities:

$$C_{ij} = \text{cost of traveling from city } i \text{ to city } j.$$

$$x_{ijk} = \begin{cases} 1, & \text{if the salesman travels from city } i \text{ to city } j \text{ and city } j \text{ is the } k^{\text{th}} \text{ city visited} \\ 0, & \text{otherwise} \end{cases}$$

Thus, for example, if there are a total of 5 cities, and the salesman's route is from city 1 to cities 3, 5, 4, 2, and back to city 1, then $x_{131} = x_{352} = x_{543} = x_{424} = x_{215} = 1$, all other x_{ijk}'s $= 0$.

Now, let us formulate the constraints. Since a city j must be visited once and only once, it must be true that

$$x_{1j1} + \sum_{i=1}^{N} \sum_{k=2}^{N-1} x_{ijk} = 1 \qquad j = 2, 3, \ldots, N \tag{9-4}$$

Also, since the last city on the route is city 1, it must be true that

$$\sum_{i=2}^{N} x_{i1N} = 1 \tag{9-5}$$

In addition, the first city on the route is also city 1; thus,

$$\sum_{j=2}^{N} x_{1j1} = 1 \tag{9-6}$$

the constraint (9-6) merely states that exactly one of the $N - 1$ cities $j = 2, 3, \ldots, N$ is visited from city 1. Since each city i is visited once and only once, we must also require that

$$x_{i1N} + \sum_{j=2}^{N} \sum_{k=2}^{N-1} x_{ijk} = 1 \qquad i = 2, 3, \ldots, N \tag{9-7}$$

Moreover, each city (except city 1) is a candidate for being the kth city visited. Consequently,

$$\sum_{i=1}^{N} \sum_{j=2}^{N} x_{ijk} = 1 \qquad k = 2, \ldots, N - 1 \tag{9-8}$$

(City 1 must be the Nth city visited; this fact is ensured by constraint (9-5).)

Thus far, we have constructed constraints which ensure that each city will be visited exactly once and that city 1 will be the first and last city on the route. However, these constraints allow the possibility that a *subtour* can occur; a subtour is a disconnected trip. In Figure 9-2, a disconnected trip is shown for a six-city problem. In this example, the "route" is from city 1 to city 2, then from city 3 to city 4, from city 2 to city 6, from city 4 to city 5, from city 5 to city 3, and finally, from city 6 to city 1 (Figure 9-2). The corresponding values of the variables x_{ijk} are

$$x_{121} = x_{262} = x_{616} = x_{343} = x_{454} = x_{535} = 1$$

$$\text{all other } x_{ijk}\text{'s} = 0$$

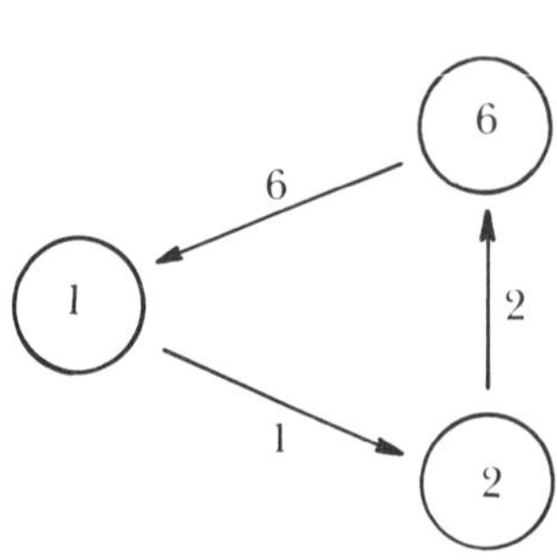

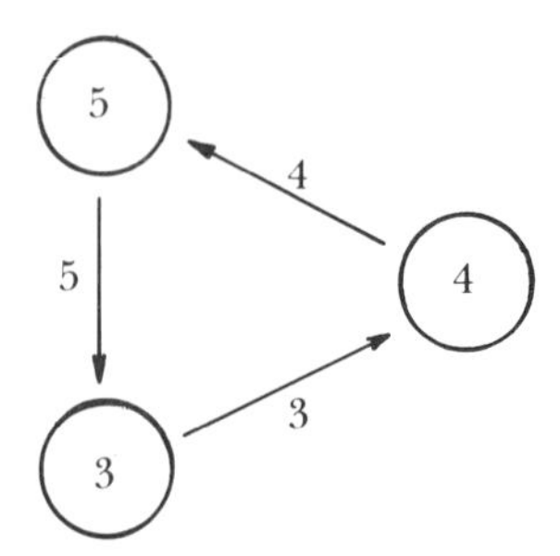

Figure 9-2

These variables satisfy the constraints (9-4) through (9-8); however, it is evident that we wish to exclude such cases from consideration. This can be done by formulating the following observation as a set of constraints: If city j is visited as the kth city on the route, then the $(k+1)^{\text{st}}$ city to be visited must be visited from city j. This implies that

$$\sum_{i=1}^{N} x_{ijk} = \sum_{p=1}^{N} x_{jp,k+1} \qquad j = 2, \ldots, N \qquad k = 2, \ldots, N-1 \tag{9-9}$$

$$x_{1j1} = \sum_{p=1}^{N} x_{jp2} \qquad j = 2, \ldots, N \tag{9-10}$$

$$x_{j1N} = \sum_{i=1}^{N} x_{ij,N-1} \qquad j = 2, \ldots, N \tag{9-11}$$

The constraints (9-4) through (9-11) completely describe the model, as long as it is also required that each x_{ijk} be a nonnegative integer. The objective function, of course, is

$$\text{Minimize} \quad z = \sum_{i=1}^{N} \sum_{j=1}^{N} \sum_{k=1}^{N} C_{ij} x_{ijk} \tag{9-12}$$

These are a total of $N^2 + 2N - 4$ constraints. For this reason, as indicated earlier, the traveling salesman problem cannot be solved by conventional integer programming algorithms for N larger than about 5 or 6.

Model 4: "Either-Or" Constraints

A bakery sells eight varieties of doughnuts. The preparation of varieties 1, 2, and 3 involves a rather complicated process, and so the bakery has decided that it would rather not bake these varieties unless it can bake and sell at least 10 dozen doughnuts of varieties 1, 2, and 3, combined. Suppose also that the capacity of the bakery prohibits the total number of doughnuts baked from exceeding 30 dozen, and that the per unit profit for a variety j doughnut is P_j dollars. If we let x_j, $j = 1, 2, \ldots, 8$, denote the number of dozens of doughnut variety j to be baked, then the maximum profit can be found by solving the following problem (assuming the bakery can sell everything it bakes):

$$\text{Maximize} \quad z = \sum_{j=1}^{8} P_j x_j$$

$$\text{Subject to} \quad \sum_{j=1}^{8} x_j \leq 30 \tag{9-13}$$

$$x_1 + x_2 + x_3 = 0 \quad \text{or} \quad x_1 + x_2 + x_3 \geq 10 \tag{9-14}$$

$$x_j \geq 0 \qquad j = 1, 2, \ldots, 8$$

The "either-or" constraint (9-14) may be handled in the following manner. Let y be an integer variable. Then consider the two constraints

$$x_1 + x_2 + x_3 \leq 30y \tag{9-15}$$

$$x_1 + x_2 + x_3 \geq 10y \tag{9-16}$$

If $y = 1$, then constraints (9-14) and (9-15) become, respectively,

$$x_1 + x_2 + x_3 \leq 30 \tag{9-17}$$

$$x_1 + x_2 + x_3 \geq 10 \tag{9-18}$$

but (9-17) is redundant, by virtue of (9-13). On the other hand, if $y = 0$, then the constraints (9-14) and (9-15) combine to yield $x_1 + x_2 + x_3 = 0$. Thus, we can now formulate the problem as

$$\text{Maximize} \quad z = \sum_{j=1}^{8} P_j x_j$$

$$\text{Subject to} \quad \sum_{j=1}^{8} x_j \leq 30$$

$$x_1 + x_2 + x_3 - 30y \leq 0$$

$$x_1 + x_2 + x_3 - 10y \geq 0$$

$$0 \leq y \leq 1$$

$$y \text{ an integer}$$

$$x_j \geq 0 \qquad j = 1, 2, \ldots, 8$$

9.3 GOMORY'S ALGORITHM

Let us now investigate several algorithms which have been developed to solve integer programming problems. Much of the early work in this area was done by Gomory.[6,7,8] In this section we shall examine one version of Gomory's algorithm for the all-integer problem. In particular, we shall be considering the following problem:

$$\text{Maximize} \quad z = \bar{c}'\bar{x} \tag{9-19}$$

$$\text{Subject to} \quad \begin{Bmatrix} A\bar{x} = \bar{b} \\ \bar{x} \geq \bar{0} \end{Bmatrix} \tag{9-20}$$

$$x_j \text{ an integer} \qquad j = 1, 2, \ldots, n \tag{9-21}$$

Without the constraints (9-21), this problem is a linear programming problem; we shall refer to this linear problem as "the related LP problem" to the integer linear programming problem (ILP).

The basic idea in Gomory's algorithm is the following: Given an integer programming problem, solve the related LP problem. If the optimal solution contains only integer-valued variables, then clearly it must be the optimal ILP solution. If some variables in the optimal LP solution are not integers, then a new constraint (to be discussed later) is added to the problem and the new related LP problem is solved. Again, the optimal solution to this new LP problem is checked to see if all its components are integers; if so, it is the optimal ILP solution for the *original* problem. If some of the variables are not integers, then another new constraint is added and the process is repeated, until one of the related LP problems has an all-integer optimal solution; the first such all-integer solution obtained is the optimal solution for the original problem.

In order for this procedure to be valid, the new constraint which is added at each stage must possess the following properties:

1. Every *integer* feasible solution to the original problem must also be a feasible solution to the new problem after the addition of the constraint.
2. The optimal solution to the LP problem solved at each stage must become infeasible after the new constraint is added.

Property (1) states that the addition of the new constraint does not eliminate any of the integer solutions from the set of feasible solutions, thus ensuring that the optimal integer solution will not become infeasible.

Property (2) is necessary, because if the addition of the constraint did not eliminate the optimal LP solution, then the latter would also be optimal for the new LP problem. Since this solution is not an all-integer solution, we would have gained nothing by adding that particular constraint.

We noted in Section 9.1 that the optimal solution to an integer programming problem might be an interior point of the convex set of feasible continuous solutions. Thus, if we were only investigating extreme points of this set, we would not be very likely to find the optimal solution. However, the addition of constraints to this set of solutions creates a new convex set of solutions. In effect, we are trying to add constraints, one at a time, in such a manner that eventually the optimal integer solution will become an extreme point of the resulting new set of feasible solutions. Note that with the addition of each constraint, the basis size for the resulting LP problem increases by one and hence the number of variables which are allowed to be positive at the same time also increases by one. This is, then, a systematic procedure for examining interior points of the original set of solutions.†

† Under certain conditions it is possible to delete a constraint which had been added previously. In Hadley[9] it is shown that it is never necessary to have more than $n + 1$ constraints at one time. This fact makes sense intuitively since the original problem only had n variables, all of which could be positive in a basic feasible solution for a set of $n + 1$ constraints.

Let us now investigate the nature of these constraints which are to be added. Each such constraint is often referred to as a "Gomory cut," since it is essentially a hyperplane which "cuts off" part of the convex set of feasible solutions, forming a new smaller convex set. Let us assume for simplicity that the optimal solution to the related LP problem contains the first m columns of A in the basis; partitioning A into two submatrices yields

$$A = [B, R] \tag{9-22}$$

where B contains the first m columns of A (the basis matrix) and R contains the remaining $(n - m)$ columns of A. The corresponding optimal solution, then, is $\bar{x}^* = [\bar{x}_B^*, \bar{x}_R^*]$ and $\bar{x}_R^* = \bar{0}$.

Now, since $A\bar{x} = \bar{b}$, we have

$$[B, R]\begin{bmatrix} \bar{x}_B^* \\ \bar{x}_R^* \end{bmatrix} = \bar{b}$$

or

$$B\bar{x}_B^* + R\bar{x}_R^* = \bar{b} \tag{9-23}$$

Since B^{-1} exists, we can rewrite equation (9-23) as

$$\bar{x}_B^* = B^{-1}\bar{b} - (B^{-1}R)\bar{x}_R^* \tag{9-24}$$

Furthermore, since R consists of the columns of A, we can express equation (9-24) in terms of the columns of the simplex tableau; namely, letting $\bar{y}_j = B^{-1}\bar{a}_j, j = 1, 2, \ldots, n$, and $\bar{y}_0 = B^{-1}\bar{b}$, we have

$$\bar{x}_B^* = \bar{y}_0 - \sum_{j \in R} \bar{y}_j x_j \tag{9-25}$$

where the summation in equation (9-25) is taken over those j whose corresponding $\bar{a}_j$ is a column of R (i.e., those x_j which are nonbasic variables).

Let us suppose, now, that the rth basic variable of $\bar{x}_B^*$ is not an integer. From (9-25), we see that the equation which contains this variable is

$$x_{Br} = y_{r0} - \sum_{j \in R} y_{rj} x_j \tag{9-26}$$

In the current solution, all $x_j, j \in R$, are zero. Thus, y_{r0} is not an integer, since currently $x_{Br} = y_{r0}$. Let us express y_{r0} as the sum of its integer part and its fractional part. Let w_{r0} be the integer (or "whole") part of y_{r0} (i.e., the greatest integer less than or equal to y_{r0}) and let f_{r0} be the remaining fractional part. Thus,

$$y_{r0} = w_{r0} + f_{r0} \tag{9-27}$$

and

$$w_{r0} \geq 0 \tag{9-28}$$

$$0 < f_{r0} < 1 \tag{9-29}$$

Let us similarly represent the y_{rj}, $j \in R$:

$$y_{rj} = w_{rj} + f_{rj} \tag{9-30}$$

$$0 \leq f_{rj} < 1 \tag{9-31}$$

If we substitute equations (9-27) and (9-30) into equation (9-26), we obtain

$$\begin{aligned} x_{Br} &= w_{r0} + f_{r0} - \sum_{j \in R} (w_{rj} + f_{rj}) x_j \\ &= \left(w_{r0} - \sum_{j \in R} w_{rj} x_j\right) + \left(f_{r0} - \sum_{j \in R} f_{rj} x_j\right) \end{aligned} \tag{9-32}$$

Now, if we wish to modify the current solution, so that the new solution is all-integer, then at least one of the x_j, $j \in R$, must become positive (since there is a unique solution with all these $x_j = 0$, and it is $\bar{x}_B^*$). In such an all-integer solution, the quantity contained in the first pair of parentheses in equation (9-32) must be an integer (positive or negative), since each x_j must be an integer and each w_{rj} is, by definition, an integer. Thus, the new value of x_{Br}, which we also desire to be an integer, can be expressed as

$$x_{Br} = \text{integer} + \left(f_{r0} - \sum_{j \in R} f_{rj} x_j\right) \tag{9-33}$$

If x_{Br} is to be an integer, then, the quantity

$$f_{r0} - \sum_{j \in R} f_{rj} x_j \tag{9-34}$$

must be an integer. However, each f_{rj} is nonnegative and each x_j is also nonnegative; thus

$$\sum_{j \in R} f_{rj} x_j \geq 0 \tag{9-35}$$

Moreover, from (9-29) and (9-34), we see subtraction of the nonnegative quantity (9-35) from the positive fraction f_{r0} will yield an integer quantity only if

$$f_{r0} - \sum_{j \in R} f_{rj} x_j \leq 0 \tag{9-36}$$

Let us examine (9-36) more closely. Firstly, recall that the current optimal LP solution $[\bar{x}_B^*, \bar{x}_R^*]$ has $\bar{x}_R^* = \bar{0}$, and hence this solution does not satisfy (9-36). Secondly, by our derivation of (9-36), every integer feasible solution to the original problem will satisfy (9-36). Thus, (9-36) possesses the two properties mentioned earlier in this section. Constraint (9-36), then, is the "Gomory cut."

In our discussion, we did not consider how to determine the Gomory cut if more than one basis variable is not an integer. There are many ways to do this; in the most frequently employed technique, the x_{Br} whose fractional part f_{r0} is the largest is used to form the constraint (9-36).

Gomory has proven that this algorithm (when the cuts (9-36) are chosen in a way slightly different from that described in the preceding paragraph) will converge in a finite number of iterations to the optimal integer solution.

Computational experience with the method indicates that it is capable of solving some reasonably large ILP problems; however, it has also been known (see Hadley[9]) to run for over 2000 iterations without converging to the optimal ILP solution on problems which require less than twenty iterations to obtain the optimal solution for the related LP problem.

Gomory[8] has extended the algorithm to the case of the mixed integer-continuous variable problem. The primary computational difference is in the definition of the cut, which is modified slightly.

Many researchers have developed different "cuts" from (9-36) which are "stronger" than Gomory's cut, in the sense that they cut out a larger portion of the feasible region. However, none has proved to be of any particular computational advantage.

We shall close this section by graphically illustrating Gomory's algorithm. Let us consider the following problem, illustrated in Figure 9-3:

$$\text{Maximize} \quad z = x_1 - x_2$$

$$\text{Subject to} \quad x_1 + 2x_2 \leq 4 \tag{9-37}$$

$$6x_1 + 2x_2 \leq 9 \tag{9-38}$$

$$x_1, x_2 \geq 0$$

$$x_1, x_2 \text{ integer}$$

The optimal solution to the related LP problem occurs at $(x_1, x_2) = (\frac{3}{2}, 0)$. The optimal simplex tableau is given in Table 9.1

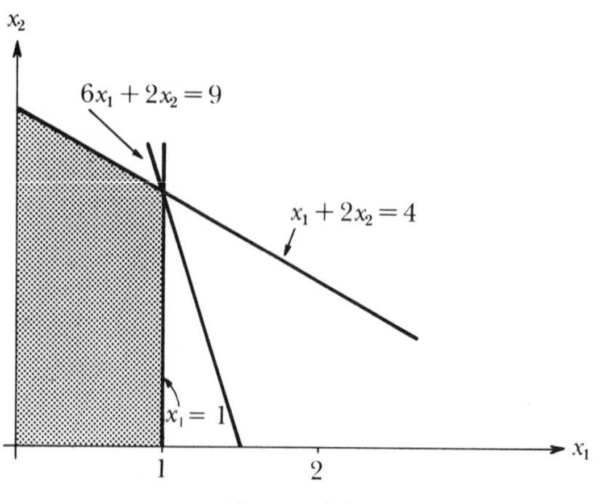

Figure 9-3

Table 9.1

Variables in Basis	$\bar{a}_1$	$\bar{a}_2$	$\bar{a}_3$	$\bar{a}_4$
$x_{B1} = x_1 = \frac{3}{2}$	1	$\frac{2}{6}$	0	$\frac{1}{6}$
$x_{B2} = x_3 = \frac{5}{2}$	0	$\frac{10}{6}$	1	$-\frac{1}{6}$
$z = \frac{3}{2}$	0	$\frac{8}{6}$	0	$\frac{1}{6}$

(where x_3 and x_4 are the slack variables added to constraints (9-37) and (9-38), respectively). Since both x_{B1} and x_{B2} are noninteger, we may use either in the calculation of the Gomory cut. We shall choose x_{B1}. From Table 9.1, we see that $y_{12} = \frac{2}{6}$ and $y_{14} = \frac{1}{6}$. Thus, equation (9-36) becomes

$$\tfrac{1}{2} - [(\tfrac{2}{6})x_2 + (\tfrac{1}{6})x_4] \leq 0$$

or

$$2x_2 + x_4 \geq 3 \tag{9-39}$$

However, since x_4 is the slack variable for (9-38), we have

$$x_4 = 9 - 6x_1 - 2x_2 \tag{9-40}$$

Substitution of equation (9-40) into (9-39) yields

$$2x_2 + (9 - 6x_1 - 2x_2) \geq 3$$

which reduces to

$$x_1 \leq 1 \tag{9-41}$$

It should be noted that the constraints (9-39) and (9-41) are completely equivalent; we have calculated (9-41) so that we may graph this example in terms of x_1 and x_2.

The optimal solution to the LP problem with the new constraint (9-41) is easily seen to occur at $(x_1, x_2) = (1, 0)$, which is the desired integer solution.

Suppose, instead, that we chose to calculate the Gomory cut using x_{B2}. From Table 9.1 we see that $y_{22} = \frac{10}{6}$, $y_{24} = -\frac{1}{6}$. Thus, $f_{22} = \frac{4}{6}$, and $f_{24} = \frac{5}{6}$ (since f_{rj} must be nonnegative, and $y_{24} = -\frac{1}{6} = -1 + \frac{5}{6}$). The new constraint would then have been

$$\tfrac{1}{2} - [(\tfrac{4}{6})x_2 + (\tfrac{5}{6})x_4] \leq 0$$

or

$$4x_2 + 5x_4 \geq 3 \tag{9-42}$$

Upon substituting equation (9-40) into (9-42) we obtain

$$4x_2 + 5(9 - 6x_1 - 2x_2) \geq 3$$

$$-30x_1 - 6x_2 \geq -42$$

$$5x_1 + x_2 \leq 7 \tag{9-43}$$

If the foregoing constraint is added to the original problem (instead of (9-41)), the resulting LP problem has an optimal solution of $(x_1, x_2) = (\frac{7}{5}, 0)$, which is not all integer. Thus, the choice of the constraints to be added affects the efficiency of the algorithm. Unfortunately, there is no *a priori* method of determining the best cut. The fact that this example was solved utilizing only one Gomory cut should not tempt the reader into thinking that the method will always work so well. A reader of the optimistic variety is invited to try solving the example of Section 9.1 with this algorithm.

There are several computational details of the algorithm which we will now mention: Recall that each time a new constraint of the form (9-36) is added to create a new problem, the current optimal solution becomes infeasible. However, only one variable is infeasible, and it is the surplus variable corresponding to the constraint (9-36). Initially, this surplus variable, call it y_1, is equal to the negative of f_{r0} (since all x_j, $j \in R$, are zero). Moreover, none of the $z_j - c_j$ are changed, since the cost corresponding to y_1 is zero. Hence, the solution still has all $z_j - c_j \geq 0$. The dual simplex method can thus be used to determine the new optimal solution. The initial tableau for this new problem is easy to construct.

9.4 THE BOUNDED VARIABLE ALGORITHM

In the previous section we mentioned several disadvantages of Gomory's algorithm. However, its main disadvantage is that until it terminates with the optimal integer solution, it does not yield *any* integer feasible solutions. Thus, in a problem in which the algorithm has performed a large number of iterations and still has not converged to the optimal solution, one cannot prematurely stop the calculations and settle for having a "reasonably good," but not necessarily optimal, integer solution. Computational experience with numerous integer programming algorithms indicates that the amount of time required to completely solve a given problem cannot be predicted with any reasonable degree of accuracy; rather, this time varies substantially from problem to problem, even for problems involving the same numbers of variables and constraints.

Thus, a desirable feature of an integer programming algorithm is the ability to rapidly obtain a feasible integer solution with a relatively good value of the objective function, even if this solution is not necessarily the optimal solution. Such algorithms have been developed by, among others, Balas,[2] Glover,[5], Krolak,[11] and Land and Doig.[12] In this section, we shall discuss Krolak's algorithm, called the Bounded Variable Algorithm. This algorithm is conceptually quite simple, and computational experience indicates that it is at least as good as the other generally used algorithms.

Consider the following integer programming problem:

$$\text{Maximize} \quad z = \bar{c}'\bar{x} \tag{9-44}$$

$$\text{Subject to} \quad A\bar{x} \leq \bar{b} \tag{9-45}$$

$$0 \leq L_j \leq x_j \leq U_j \qquad j = 1, 2, \ldots, n \tag{9-46}$$

$$x_j \text{ integer}, \qquad j = 1, 2, \ldots, n \tag{9-47}$$

(As usual, A is an $m \times n$ matrix, and $\bar{c}$ and $\bar{b}$ are n and m component vectors, respectively.) We shall also assume that every component of $\bar{c}$ is an integer.

Suppose, now, that we solve two linear programming problems:

1. Maximize $z = \bar{c}'\bar{x}$

   Subject to (9-45) and (9-46)

2. Minimize $z = \bar{c}'\bar{x}$

   Subject to (9-45) and (9-46)

The optimal values of z for these two problems are not necessarily integers, since the optimal solutions are not necessarily integers. However, the optimal value of z for the integer programming problem (9-44) through (9-47), z^*, will be an integer, since every c_j and x_j are integers. If we let z_U be the value of the optimal z to problem 1, rounded down to the nearest integer, and let z_L be the value of the optimal z to problem 2, rounded up to the nearest integer, then clearly

$$z_L \leq z^* \leq z_U \tag{9-48}$$

In fact, given any feasible integer solution $\hat{\bar{x}}$, with corresponding $\hat{z}$, it must be true that

$$z_L \leq \hat{z} \leq z^* \leq z_U \tag{9-49}$$

Thus, if we have somehow obtained such an integer solution $\hat{\bar{x}}$, we would then like to be able to answer:

1. Does there exist an integer solution with a larger value of z than $\hat{z}$? If not, then $\hat{\bar{x}}$ is the optimal solution.
2. If such an integer solution does exist, then what is it?

Any algorithm which is capable of answering these questions can solve integer programming problems. To see this, let us formulate the structure of such an algorithm.

Observe that once we have obtained an integer solution $\hat{\bar{x}}$ with corresponding $\hat{z}$, then we are no longer interested in integer solutions with values of z smaller than $\hat{z}$. In order to eliminate such solutions from consideration, we add the following constraint to the problem

$$\bar{c}'\bar{x} \geq \hat{z} \tag{9-50}$$

Each time we obtain a better integer solution, we replace the right-hand side of (9-50) with the new value of z. Initially, we may use (z_L) as the right-hand side of (9-50).

The step-by-step procedure is given below.

Algorithm

1. Find a feasible integer $\hat{\bar{x}}$ to the integer programming problem (9-44) through (9-47) and (9-50) or show that none exists. If such an $\hat{\bar{x}}$ is found, with corresponding $\hat{z}$, go to Step 2. If no such $\hat{\bar{x}}$ exists, go to Step 3.

2. Replace the right-hand side of (9-50) with $(\hat{z} + 1)$. If $\hat{z} = z_U$, then $\hat{\bar{x}}$ is the optimal solution; go to Step 4. If $\hat{z} < z_U$, we don't yet know whether or not $\hat{\bar{x}}$ is optimal. Return to Step 1 (note that the new constraint $\bar{c}'\bar{x} \geq (\hat{z} + 1)$ forces the current best found solution, $\hat{\bar{x}}$, to be infeasible; thus, in re-returning to Step 1, we are attempting to find a better solution; since z will be an integer for any integer solution, we know that any such better solution must have a value of z which is at least one unit greater than $\hat{z}$).

3. No feasible integer solution exists which satisfies all the constraints; if we have not yet found any feasible integer solution, then there is no integer solution to the original problem. Go to Step 5. If, on the other hand, we had previously found a feasible integer $\hat{\bar{x}}$, then the constraint which now cannot be satisfied must be (9-50): $\bar{c}'\bar{x} \geq \hat{z} + 1$. Thus, $\hat{\bar{x}}$ is the optimal solution. Go to Step 4.

4. The optimal solution has been found.

5. No feasible integer solution exists for the original problem.

Observe that this algorithm requires only a finite number of operations, provided that the operations required to carry out Step 1 are finite; every time a new feasible solution is found, the right-hand side of (9-50) is increased a positive amount; sooner or later the process must terminate because we have determined that no feasible solution exists with a value of z larger than the current $\hat{z}$, or because $\hat{z} = z_U$.

We have not yet discussed how we are to carry out Step 1; this is the crucial step in the algorithm. The Bounded Variable Algorithm provides one method for doing so.

Since Step 1 calls for the determination of whether or not a feasible solution exists, it might appear that an exhaustive search of all possible combinations of integer values of the x_j's, within the bounds (9-46) would be necessary in order to determine whether or not a feasible solution does in fact exist. Such a procedure is of course finite. However, it is obvious that it would be computationally impossible to calculate all such combinations, unless the bounds (9-46) yield sufficiently small intervals for each x_j. In the Bounded Variable Algorithm a method is given for successively reducing these intervals by periodic recalculation of the bounds.

To see how this is done, let us consider the ith constraint of (9-45):

$$\sum_{k=1}^{n} a_{ik}x_k \leq b_i \tag{9-51}$$

For at least one j, $j = 1, 2, \ldots, n$, $a_{ij} \neq 0$. There are two cases; $a_{ij} < 0$ or $a_{ij} > 0$. Let us rewrite (9-51) as follows:

$$\sum_{\substack{k=1 \\ k \neq j}}^{n} a_{ik}x_k + a_{ij}x_j \leq b_i \tag{9-52}$$

We now define the following two sets:

$$J_i^P = \{k \mid a_{ik} > 0, k \neq j\}$$

$$J_i^N = \{k \mid a_{ik} < 0, k \neq j\}$$

then, (9-52) may be written as

$$\sum_{k \in J_i^P} a_{ik}x_k + \sum_{k \in J_i^N} a_{ik}x_k + a_{ij}x_j \leq b_i \tag{9-53}$$

Case 1: $a_{ij} > 0$

Solving (9-53) for x_j yields

$$x_j \leq \left[b_i - \sum_{k \in J_i^P}^{n} a_{ik}x_k - \sum_{k \in J_i^N} a_{ik}x_k \right] \Big/ a_{ij} \tag{9-54}$$

Since all the terms in the first summation are nonnegative, and since $L_j \leq x_j$, $j = 1, 2, \ldots, n$, the following inequality must hold:

$$-\sum_{k \in J_i^P} a_{ik}x_k \leq -\sum_{k \in J_i^P} a_{ik}L_k \tag{9-55}$$

Similarly, we obtain

$$-\sum_{k \in J_i^N} a_{ik}x_k \leq -\sum_{k \in J_i^N} a_{ik}U_k \tag{9-56}$$

Substitution of the inequalities (9-55) and (9-56) into (9-54) yields

$$x_j \leq \left[b_i - \sum_{k \in J_i^P} a_{ik}L_k - \sum_{k \in J_i^N} a_{ik}U_k \right] \Big/ a_{ij} \tag{9-57}$$

Thus, if $a_{ij} > 0$, the ith constraint provides an upper bound for x_j. Let us denote the right-hand side of (9-57) by U_{ij}. U_{ij}, then, is an upper bound on x_j obtained from the ith constraint. For every constraint for which $a_{ij} > 0$, we can obtain such an upper bound; x_j must simultaneously satisfy each of the resulting inequalities (9-57). In addition, of course, x_j must be less than or equal to U_j. Therefore, a new upper bound on x_j, denoted by $\hat{U}_j$, can be defined by

$$\hat{U}_j = \text{Min}\ \{\text{Min}\ [U_{ij}],\ U_j\} \tag{9-58}$$

where the Min $[U_{ij}]$ is taken over those i for which $a_{ij} > 0$.

Case 2: $a_{ij} < 0$

If $a_{ij} < 0$, when (9-53) is solved for x_j, the sense of the inequality sign becomes reversed. Thus, we obtain

$$x_j \geq \left[b_i - \sum_{k \in J_i^P} a_{ik}x_k - \sum_{k \in J_i^N} a_{ik}x_k\right] \Big/ a_{ij} \tag{9-59}$$

Also, it must be true that

$$-\left(\frac{1}{a_{ik}}\right) \sum_{k \in J_i^P} a_{ik}x_k \geq -\left(\frac{1}{a_{ij}}\right) \sum_{k \in J_i^P} a_{ik}L_k$$

$$-\left(\frac{1}{a_{ij}}\right) \sum_{k \in J_i^N} a_{ik}x_k \geq -\left(\frac{1}{a_{ij}}\right) \sum_{k \in J_i^N} a_{ik}U_k$$

Substitution of these inequalities into (9-59) yields

$$x_j \geq \left(b_i - \sum_{k \in J_i^P} a_{ik}L_k - \sum_{k \in J_i^N} a_{ik}U_k\right) \Big/ a_{ij} \tag{9-60}$$

As before, an inequality of the form (9-60) is obtained for each $a_{ij} < 0$. Thus, a new lower bound $\hat{L}_j$ for x_j can be obtained from

$$\hat{L}_j = \text{Max}\ \{\text{Max}\ [L_{ij}],\ L_j\} \tag{9-61}$$

where each L_{ij} represents the right-hand side of (9-60) and the Max $[L_{ij}]$ is taken over those i for which $a_{ij} < 0$.

We state now the procedure employed by the Bounded Variable Algorithm for performing the calculations required in Step 1 of the algorithm described earlier in this section.

Procedure for Step 1

(a) For each variable x_j, let F_j be the set of all possible feasible values for x_j; thus,

$$F_j = \{k \mid k = L_j, L_j + 1, L_j + 2, \ldots, U_j\}$$

(b) Pick an integer from F_1, say k_1. Set $x_1 = k_1$.

(c) Setting $\hat{L}_1 = \hat{U}_1 = k_1$, use (9-58) and (9-61) to obtain new upper and lower bounds for x_2. If $L_2 > U_2$, then no feasible solution exists with $x_1 = k_1$ (i.e., there are no feasible values for x_2). Go to

Step (e). If $\hat{L}_2 \leq \hat{U}_2$, then select some value from F_2, say k_2, and set $x_2 = k_2$. Go to Step (d).

(d) Continue as in Step (c), generating, in succession, a set F_j for each variable x_j, setting x_j equal to some integer from the set F_j, and calculating new bounds for x_{j+1}. These new bounds, $\hat{L}_{j+1}$ and $\hat{U}_{j+1}$, are calculated by setting both the lower and upper bounds for variables $x_1, \ldots, x_j$, equal to the currently set values of these variables. Eventually, either a feasible solution is found (if all n variables have been set) or for some variable, say x_k, the calculation of new bounds yields $\hat{L}_k > \hat{U}_k$. This latter case implies that no feasible solution exists with variables $x_1, \ldots, x_{k-1}$ set at their current values. In this case, go to Step (e). If a feasible solution has been found, go to Step (g).

(e) Since $\hat{L}_k > \hat{U}_k$ for some variable x_k, we must change the value of at least one other variable if we are to find a feasible solution. We shall change the $(k - 1)^{\text{st}}$ variable. First, we shall delete from F_{k-1} the current value of x_{k-1}; then, set x_{k-1} equal to another integer from F_{k-1}, provided that F_{k-1} still contains at least one integer. If it does not, go to Step (f); otherwise, return to Step (d).

(f) Some set F_{k-1} was found to be empty. If $k > 2$, then there is no feasible solution having the currently assigned values of $x_1, x_2, \ldots, x_{k-2}$, since this combination of values has been tried for all possible values of x_{k-1} and has failed to yield a solution. Reduce k by one and return to Step (e). If $k \leq 2$, however, then no possible solution exists, since for all possible values for x_1, no feasible solution could be found. In this case, go to Step (h).

(g) We have found a feasible solution.

(h) We have determined that no feasible solution exists.

We have now completely described the basic structure of the Bounded Variable Algorithm. However, a few additional remarks are desirable.

First, note that nothing has been said about the *order* in which elements from each set F_j are selected in Steps (b), (c), and (d). Obviously, this order can have an enormous effect on the efficiency of the algorithm. Krolak[11] found that the best selection rule was as follows: If x_j^* denotes the optimal value of x_j for the *continuous* linear programming problem, then pick either the next highest integer or the next lowest integer to x_j^* (if x_j^* is not an integer) as the first choice for x_j. Choose the other of these two values second. For the succeeding choices, alternate between the largest integer in F_j less than x_j^* and the smallest integer in F_j greater than x_j^*.

For the last variable, x_n, a better choice can be made. Since all of the variables $x_1, \ldots, x_{n-1}$ have been set, the newly calculated upper and lower bounds for x_n are exact. In other words, any integer value for x_n, $\hat{L}_n \leq x_n \leq \hat{U}_n$, will yield a feasible solution. Thus, we choose $x_n = \hat{U}_n$ if $c_n > 0$ and $x_n = \hat{L}_n$ if $c_n < 0$; this will yield the largest possible value of z, given the current values for $x_1, \ldots, x_{n-1}$.

These selection rules are designed to obtain a "good" feasible integer solution as soon as possible. This is important, because if this solution is near optimal (its value of z is close to z^*), then the addition of the constraint (9-50) will result in a relatively small convex set; hence, the sets F_j should become relatively small. Moreover, if we wish to terminate the algorithm before it has completed its calculations, we will at least have obtained a reasonably good feasible integer solution; and, since we have an upper bound on z^*, we can estimate how good our solution is. It is likely, in fact, that the true optimal solution will be found in a reasonably short amount of time. The most time-consuming aspect of the algorithm occurs once the optimal solution has been found, when the algorithm is attempting to find a better solution.

9.5 BRANCH AND BOUND METHODS

Many of the most recently developed algorithms for solving integer programming problems employ a technique that has come to be called a "branch and bound method," or a "truncated enumeration method." A branch and bound approach can be applied to any optimization problem in which there are a finite number of feasible solutions which must be examined in order to determine the optimal solution. Thus, any bounded integer programming problem may be solved by this technique.

As the reader should by now be fully aware, the fact that, in some given optimization problem, we need examine a mere finite number of solutions to find the optimum is often of little help in solving the problem, since the number of such solutions usually increases combinatorially or exponentially as the number of variables increases. For example, we have seen that the total possible number of basic feasible solutions to a linear programming problem is $\binom{n}{m}$, where n is the number of variables and m the number of constraints. Consider now an integer programming problem in which all the n variables must be either 0 or 1; the total possible number of feasible integer solutions is 2^n, which increases exponentially with n.

In a branch and bound method, the total set of solutions under consideration is systematically subdivided into smaller sets, in such a way that large subsets of solutions may be discarded from consideration without each solution in these subsets having to be examined. If the method is to be successful, only a very small fraction of the total number of solutions will have to be examined before the optimal solution is determined.

The branch and bound approach has proven to be most successful in solving very special types of integer programming problems, such as the traveling salesman problem. In fact, the only known methods for solving large traveling salesman problems (with as many as 70 cities) exactly are two quite different branch and bound algorithms, one developed by Shapiro,[15] the other by Little, Murty, Sweeney, and Karel.[13] These algorithms, along with several others, are discussed and compared in Bellmore and Nemhauser.[4] The paper by Little *et al.* and a branch and bound algorithm for general integer linear programming problems published by Land and Doig[12] are generally credited with demonstrating the potential of the branch and bound approach and with popularizing this technique.

The general branch and bound approach is discussed by Agin.[1] It is sufficient, however, to describe the technique by considering a branch and bound algorithm for a particular type of problem. In this section we shall present a branch and bound algorithm for solving the simple assignment problem. Before doing so, we shall briefly describe the assignment problem and compare it with the traveling salesman problem.

The Simple Assignment Problem

A small wine-tasting company with five employees has received contracts to test three wines. One employee is required for the testing of each wine. In addition, one employee is required for report-writing and one for administrative duties. Each of the five employees is capable of performing any of the five tasks, and the company likes to vary the assignments of the men to the tasks, partly to maintain objectivity in the testing and partly out of necessity! At any rate, the five employees currently on hand perform the tasks with varying degrees of efficiency. From past experience, the company has been able to construct a table showing each employee's relative efficiency for each of the five tasks. In this table (Table 9.2) tasks A, B, and C represent the three wines to be tested, task D represents reporting-writing, and task E represents administrative duties. The larger the number, the more efficient the employee is at a given job. Thus, the most efficient overall assignment of men to tasks is the assignment which maximizes the sum of the efficiency numbers. If we circle the numbers in each box of Table 9.2 corresponding to an assignment (e.g., the second box in the first row if employee II is assigned to task A), then a complete assignment will consist of one circled number in each row and in each column of the table.

Table 9.2

Task \ Employee	I	II	III	IV	V
A	5	7	3	6	2
B	6	3	5	5	5
C	1	2	3	2	3
D	4	6	7	5	4
E	3	4	3	4	4

Let us formalize this model mathematically. Suppose there are N men to be assigned to N jobs. The table of efficiency numbers (or costs, or profits) may be considered to be a square $N \times N$ matrix. Let us denote this matrix by $P = [p_{ij}]$, where p_{ij} is the profit to be made by assigning the jth man to the ith job. Define, now, an integer variable x_{ij}:

$$x_{ij} = \begin{cases} 1, & \text{if the } j^{\text{th}} \text{ man is assigned to the } i^{\text{th}} \text{ job} \\ 0, & \text{otherwise} \end{cases}$$

then, we wish to

$$\text{Maximize} \quad z = \sum_{j=1}^{N} \sum_{i=1}^{N} p_{ij} x_{ij} \tag{9-62}$$

$$\text{Subject to} \quad \sum_{j=1}^{N} x_{ij} = 1, \qquad i = 1, 2, \ldots, N \tag{9-63}$$

$$\sum_{i=1}^{N} x_{ij} = 1, \qquad j = 1, 2, \ldots, N \tag{9-64}$$

$$x_{ij} \geq 0, \text{ all } i \text{ and } j \tag{9-65}$$

$$x_{ij} \text{ integer, all } i \text{ and } j \tag{9-66}$$

The constraints (9-63) and (9-64) ensure that only one x_{ij} from each row and column will be equal to 1, the rest 0. Notice that this formulation is identical with a transportation problem, except for the integrality requirement (9-66). Recall, however, that the optimal solution to such a problem, using a transportation problem algorithm, will always be all integer. Thus, the assignment problem can always be solved by such methods.†

Let us return for a moment to the traveling salesman problem. Note that if we define

$$x_{ij} = \begin{cases} 1, & \text{if city } j \text{ is visited from city } i \\ 0, & \text{otherwise} \end{cases}$$

† In fact, there is an even more efficient algorithm for solving assignment problems. See Hillier and Lieberman[10] for a description of this algorithm.

and if we let $-p_{ij}$ denote the cost of going from city i to city j, then *any solution to the traveling salesman problem is also a solution to the assignment problem.* The converse, however, is not true since, for example, $x_{11} = x_{22} = \ldots = x_{NN} = 1$, all other $x_{ij} = 0$, is a feasible solution for the assignment problem, but not for the traveling salesman problem.

To illustrate the branch and bound approach let us develop an algorithm for solving the assignment problem. This algorithm is very similar to the traveling salesman algorithm of Little, *et al.*[13] However, the assignment problem branch and bound algorithm is somewhat less complicated. First, let us define the following terms:

Node—a specified subset of the set of all possible solutions (node 0 represents the set of all solutions)

Branch—a directed line segment connecting two nodes

Tree of Solutions—a graph containing all the nodes and branches, in a particular problem

Path to Node k—a sequence of nodes and branches from node 0 to node k

Terminal Node—a node representing one feasible solution

For example, suppose we wish to solve an assignment problem with 3 men and 3 jobs. Then there are six possible assignments, as indicated in Table 9.3. The tree of solutions for this problem is shown in Figure 9-4. The large circles represent nodes, with the smaller circles containing the corresponding node number (for reference purposes). The equations contained within a given node are constraints imposed at that node; thus, at node 1, $x_{11} = 1$, and so node 1 represents the set of all solutions for which $x_{11} = 1$. Node 0 represents the set of all possible solutions. A solution at any given node must satisfy all constraints imposed at that node, and in addition, *it must satisfy all constraints imposed by every node which is on the path from node* 0 *to that node.* Node 4, then, represents the set of all solutions such that $x_{11} = x_{22} = 1$.

Table 9.3

Node Number	Assignment Number	Assignment		
		Man #1	*Man #2*	*Man #3*
10	1	1	2	3
11	2	1	3	2
12	3	2	1	3
13	4	2	3	1
14	5	3	1	2
15	6	3	2	1

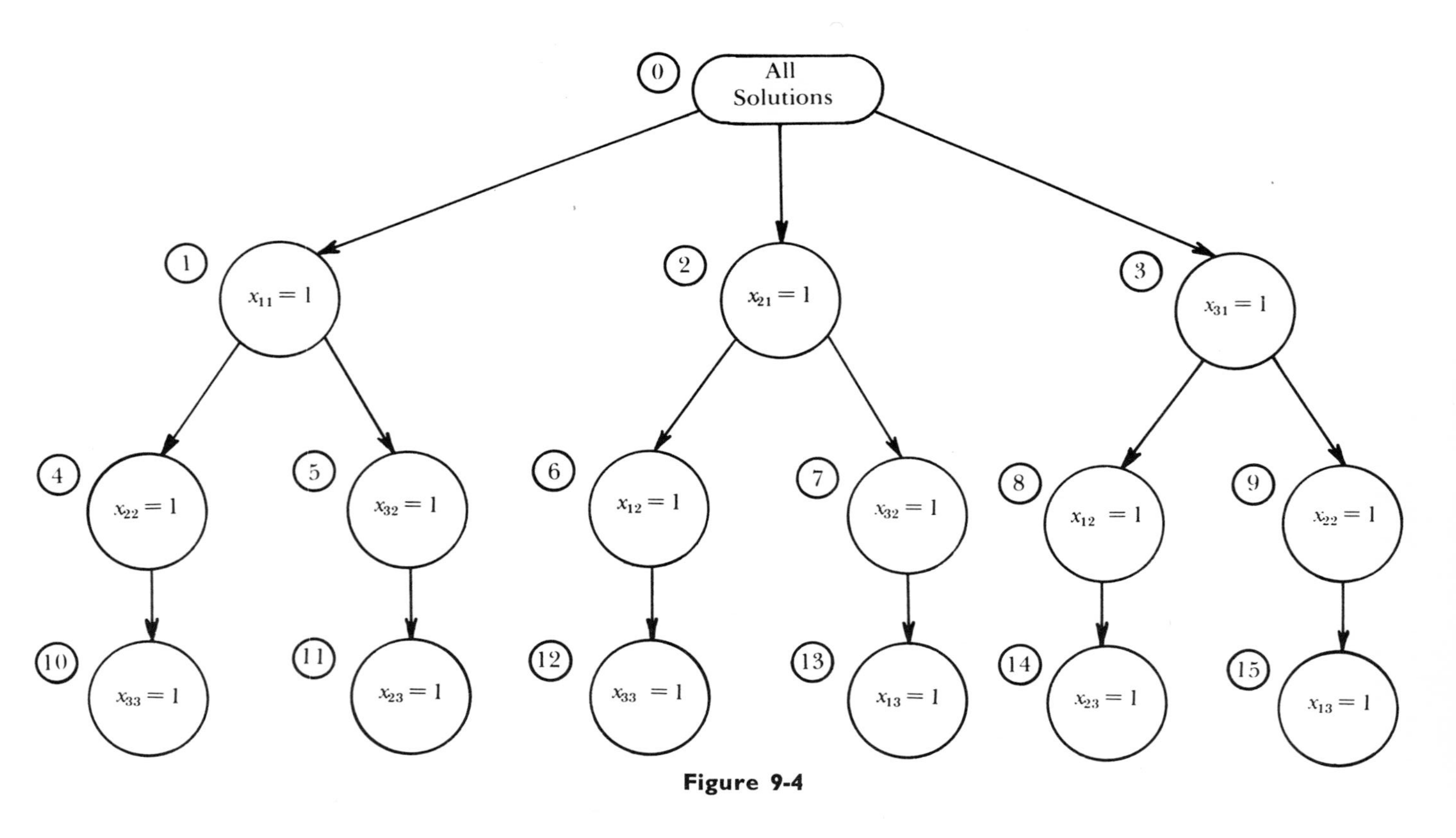
0
All Solutions
1
$x_{11} = 1$
2
$x_{21} = 1$
3
$x_{31} = 1$
4
$x_{22} = 1$
5
$x_{32} = 1$
6
$x_{12} = 1$
7
$x_{32} = 1$
8
$x_{12} = 1$
9
$x_{22} = 1$
10
$x_{33} = 1$
11
$x_{23} = 1$
12
$x_{33} = 1$
13
$x_{13} = 1$
14
$x_{23} = 1$
15
$x_{13} = 1$

Figure 9-4

Let us examine in a little more detail how the tree of Figure 9-4 was generated. Beginning with node 0, we wish to subdivide the set of all solutions. In this example, we have subdivided this set (represented by node 0) to three subsets: the set of all solutions in which man 1 is assigned to job 1 (represented by node 1), the set of all solutions in which man 1 is assigned to job 2 (represented by node 2), and the set of all solutions in which man 1 is assigned to job 3 (represented by node 3).

Next, each of these three nodes must be subdivided. Node 1, for example, is subdivided into two subsets, represented by nodes 4 and 5. Node 4 represents the set of all solutions in which $x_{22} = 1$ *and* $x_{11} = 1$; there is only one such solution: $x_{11} = x_{22} = x_{33} = 1$. Thus, node 4 is "subdivided" into one subset, representing one feasible solution, resulting in the terminal node, 10. Note that nodes 4 and 5 represent the only possible subdivision of node 1 which would lead to feasible solutions (i.e., since node 1 implies that man 1 is assigned to job 1, man 2 can only be assigned to job 2 or to job 3). Similar reasoning leads to the generation of the other nodes. The six terminal nodes represent the six possible solutions (see Table 9.3).

Let us now turn to the problem of efficiently utilizing such a tree of solutions in solving the assignment problem. For each terminal node, we can calculate the corresponding value of the objective function, and the optimal solution could thus be obtained by finding the maximum of these values. However, this would amount to total enumeration of all feasible solutions. Consider, instead, the following scheme:

1. Find one feasible solution and calculate the corresponding value of the objective function, $\hat{z}$. Note that, if the optimal value of z is z^*, then $\hat{z} \leq z^*$. Thus, $\hat{z}$ is a *lower bound* for z^*; let $\hat{z} = z_L$.

2. Subdivide each node which currently is the last node in a path (initially, just node 0) into further subdivisions (nodes); for each such newly created node, calculate an *upper bound* for z. Thus, if z_U^k denotes the upper bound on z at node k, then z_U^k is the largest possible value that z could attain for any solution which is in the set of solutions represented by node k. If, for any of these values z_U^k, it is true that

$$z_U^k < z_L \qquad \textbf{(9-67)}$$

then no solution which could be generated by a path through node k can have a value of z as large as the currently known best value of z, z_L. Thus, whenever inequality (9-67) holds, we may eliminate the entire set of solutions represented by node k from consideration; no subdivisions of node k are made. Instead, we move back up the tree to the node from which node k was generated, say node p. If there is another subdivision of node p to consider, then this subdivision (node) is considered in the same manner as node k. If all subdivisions of node p have been considered and for all these subdivisions the inequality (9-67) holds, then we again move back up the tree to the node from which node p was generated and repeat the process.

3. If, on the other hand, the upper bound at node k, z_U^k, violates inequality (9-67), then we wish to continue subdividing the subset of solutions represented by this node. Thus, we return to Step 2.

4. Whenever a stage is reached in the subdivision of the nodes in which a node represents a single feasible solution, then the corresponding value of z, $\hat{z}$, is calculated and compared with the current lower bound, z_L. If

$$\hat{z} > z_L \tag{9-68}$$

then this new feasible solution is better than the previously found best solution; thus, $\hat{z}$ is a new lower bound on z^* (the true optimal value of z). In this case, let $z_L = \hat{z}$ and save the feasible solution corresponding to $\hat{z}$. If, however, inequality (9-68) is not satisfied, this new feasible solution cannot be the optimal solution. In either case, we have reached a terminal node and must move back up the tree until we reach a node from which other subdivisions exist which have not yet been considered. Having reached such a node, we return to Step 2.

The scheme described in Steps 1, 2, 3, and 4 is essentially the basic branch and bound method. There is one major variation of the foregoing method, which will be discussed later in this section.

In order to apply the approach just listed we must be able to calculate the upper bounds z_U^k. The method of calculating these z_U^k differs with the type of problem being solved. We shall describe a procedure for calculating these upper bounds for the assignment problem.

Let us assume that node k, for example, represents the set of all solutions to the assignment problem in which $x_{12} = x_{35} = 1$ (that is, man 2 is assigned to job 1 and man 5 is assigned to job 3). Since man 2 and man 5 have already been assigned, node k corresponds to a smaller assignment problem in which rows 1 and 3 and columns 2 and 5 have been deleted. Thus, we need to determine a method for obtaining an upper bound for the solution to *any* assignment problem in order to be able to determine an upper bound for node k. But, this is an easy matter; we need only note that since any solution to an assignment column contains exactly one element from each row, an upper bound on the solution can be obtained by adding the maximum elements of each row of P together. Clearly, the optimal solution can be no larger than this sum; however, the solution corresponding to this value of z may not be feasible, since we have ignored the fact that the solution must also contain exactly one element from each *column*.

In calculating an upper bound for node k, we first delete the rows and columns corresponding to the constraints imposed at node k. For every constraint of the form $x_{qr} = 1$ which is imposed at node k, row q and column r are deleted. In the resulting smaller assignment problem, we calculate the upper bound, as described in the preceding paragraph. To this number, we add the sum of the elements of P corresponding to each of the constraints imposed at node k (i.e., the profit obtained from the assignments specified by the constraints imposed at node k).

Symbolically, if we define the following sets:

$$I_k = \{i \mid \text{row } i \text{ is not deleted at node } k\}$$
$$J_k = \{j \mid \text{column } j \text{ is not deleted at node } k\}$$
$$D_k = \{(i, j) \mid \text{row } i \text{ and column } j \text{ are deleted at node } k\}$$

then, the upper bound at node k, z_U^k, is

$$z_U^k = \sum_{(i,j) \in D_k} p_{ij} + \sum_{i \in I_k} \operatorname*{Max}_{j \in J_k} p_{ij} \tag{9-69}$$

Let us solve the 3×3 assignment problem with the following profit matrix:

$$P = \begin{bmatrix} 7 & 3 & 4 \\ 5 & 3 & 6 \\ 2 & 6 & 4 \end{bmatrix}$$

To begin, we need an initial solution to obtain a lower bound; for $x_{11} = x_{22} = x_{33}$, the corresponding value of z is $7 + 3 + 4 = 14$. We shall let $z_L = 14$, initially. This example is illustrated in Figure 9-5; the nodes are numbered in the order in which they are generated by the branch and bound procedure.

We first subdivide node 0 into three subdivisions, as shown in Figure 9-5. Next, we calculate the upper bound at node 1, yielding $z_U^1 = p_{11} + \text{Max}\,\{3, 6\} + \text{Max}\,\{6, 4\} = 7 + 6 + 6 = 19$. Thus, since $z_U^1 > z_L$, we then subdivide node 1 into node 4 and node 5. The upper bound for node 4 is then calculated:

$$\begin{aligned} z_U^4 &= p_{11} + p_{22} + \text{Max}\,\{4\} \\ &= 7 + 3 + 4 \\ &= 14 \end{aligned}$$

Since $z_U^4 \not> z_L$, we terminate this path and go to the next subdivision of node 1, which is node 5. Thus,

$$\begin{aligned} z_U^5 &= p_{11} + p_{23} + \text{Max}\,\{6\} \\ &= 7 + 6 + 6 \\ &= 19 \end{aligned}$$

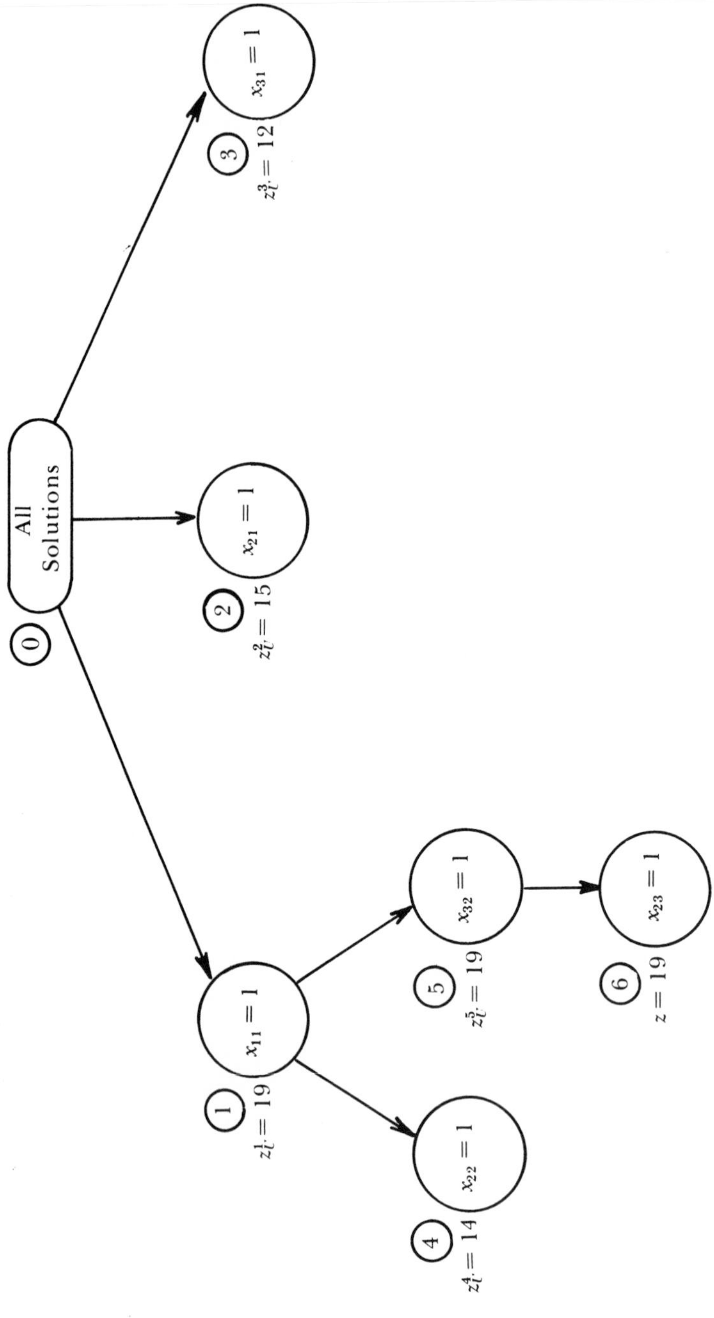

Figure 9-5

Now, $z_U^5 > z_L$, so we wish to subdivide node 5. However, as we have noted previously, this node can only be "subdivided" into one node (node 6), yielding a feasible solution: $x_{11} = x_{32} = x_{23} = 1$, $\hat{z} = 19$. Since $\hat{z} > z_L$, we have obtained a new lower bound. Thus, we now set $z_L = 19$.

Moving back up the tree now, the next node to be examined is node 2, the second subdivision of node 0. Calculating the upper bound for node 2, we find that

$$\begin{aligned} z_U^2 &= p_{21} + \text{Max}\,\{3, 4\} + \text{Max}\,\{6, 4\} \\ &= 5 + 4 + 6 \\ &= 15 \end{aligned}$$

Since $z_U^2 < z_L$, we do not need to subdivide node 2. We then go to the next node to be examined, node 3, and find that

$$\begin{aligned} z_U^3 &= p_{31} + \text{Max}\,\{3, 4\} + \text{Max}\,\{3, 6\} \\ &= 2 + 4 + 6 \\ &= 12 \end{aligned}$$

Here again, the upper bound is smaller than our current best solution, $z_L = 19$, and so there is no reason to subdivide node 3.

Moving back up the tree, we again return to node 0. Since we have examined all the subdivisions of node 0, we have exhausted all possibilities for finding a solution which is better than $z_L = 19$. The optimal solution, then, is $x_{11} = x_{32} = x_{23} = 1$, with $z^* = 19$.

Notice that only six nodes (besides node 0) were generated, whereas, from Figure 9-4, we see that a total of 15 nodes would have been generated, if we had had to examine all possible solutions.

Obviously, the efficiency of a branch and bound method depends heavily on how good the bounds z_L and z_U^k are. For this reason, it is desirable to obtain a good initial solution as rapidly as possible, in order to have as large a lower bound as possible. Often, certain other methods, known as *heuristics* or *suboptimal methods*, are employed to this end. These methods have the feature that they are very fast and usually provide a very good, if not optimal, solution.

The branch and bound algorithm described in this section used the "branching rule" that the next node to be subdivided was the node whose z_U^k was the most recently calculated upper bound, provided that $z_U^k > z_L$. Another commonly used branching rule is to always subdivide that node whose z_U^k is the largest. With this rule, it is very likely that the optimal solution, or a near-optimal solution, will be obtained early in the generation of the nodes, and hence a very good z_L will be available. In this manner, a smaller number of nodes may have to be examined than with the first branching rule.

However, the latter method requires that all the nodes which are currently at the end of a path must be saved, because eventually each of these may have the largest z_U^k. Thus, a large amount of computer storage is necessary if such a rule is to be used. Furthermore, it is impossible to obtain a good estimate of the amount of storage required, in advance of solving a given problem. The original branching rule given in this section has the advantage of requiring a comparatively small amount of computer storage. In practice, then, some combination of these two rules is the most desirable type of branching rule. Such a rule would limit the total storage required but would not necessarily branch from the most recently examined node.

9.6 DYNAMIC PROGRAMMING: INTRODUCTION

In this section we wish to briefly introduce the reader to a technique for solving certain types of optimization problems which is philosophically different from most of the methods we have thus far considered. The technique to which we refer is generally known as *dynamic programming*. The term dynamic programming does not refer to a particular type of problem (as does linear programming) but rather to a computational method for solving optimization problems. Much of the credit for the development of the technique, as well as the name dynamic programming, belongs to Richard Bellman.

Dynamic programming can be used to solve a wide variety of problems. Such problems include those in which the variables are continuous, as well as discrete. Another class of problems which is usually solved by dynamic programming involves the optimization of a definite integral. In these problems the unknown is a *function*, rather than a variable. Such problems are called problems in *the calculus of variations*.

In this section we shall only consider optimization problems in integer variables. For more complete discussions of dynamic programming the reader should consult Bellman and Dreyfus,[3] Nemhauser,[14] and Hillier and Lieberman.[10]

The dynamic programming approach for solving a problem with n variables is to convert this problem to n subproblems, each of which contains only one variable. These subproblems are then solved *sequentially*, in such a way that the combined optimal solutions for the n subproblems yields the optimal solution for the original problem. Generally speaking, the amount of computation involved in solving an optimization problem increases exponentially with the number of variables, but only linearly with the number of subproblems. Thus, the dynamic programming approach will usually greatly reduce the amount of computation necessary to solve a problem with a large number of variables.

Before describing more generally the nature of the problems which can be solved by dynamic programming, let us consider the following illustrative example:

$$\text{Maximize} \quad z = (x_1^3 - x_1) + x_2^2 - x_3 \tag{9-70}$$

$$\text{Subject to} \quad 3x_1 + 4x_2 + 2x_3 \leq 16 \tag{9-71}$$

$$x_1, x_2, x_3 \geq 0$$

$$x_1, x_2, x_3 \text{ integer}$$

Note that this problem is a nonlinear integer programming problem, and that the objective function is neither convex nor concave. Thus, no previously discussed method will obtain the global optimal solution.

The objective function (9-70) is separable, however. Let us define

$$f_1(x_1) = x_1^3 - x_1 \tag{9-72}$$

$$f_2(x_2) = x_2^2 \tag{9-73}$$

$$f_3(x_3) = -x_3 \tag{9-74}$$

The objective function then becomes

$$z = f_1(x_1) + f_2(x_2) + f_3(x_3) \tag{9-75}$$

Thus, if we denote by z^* the optimal value of z, then

$$z^* = \underset{x_1, x_2, x_3}{\text{Max}} [f_1(x_1) + f_2(x_2) + f_3(x_3)] \tag{9-76}$$

where the maximization in equation (9-76) is taken over all nonnegative integers x_1, x_2, x_3 satisfying (9-71).

Let us denote the optimal solution† by x_1^*, x_2^*, x_3^*.

Suppose, for the moment, that we happen to know the values x_2^*, x_3^*. Then, the problem would be restated as a one-variable problem, as follows:

$$\underset{x_1}{\text{Max}} \{f_1(x_1) + [f_2(x_2^*) + f_3(x_3^*)]\} \tag{9-77}$$

$$\text{Subject to} \quad 3x_1 \leq 16 - 4x_2^* - 2x_3^* \tag{9-78}$$

$$x_1 \text{ a nonnegative integer}$$

Of course, we do not know x_2^*, x_3^*. However, we do know that x_2^*, x_3^* must satisfy

$$16 - 4x_2^* - 2x_3^* \geq 0 \tag{9-79}$$

† If there is more than one optimal solution, then let x_1^*, x_2^*, x_3^* denote one such solution. The technique which we are describing, however, is capable of finding alternative optima, if desired. This will be discussed later.

otherwise, there would be no feasible solution for the one-variable problem (9-77) and (9-78). Note also that, since $f_2(x_2^*)$ and $f_3(x_2^*)$ are constants, the objective function of (9-77) may be replaced by

$$\max_{x_1} [f_1(x_1)] \tag{9-80}$$

Let us now introduce a new parameter, λ, and write (9-78) as

$$3x_1 \le \lambda \tag{9-81}$$

we shall allow λ to assume any of the values 0, 1, . . . , 16.

Now suppose that we solve the one-dimensional optimization problem defined by (9-80) and (9-81) *for each value of* λ, and store the results in a table (Table 9.4):

Table 9.4

λ	$g_1(\lambda)$	$\bar{x}_1$
0	0	0
1	0	0
2	0	0
3	0	1
4	0	1
5	0	1
6	6	2
7	6	2
8	6	2
9	24	3
10	24	3
11	24	3
12	60	4
13	60	4
14	60	4
15	120	5
16	120	5

The function $g_1(\lambda)$ is defined by

$$g_1(\lambda) \equiv \max_{x_1} [f_1(x_1)] \tag{9-82}$$

since the maximization depends upon λ, by (9-81). The last column of Table 9.4 contains the value of x_1 for which the maximum is attained.† Thus, the parameter λ effectively restricts the range on x_1, just as this range would be restricted by x_2^*, x_3^*.

† In case the maximum occurs at more than one value of x_1, we could record all of these values in the table if we desired to find alternative optima. However, the amount of additional computation required (not at this stage but at later stages in the solution procedure) usually makes the determination of alternative optima prohibitive. Thus, any value for which the maximum occurs is entered in the table.

Now, let us consider the following optimization problem: Find

$$g_2(\lambda) \equiv \operatorname*{Max}_{x_2} \left\{ \operatorname*{Max}_{x_1} [f_1(x_1)] + f_2(x_2) \right\} \qquad \textbf{(9-83)}$$

$$\text{Subject to} \quad 3x_1 \leq \lambda - 4x_2 \qquad \textbf{(9-84)}$$

where the $\operatorname*{Max}_{x_1} [f_1(x_1)]$ is taken over nonnegative integers x_1 satisfying (9-84). What is the significance of this problem? First of all, any feasible solution to this problem is a feasible solution to the original problem. Secondly, for fixed λ, the problem is actually a one-variable problem. To see this, observe that, by definition of $g_1(\lambda)$, the $\operatorname*{Max}_{x_1} [f_1(x_1)]$ over x_1 satisfying (9-84) is nothing more than $g_1(\lambda - 4x_2)$. Thus, equation (9-83) may be expressed as

$$g_2(\lambda) = \operatorname*{Max}_{x_2} [g_1(\lambda - 4x_2) + f_2(x_2)] \qquad \textbf{(9-85)}$$

For fixed λ, therefore, x_2 is the only variable in the objective function. Moreover, since it must be true that

$$\lambda - 4x_2 \geq 0$$

in order to maintain feasibility, we see that the maximization in (9-85) is taken over nonnegative integers x_2 between 0 and $\lambda/4$. Note that for any such integer, we have already calculated (in Table 9.4) the value of $g_1(\lambda - 4x_2)$, since $(\lambda - 4x_2)$ is an integer between 0 and λ. For example, with $\lambda = 16$, $x_2 = 3$ we find that $g_1(16 - 12) = g_1(4) = 0$. It is a simple matter, therefore, to generate a table of values for $g_2(\lambda)$, for each value of λ, $\lambda = 0, 1, \ldots, 16$, just as we did for $g_1(\lambda)$. Table 9.5 is the result of our example.

Thus far, we have not yet made the connection between the two one-variable problems (that of finding $g_1(\lambda)$ and that of finding $g_2(\lambda)$) and our original three-variable problem. To do so, let us now consider still a third one-variable problem: Find

$$g_3(\lambda) \equiv \operatorname*{Max}_{x_3} \{g_2(\lambda - 2x_3) + f_3(x_3)\} \qquad \textbf{(9-86)}$$

where the maximization is over nonnegative integers x_3 between 0 and $\lambda/2$.

Combining equations (9-82), (9-83), and (9-86) yields

$$\begin{aligned} g_3(\lambda) &= \operatorname*{Max}_{x_3} \{g_2(\lambda - 2x_3) + f_3(x_3)\} \\ &= \operatorname*{Max}_{x_3} \left\{ \operatorname*{Max}_{x_2} [g_1(\lambda - 2x_3 - 4x_2) + f_2(x_2)] + f_3(x_3) \right\} \\ &= \operatorname*{Max}_{x_3} \left\{ \operatorname*{Max}_{x_2} [\operatorname*{Max}_{x_1} (f_1(x_1) + f_2(x_2)] + f_3(x_3) \right\} \qquad \textbf{(9-87)} \end{aligned}$$

Table 9.5

λ	$g_2(\lambda)$	$\hat{x}_2$
0	0	0
1	0	0
2	0	0
3	0	0
4	1	1
5	1	1
6	6	0
7	6	0
8	6	0
9	24	0
10	24	0
11	24	0
12	60	0
13	60	0
14	60	0
15	120	0
16	120	0

The ranges over which the maximizations in (9-87) are taken are, respectively

$$0 \le x_3 \le \tfrac{1}{2}\lambda \tag{9-88a}$$

$$0 \le x_2 \le \tfrac{1}{4}(\lambda - 2x_3) \tag{9-88b}$$

$$0 \le x_1 \le \tfrac{1}{3}(\lambda - 4x_2 - 2x_3) \tag{9-88c}$$

Upon multiplying inequalities (9-88c) by 3, noting that x_1, x_2, x_3 are non-negative, and summing the resulting inequalities, we obtain

$$0 \le 3x_1 + 4x_2 + 2x_3 \le \lambda \tag{9-89}$$

The relations (9-89) imply that, for $\lambda = 16$, a feasible solution to our original problem is also a feasible solution to (9-88), and vice-versa. Thus, it must be true that the maximizations in (9-87) are equivalent to

$$z^* = \operatorname*{Max}_{x_1, x_2, x_3} [f_1(x_1) + f_2(x_2) + f_3(x_3)]$$

In other words,

$$g_3(16) = z^* \tag{9-90}$$

Thus, in order to solve the original three-variable problem, we need only calculate $g_3(16)$. From equation (9-86),

$$g_3(16) = \operatorname*{Max}_{x_3} [g_2(16 - 2x_3) + f_3(x_3)]$$

where $x_3 = 0, 1, \ldots, 8$. For all possible values of x_3, then, we have $g_2(16 - 2x_3)$ in Table 9.5. Upon performing this maximization, we find that $g_3(16) = z^* = 120$, and the corresponding value of x_3 is

$x_3^* = 0$. Now, we must find x_2^* and x_1^*. But, from (9-87), with $\lambda = 16$, we see that

$$\begin{aligned} z^* &= g_3(16) = g_2(16 - 2x_3^*) + f_3(x_3^*) \\ &= g_2(16) + f_3(0) \end{aligned}$$

From the $g_2(\lambda)$ table (Table 9.5), we find that the value of x_2 corresponding to $g_2(16)$ is $x_2^* = 0$. To find x_1^*, we return to the second line of equation (9-87) and obtain

$$\begin{aligned} z^* &= g_3(16) = g_1(\lambda - 4x_2^* - 2x_3^*) + f_2(x_2^*) + f_3(x_3^*) \\ &= g_1(16 - 0 - 0) = g_1(16) \end{aligned}$$

The value of x_1 corresponding to $g_1(16 - 4x_2^* - 2x_3^*) = g_1(16)$ is found from Table 9.4 to be $x_1^* = 5$.

The optimal solution to the original problem is, therefore, $z^* = 120$, with $x_1^* = 5$, $x_2^* = 0$, $x_3^* = 0$.

Let us summarize this procedure:

1. A parameter λ was introduced, which could assume any of the values $\lambda = 0, 1, \ldots, 16$.
2. For each value of λ, a one-variable optimization problem was solved, in each of the variables x_1, x_2.
3. For $\lambda = 16$, a one-variable optimization problem was solved, yielding z^* and x_3^*.
4. The optimal values of x_2 and x_1 were obtained from the tables generated in Step 2.

Thus, we have reduced the problem of solving a three-variable optimization problem to that of solving a sequence of one-variable problems. In general, as the number of variables increases, the computational savings afforded by this approach is significant.

The dynamic programming technique just described is easily generalized to a problem involving n variables, as follows:

Consider the problem

$$\text{Maximize} \quad z = \sum_{j=1}^{n} f_j(x_j) \tag{9-91}$$

$$\sum_{j=1}^{n} a_j x_j \leq b \tag{9-92}$$

all x_j nonnegative integers

We shall assume that the a_j's and b are all strictly positive integers.

We now define a sequence of functions

$$g_k(\lambda) = \underset{x_1, \ldots, x_k}{\text{Max}} \left\{ \sum_{j=1}^{k} f_j(x_j), \right\} \qquad k = 1, \ldots, n \tag{9-93}$$

where the maximization is taken over nonnegative integers $x_1, \ldots, x_k$ satisfying

$$\sum_{j=1}^{k} a_j x_j \le \lambda \tag{9-94}$$

Observe that

$$g_1(\lambda) = \underset{x_1}{\text{Max}}\,[f_1(x_1)]$$

where x_1 is an integer between 0 and λ/a_1.

The parameter λ is allowed to assume the values $\lambda = 0, 1, \ldots, b$. Thus, we construct a table for the values $g_1(\lambda)$ (with the corresponding values of x_1, as in our example).

Next, we calculate sequentially the other $g_k(\lambda)$, $k = 2, \ldots, (n-1)$, from the relation

$$g_k(\lambda) = \underset{x_k}{\text{Max}}\,\{f_k(x_k) + g_k(\lambda - a_k x_k)\} \tag{9-95}$$

where the maximization is over nonnegative integers x_k between 0 and λ/a_k. Finally, after a table has been generated for $g_{n-1}(\lambda)$, we then calculate

$$g_n(b) = \underset{x_n}{\text{Max}}\,\{f_n(x_n) + g_{n-1}(\lambda - a_n x_n)\} \tag{9-96}$$

From (9-96), we obtain $z^* = g_n(b)$ and the corresponding x_n^*. Next we obtain x_{n-1}^* from the $g_{n-1}(\lambda)$ table; x_{n-1}^* is the value of x_{n-1} corresponding to $g_{n-1}(b - a_n x_n^*)$. Similarly, x_{n-2}^* is the value of x_{n-2} corresponding to $g_{n-2}(b - a_n x_n^* - a_{n-1} x_{n-1}^*)$ in the $g_{n-2}(\lambda)$ table. Continuing in this manner, we obtain sequentially the optimal values of $x_n, x_{n-1}, \ldots, x_2, x_1$.

Observe from equation (9-95) that the calculation of the table for each $g_k(\lambda)$ involves the solving of only a one variable optimization problem. Thus, the amount of computation necessary to solve the original problem is proportional to the number of variables and therefore does not increase exponentially (although the total number of feasible solutions does in fact increase exponentially: if each of the variables in a given problem, for example, were allowed to assume only the values 0 or 1, the total possible number of feasible solutions might be as large as 2^n).

9.7 DYNAMIC PROGRAMMING: GENERAL DISCUSSION; AN EXAMPLE

In the previous section, we solved an optimization problem by dynamic programming. Although the problem we considered had a very special structure, the use of dynamic programming is not restricted to problems of such structure. There are, however, several criteria which must be satisfied by an optimization problem in order for it to be solvable by dynamic programming.

Before specifically examining such criteria, let us observe that there is often a variety of ways of mathematically formulating a given problem. For example, consider the traveling salesman problem. Ignoring the objective function for the moment, a verbal definition of a feasible solution for the traveling salesman problem is: any route which begins at city 1 and goes to each of the remaining given $n - 1$ cities exactly once and then returns to city 1 is a feasible solution. Mathematically, we have seen, in Section 9.2, an integer linear programming formulation of this problem. Alternatively, we might have formulated the problem as a network flow problem, in which each city is a node (city 1 being the source and the sink). Still another method of describing a feasible solution is by means of a permutation matrix:† Given the permutation matrix $P = [p_{ij}]$, if the salesman travels from city i to city j, then $p_{ij} = 1$; otherwise $p_{ij} = 0$.

To solve an optimization problem by dynamic programming, one must first formulate the problem in a dynamic programming framework. Let us now discuss the characteristics of this framework.

The n variable optimization problem defined by (9-91) and (9-92) was solved by solving a sequence of one variable problems. Each of these subproblems involved a parameter λ, in addition to one of the variables x_j. The parameter λ is called a *state parameter.* The subproblems were linked successively by the equations (9-95) and (9-96). These are called the *recursion relations.* The kth subproblem,

$$\text{find} \quad g_k(\lambda) = \operatorname*{Max}_{x_k} \{f_k(x_k) + g_{k-1}(\lambda - a_k x_k)\}$$

where x_k is an integer between 0 and λ/a_k, is dependent only upon the $(k - 1)^{\text{st}}$ subproblem (for each λ). The kth subproblem is often referred to as the kth *stage* of the optimization procedure.

In general terms, an optimization problem can be solved by dynamic programming if it can be formulated in such a way that the following conditions are satisfied:

1. The problem can be divided into n stages, each of which involves the solving of a one variable subproblem.‡

---

† A *permutation matrix* is a square matrix whose colums are columns of an identity matrix, but which may be reordered. Thus, for example, the matrices

$$\begin{bmatrix} 1 & 0 & 0 \\ 0 & 0 & 1 \\ 0 & 1 & 0 \end{bmatrix}, \quad \begin{bmatrix} 0 & 1 & 0 \\ 1 & 0 & 0 \\ 0 & 0 & 1 \end{bmatrix}, \quad \begin{bmatrix} 0 & 0 & 1 \\ 0 & 1 & 0 \\ 1 & 0 & 0 \end{bmatrix}$$

are all permutation matrices. Every row and every column of a permutation matrix contains exactly one 1 and the rest 0's.

‡ Occasionally, a state might consist of solving a problem of more than one variable. However, this is to be avoided if possible. We shall not consider such cases in this text.

Table 9.6

λ_1 \ λ_2	0	1	2	...
0	$g_k(0, 0)$	$g_k(0, 1)$	$g_k(0, 2)$	...
1	$g_k(1, 0)$	$g_k(1, 1)$	$g_k(1, 2)$	...
2	$g_k(2, 0)$	$g_k(2, 1)$	$g_k(2, 2)$	...
.	.	.	.	.
.	.	.	.	.
.	.	.	.	.

2. A recursion relation between the $(k - 1)^{st}$ stage and the kth stage can be established.
3. The solution to the kth stage problem depends only upon the $(k - 1)^{st}$ stage and on the state parameters.

It should be noted that it is possible that more than one state parameter will be required in order to formulate the recursion relations. However, the efficiency and power of dynamic programming decreases rapidly as the number of state parameters increases. To see this, consider one of the functions $g_k(\lambda)$. If g_k were a function of two parameters, λ_1, λ_2, then instead of generating one column of values for $g_k(\lambda)$, we would need to calculate a matrix of values, $g_k(\lambda_1, \lambda_2)$, as illustrated in Table 9.6.

If g_k were a function of three state parameters, a three-dimensional array would have to be calculated, for each g_k.

Thus, one of the main limitations on the types of problems solvable by dynamic programming is the number of state parameters involved. This is often referred to as the *dimensionality restriction* (not the number of variables in the original problem). Currently, it is usually not possible to use dynamic programming to solve a problem containing more than two or three state parameters.

Another factor which restricts the size of problems solvable by dynamic programming is the amount of storage required by the tables of values for the $g_k(\lambda)$'s. The size of these tables depends upon the allowable range for the state parameter and the number of tables depends upon the number of stages. However, with the high-speed disc storage and also the large core memory storage currently available with some digital computer systems, such restrictions are not as crucial as they once were.

One of the main reasons that dynamic programming is particularly well-suited for integer variable problems is that such restrictions will help to reduce the size of the tables. If a variable were allowed to be continuous, the state parameter λ would also be continuous; in order to generate a table of values for a continuous parameter, it is necessary to make discrete approximations for λ, and (depending upon how good an approximation is desired) this will generally increase the size of each table substantially.

Moreover, if upper and lower bounds for the variables are given, these would tend to reduce the size of the tables (see Exercise 17, page 366) and hence are a computational advantage for dynamic programming, whereas in most other optimization techniques (except perhaps search methods) the inclusion of upper bound constraints in an optimization problem increases the computational effort required to solve the problem.

We shall conclude our discussion of dynamic programming with an example.

An Airplane Route Problem

A manufacturing company executive wishes to make a random inspection of the company's plants. The plants are located in two districts. Instead of inspecting each plant, the executive has decided to inspect exactly one plant in each district. The executive will begin his inspection trip in Los Angeles and conclude it in New York. Table 9.7 below lists the locations of the plants in each of the two districts and the relevant travel costs are given in Table 9.8. The executive wishes to determine which plants to inspect so that travel costs are minimized.

There are three stages in this problem: The solution to the first stage is one of the Western District cities, and the solution to the second stage is one of the Eastern District cities. The third stage has only one solution, New York. However, as we shall see, the cost corresponding to this stage will vary depending on the previous stage (from which city the executive travels to New York). Once these two cities are specified, the complete route is determined. Thus, there are actually two variables in the problem. Let us denote by x_1 the city visited in stage one and by x_2 the city visited in stage two. In order to quantify the values these variables can assume (rather than, for example, letting x_1 = Denver) we shall assign the values 1, 2, 3 to

Table 9.7

Western District Plants	Eastern District Plants
Denver	Chicago
Santa Fe	Cincinnati
Kansas City	Pittsburgh
	Wheeling

Table 9.8

TRAVEL COSTS

LOS ANGELES TO:	
Denver	30
Santa Fe	45
Kansas City	60

To \ From	DENVER	SANTA FE	KANSAS CITY
Chicago	35	55	42
Cincinnati	60	80	70
Pittsburgh	75	90	83
Wheeling	80	95	90

TO N.Y. FROM:	
Chicago	90
Cincinnati	75
Pittsburgh	40
Wheeling	40

Denver, Sante Fe, and Kansas City, respectively, and the values 1, 2, 3, 4, to Chicago, Cincinnati, Pittsburgh, and Wheeling, respectively. Thus, x_1 will equal either 1, 2, or 3, and x_2 will equal 1, 2, 3, or 4.

We must now determine the recursion relations. Let us define the following cost functions:

$$f_1(x_1) = \text{cost of traveling from Los Angeles to city } x_1$$

$$f_2(x_1, x_2) = \text{cost of traveling from } x_1 \text{ to } x_2$$

$$f_3(x_2) = \text{cost of traveling from } x_2 \text{ to New York}$$

All relevant values of $f_1(x_1)$, $f_2(x_1, x_2)$, and $f_3(x_2)$ are given in Table 9.8. Thus, we wish to

$$\text{Minimize} \quad z = f_1(x_1) + f_2(x_1, x_2) + f_3(x_2)$$

We now introduce the state parameter λ. For this problem, it is convenient to let λ represent the city the executive is currently traveling to.

Define

$$g_j(\lambda) = \text{minimum cost of traveling to city } \lambda \text{ in the } j^{\text{th}} \text{ stage of the trip, } j = 1, 2, 3$$

For the third stage, there is only one possible city, New York, and hence only one value for λ; we arbitrarily assign a value of $\lambda = 1$ for the third stage.

From our definitions of the f_j's and g_j's, we therefore find that

$$z^* = \min_{x_1, x} \{f_1(x_1) + f_2(x_1, x_2) + f_3(x_2)\} = g_3(1)$$

and, moreover, that

$$\begin{aligned} g_1(\lambda) &= f_1(\lambda) \\ g_2(\lambda) &= \operatorname*{Min}_{x_1} \{g_1(x_1) + f_2(x_1, \lambda)\} \\ g_3(1) &= \operatorname*{Min}_{x_2} \{g_2(x_2) + f_3(x_2)\} \end{aligned}$$

Therefore, we calculate Table 9.9 for $g_1(\lambda)$:

Table 9.9

λ	$g_1(\lambda)$	x_1
1	30	1
2	45	2
3	60	3

Next, we calculate Table 9.10 for $g_2(\lambda)$:

Table 9.10

λ	$g_2(\lambda)$	x_1
1	65	1
2	90	1
3	105	1
4	110	1

In order to obtain $g_2(2)$, for example, we calculate

$$\begin{aligned} g_2(2) &= \operatorname*{Min}_{x_1} \{g_1(x_1) + f_2(x_1, 2)\} \\ &= \operatorname{Min} \{[g_1(1) + f_2(1, 2)] + [g_1(2) + f_2(2, 2)] \\ &\quad + [g_1(3) + f_2(3) + f_2(3, 2)]\} \\ &= \operatorname{Min} \{[30 + 60] + [45 + 80] + [60 + 70]\} \\ &= 90 \end{aligned}$$

Finally, we compute

$$\begin{aligned} g_3(1) &= \operatorname*{Min}_{x_2} \{g_2(x_2) + f_3(x_2)\} \\ &= \operatorname{Min} \{[g_2(1) + f_3(1)] + [g_2(2) + f_3(2)] \\ &\quad + [g_2(3) + f_3(3)] + [g_2(4) + f_3(4)]\} \\ &= \operatorname{Min} \{[65 + 90] + [90 + 75] + [105 + 40] + [110 + 40]\} \\ &= \operatorname{Min} \{155 + 165 + 145 + 150\} \\ &= 145, \end{aligned}$$

corresponding to $x_2 = 3$.

Therefore, the total minimum cost for the trip is $z^* = 145$, and $x_2^* = 3$, which means that the executive travels from the third city in the Eastern District to New York; i.e., from Pittsburgh. In order to complete the optimal solution, we must return to Table 9.10, with $\lambda = 3$, and we find that $x_1^* = 1$, which corresponds to Denver.

The optimal route, then is Los Angeles to Denver to Pittsburgh to New York.

We note in conclusion that there are other possible dynamic programming formulations of the airplane route problem. Many of the decisions made in the preceding formulation were completely arbitrary (such as the values assigned to x_1, x_2). We were merely attempting to structure the problem mathematically so that it could be conveniently solved by dynamic programming.

In many cases, a dynamic programming formulation for a given optimization problem may not be readily apparent. Often, a good deal of cleverness is required to find the right formulation. Unfortunately, the only remedy for this situation is experience and perseverance. When attempting to formulate a given problem in terms of dynamic programming, it is often useful to examine what others have done with similar problems, by consulting such sources as Bellman and Dreyfus,[3] Hillier and Lieberman,[10] or Nemhauser.[14]

EXERCISES

Formulation of Integer Problems

1. A bank has set up an investment division, whose task it is to invest the bank's surplus cash. The investment manager of the bank would like to determine an investment plan for the next twelve months, with the following features:

   (a) There are a total of N projects which may be invested in.

   (b) Investment in any of the N projects may be begun at the beginning of each month.

   (c) Any revenue earned by any of the projects in the kth month may be reinvested in any of the N projects, beginning with the $(k+1)^{\text{st}}$ month.

   The following facts must also be incorporated into the investment plan:

   (d) In the kth month, a total of D_k dollars will be provided for investment by the bank.

   (e) A total of d_{jk} dollars is required for investment in project j in the kth month.

(f) If investment is made in project j in the kth month, a total of p_{jk} dollars will be earned (which can be later reinvested, as per (c)).

If p_j is the present worth of all future profits earned by investment in project j, formulate an integer programming problem whose optimal solution will tell the investment manager in which month, if any, he should invest in each of the projects, so that the total present worth of all future projects is maximized. Be sure to incorporate feature (c).

2. Suppose that in the investment plan of exercise 1, it is also required that if investment is made in project j at all, then investment must be made in project j for at least m_j months (not necessarily consecutively). Incorporate this restriction into the integer programming formulation of exercise 1.

3. In Model 6 of Section 9.2, a procedure for formulating "either/or" constraints in terms of a set of linear constraints and an integer variable was discussed. Generalize this procedure to the case of any two "either/or" constraints. That is, if only one of the two constraints

$$\begin{Bmatrix} \bar{a}_1'\bar{x} \le b_1 \\ \bar{a}_2'\bar{x} \le b_2 \end{Bmatrix}$$

must be satisfied, show how the introduction of an integer variable allows us to include both possibilities in one mixed integer-continuous variable problem. (Hint: Assume that upper bounds L_1, L_2 have been found for each constraint, such that, for all feasible $\bar{x}$,

$$\begin{Bmatrix} \bar{a}_1'\bar{x} - b_1 \le L_1 \\ \bar{a}_1'\bar{x} - b_2 \le L_2 \end{Bmatrix}$$

Explain how these bounds may be obtained via linear programming. Next, consider the following constraints

$$\begin{Bmatrix} \bar{a}_1'\bar{x} - yL_1 \le b_1 \\ \bar{a}_1'\bar{x} + yL_2 \le b_2 + L_2 \\ 0 \le y \le 1 \\ y \text{ integer} \end{Bmatrix}$$

Do these constraints force at least one of the original constraints to be satisfied?)

4. Explain how constraints of the form $|x_j - a| \geq b$ can be represented in terms of linear constraints.

5. Explain how constraints of the form

$$\sum_{j=1}^{n} |x_j - a_j| \geq b$$

can be represented in terms of linear constraints.

Solution of Integer Problems

6. Solve the example of Section 9.1 by Gomory's Algorithm.

7. Solve the example of Section 9.1 by the Bounded Variable Algorithm.

8. Solve the problem of Section 9.3 by the Bounded Variable Algorithm.

9. Solve the following problem by Gomory's Algorithm:

$$\text{Minimize} \quad z = x_2 - 2x_1$$

$$\text{Subject to} \quad \begin{pmatrix} 2x_1 + 6x_2 \leq 9 \\ 2x_2 + x_2 \leq 4 \\ x_1, x_2 \geq 0 \\ x_1, x_2 \text{ integer} \end{pmatrix}$$

10. Solve the problem of exercise 9 by the Bounded Variable Algorithm.

11. Use a branch and bound method to solve the assignment problem with the following profit matrix:

$$P = \begin{bmatrix} 9 & 8 & 3 & 6 \\ 7 & 2 & 6 & 5 \\ 5 & 4 & 7 & 3 \\ 3 & 6 & 5 & 4 \end{bmatrix}$$

12. (a) Develop a branch and bound procedure for solving an ordinary integer programming problem.

(b) Apply your procedure to the following problem:

$$\text{Maximize } z = 11x_1 + 4x_2$$

$$\text{Subject to } \quad x_1 + 2x_2 \leq 4$$

$$5x_1 + 2x_2 \leq 16$$

$$2x_1 - xx_2 \leq 4$$

$$x_1, x_2 \geq 0$$

$$x_1, x_2 \text{ integers}$$

Dynamic Programming

13. Solve by dynamic programming:

$$\text{Maximize } \quad z = 3x_1(2 - x_1) + 2x_2(2 - x_2)$$

$$\text{Subject to } \quad x_1 + x_2 \leq 3$$

$$x_1, x_2 \text{ nonnegative integers}$$

14. Formulate and solve by dynamic programming the assignment problem of exercise 11.

15. An automobile manufacturing company is concerned with sales in 3 of its divisions. The company has decided to add a total of 6 dealerships to these divisions, with at least one dealership going to each division (and thus no more than 4 dealerships will be given to any one division). The company's marketing department has reported that the expected profit that the company will receive can be summarized as follows:

No. of Dealerships Given	Profit* in Division A	Profit* in Division B	Profit* in Division C
0	100	200	150
1	200	210	160
2	280	220	170
3	330	225	180
4	340	230	200

* (in millions of dollars)

The company wishes to award the 6 dealerships so that total profit is maximized.

Formulate and *solve* this problem by means of dynamic programming.

16. A manufacturing company has decided to embark on a vigorous austerity program over the next four months. During this time, the company plans to lay off a total of 1000 men. However, in order to pacify the union local, the company has decided that it will not lay off more than 400 men in any one month.

    These men are each currently earning \$600 per month. As a result of the last union contract, the men are due for a \$100 per month raise in 2 months.

    The company's accountant, calculates that the company would realize gross sales per man of \$900, \$800, \$1000, and \$1100 in each of the next four months.

    The company wishes to minimize its net loss in profit. Assume that men will only be laid off in multiples of 100 men.

    (a) Assuming this problem were to be solved by dynamic programming, define the variables and the state parameter.
    (b) For each stage, determine the possible range of values of the state parameter and of the variable(s)—possibly as functions of the state parameter.

    Do *not* attempt to solve the problem.

17. Solve by dynamic programming:

$$\begin{aligned} \text{Maximize} \quad & z = 2x_1 + x_2 + x_3 + 3x_4^2 \\ \text{Subject to} \quad & x_1 + 2x_2 + x_3 + 4x_4 \leq 10 \\ & x_1 \geq 2 \\ & x_2 \geq 3 \\ & x_1, x_2, x_3, x_4 \text{ nonnegative integers} \end{aligned}$$

18. A plant manager has 6 handymen available and 4 tasks to be performed, each of which can be handled by as many as six men. The men's ability to work together on a given task varies with the number of men assigned to the task and is different for each of the four tasks. The plant manager has developed a measure of this ability to work together, which he calls "productivity."

The various productivity numbers are given in the table below:

No. Men Assigned	Task A	B	C	D
0	0	0	0	0
1	5	3	1	5
2	10	7	6	9
3	13	12	13	13
4	14	16	15	17
5	12	16	16	20
6	10	16	17	22

The manager wishes to determine how to assign the men to the tasks so that the total productivity is maximized. No man will be assigned to more than one task.

Formulate and solve this problem by dynamic programming.

REFERENCES

1. Agin, N.: Optimum seeking with branch and bound. *Manage. Sci.*, *13*, B-176-185 (1966).
2. Balas, E.: An additive algorithm for solving linear programs with zero-one variables. *Oper. Res.* *13*, 517–544 (1965).
3. Bellman, R., and Dreyfus, S.: *Applied Dynamic Programming*. Princeton University Press, Princeton, N.J. (1962).
4. Bellmore, M., and Nemhauser, G. L.: The traveling salesman problem: a survey. *Oper. Res.*, *16*, 538–558 (1968).
5. Glover, F.: Truncated enumeration methods for solving pure and mixed integer programs. Working Paper No. 27, Operations Research Center, Berkeley, California (1966).
6. Gomory, R. E.: Outline of an algorithm for integer solutions to linear programs. *Bull. Amer. Math. Soc.*, *64*, 275–278 (1958).
7. Gomory, R. E.: An Algorithm for Integer Solutions to Linear Programming. Princeton-IBM Mathematics Research Project, Technical Report No. 1 (1958).
8. Gomory, R. E.: An Algorithm for the Mixed Integer Problem. P-1885, The RAND Corporation (1960).
9. Hadley, G.: *Nonlinear and Dynamic Programming*. Addison-Wesley, Reading, Massachusetts (1964).
10. Hillier, F. S., and Lieberman, G. J.: *Introduction to Operations Research*. Holden-Day, San Francisco (1967).
11. Krolak, P. D.: The Bounded Variable Algorithm for Solving Integer Linear Programming Problems. Washington University, Department of Applied Mathematics and Computer Science, Report No. COO-1493-1518 (1968).
12. Land, A. H., and Doig, A.: An automatic method of solving discrete programming problems. *Econometrica*, *25*, 497–520 (1960).
13. Little, J. D. C., Murty, G., Sweeney, D. W., and Karel, C.: An algorithm for the traveling salesman problem. *Oper. Res.*, *11*, 972–989 (1963).
14. Nemhauser, G. L.: *Introduction to Dynamic Programming*. John Wiley and Sons, New York (1966).
15. Shapiro, D. M.: Algorithms for the Solution of the Optimal Cost and Bottleneck Traveling Salesman Problems. Washington University, Department of Applied Mathematics and Computer Science, Report No. AM-67-2 (1967).

ANSWERS TO SELECTED EXERCISES

CHAPTER 2

1.

(a) $\begin{bmatrix} 7 & 0 & -4 & 4 \\ 4 & 0 & -2 & 0 \\ 4 & -2 & 3 & 8 \\ 3 & -5 & 12 & 12 \end{bmatrix}$

(c) undefined

(e) $\begin{bmatrix} 2 & 2 & 18 & 4 \\ 5 & -3 & 7 & 5 \\ 0 & 3 & 8 & -2 \end{bmatrix}$

(g) same as (e)

(i) $\begin{bmatrix} 1 & 3 & -1 \\ 0 & 4 & 0 \\ -2 & -1 & 0 \\ 0 & 0 & 0 \end{bmatrix}$

(k) undefined

(m) 0

(o) undefined

3. AD does not necessarily equal DA. Example:

$$\text{Let } A = \begin{bmatrix} 1 & 2 \\ 3 & 4 \end{bmatrix} \qquad D = \begin{bmatrix} 1 & 0 \\ 0 & 2 \end{bmatrix}$$

6. $(AA')' = (A')'A' = AA'$

8. (a) 1 (c) 4 (e) −24

11. (a) 4 (c) 2

13. A^{-1} may not exist.

15. (a)
$$\left[\begin{array}{cc|cc} 1 & 0 & 7/4 & 1/2 \\ 0 & 1 & -1/4 & 1/2 \\ \hline 0 & 0 & 0 & 0 \\ 0 & 0 & 0 & 0 \end{array}\right] \left[\begin{array}{c} x_1 \\ x_2 \\ \hline x_3 \\ x_4 \end{array}\right] = \left[\begin{array}{c} 3/4 \\ -1/4 \\ \hline 0 \\ 0 \end{array}\right]$$

16. (a) $x_1 = x_2 = x_3 = 1$

17. (a) linearly dependent
(c) linearly dependent

21. (a) S_1 is a vector space.
(d) S_4 is not a vector space.
(f) S_6 is a vector space.

22. (c) dimension = 2; a basis is $\begin{bmatrix} 3 \\ 0 \\ 2 \end{bmatrix}, \begin{bmatrix} 0 \\ 1 \\ 0 \end{bmatrix}$

(e) dimension = 3; a basis is $\begin{bmatrix} 0 \\ 0 \\ 0 \\ 1 \\ 0 \end{bmatrix}, \begin{bmatrix} 0 \\ 1 \\ 1 \\ 0 \\ 0 \end{bmatrix}, \begin{bmatrix} 0 \\ 0 \\ 0 \\ 0 \\ 1 \end{bmatrix}$

(f) dimension = 2; a basis is $\begin{bmatrix} 0 \\ 1 \\ 1 \\ 0 \\ 0 \end{bmatrix}, \begin{bmatrix} 0 \\ 0 \\ 0 \\ 3 \\ 1 \end{bmatrix}$

23. (b) $f(\bar{x}) = [x_1, x_2] \begin{bmatrix} 1 & -1 \\ -1 & 1 \end{bmatrix} \begin{bmatrix} x_1 \\ x_2 \end{bmatrix}$

24. (b) $\begin{bmatrix} 1 & -1 \\ -1 & 1 \end{bmatrix}$ is positive semidefinite

CHAPTER 3

1. (a) $H = \{[x_1, x_2, x_3] \mid 2x_1 - 3x_2 - 2x_3 = 0\}$

5. (a) closed, unbounded
(c) neither open nor closed; bounded
(e) open, bounded
(i) closed, bounded

6. (a) not convex
(c) convex
(e) convex
(g) convex
(i) convex

7. (a) not convex
(c) not convex
(d) convex

9. Extreme points are: (0,0), (0,3), (1,3), (2,2), (3,0).

10. $\bar{a} = \frac{1}{2}\bar{x}_1 + \frac{1}{4}\bar{x}_2 + \frac{1}{4}\bar{x}_3$

14. (a) all of E^2

16. The hyperplane is the line through the points (1,0) and (4,3). It is unique.

18. No. The hyperplane passing through the points (4,3) and (−3,−2) is unique; however, it is not a supporting hyperplane, since it contains points lying in the interior of the convex polyhedron.

CHAPTER 4

2. $x = -2 + \sqrt{3}$ is a minimum; $x = -2 - \sqrt{3}$ is a maximum

3. $r = \frac{2B}{ze^2}$

4. $-0.829, 1.094, 4.659$

6. $\text{side} = \frac{2\sqrt{3}}{3} R$

10. $x_1 = \sqrt{\frac{3}{2}} = x_2;\; x_3 = \frac{4}{3}\sqrt{\frac{3}{2}};\; f(\bar{x}) = 9$

11. $x_1^2 = x_2^2 = x_3^2 = \frac{c^2}{3};\; f(\bar{x}) = \left(\frac{c^2}{3}\right)^3$

14. $x_1 = 2.659;\; x_2 = 1.880,\; z = 14.142$

16. $x_1 = 50,\; x_2 = 450,\; x_3 = 133.33$

CHAPTER 5

1. $x_{min} = 7.8102,\; f(x_{min}) = 703.73$

2. $x_{min} = 0.6666667,\; f(x_{min}) = -0.333334$

7. $x = 15.9744$
   $y = 24.0460$
   $z = 71.5681$

15. $x_1 = x_2 = 1$

20. $x = 3000$

CHAPTER 6

1. $x_1 = x_2 = 0;\; x_3 = x_4 = 5;\; z = 40$

4. **Final Tableau**

$\bar{c}_B$	c_j Basis vectors	$\bar{x}_B$	2 $\bar{a}_1$	3 $\bar{a}_2$	1 $\bar{a}_3$	6 $\bar{a}_4$	0 $\bar{a}_{s1}$	0 $\bar{a}_{s2}$	0 $\bar{a}_{s3}$
0	$\bar{a}_{s1}$	2.6666	6.3333	.3333	2.6666	0	1	−.33333	0
6	$\bar{a}_4$	3.3333	.66666	1.6666	.33333	1	0	.33333	0
0	$\bar{a}_{s3}$	1.3333	−.3333	−.33333	4.3333	0	0	−.66666	1
		20	2	7	1	0	0	2	0

[$\bar{a}_{s1}$, $\bar{a}_{s2}$, $\bar{a}_{s3}$ correspond to slack variables]

5. **Final Tableau**

$\bar{c}_B$	c_j Basis vectors	$\bar{x}_B$	3 $\bar{a}_1$	−4 $\bar{a}_2$	5 $\bar{a}_3$	1 $\bar{a}_4$	0 $\bar{a}_{s1}$	0 $\bar{a}_{s2}$	0 $\bar{a}_{s3}$
0	$\bar{a}_{s1}$	45	12	−1	0	21	1	0	3
0	$\bar{a}_{s2}$	63	10	−8	0	21	0	1	4
5	$\bar{a}_3$	7.5	1.5	−.5	1	2	0	0	.5
		37.5	4.5	1.5	0	9	0	0	2.5

[$\bar{a}_{s1}$, $\bar{a}_{s2}$, $\bar{a}_{s3}$ correspond to slack variables]

6. **Final Tableau**

$\bar{c}_B$	c_j Basis vectors	$\bar{x}_B$	3 $\bar{a}_1$	1 $\bar{a}_2$	0 $\bar{a}_{s1}$	0 $\bar{a}_{s2}$
0	$\bar{a}_{s1}$	13.8	0	2.8	1	−.4
3	$\bar{a}_1$	3.6	1	.6	0	.2
		10.8	0	.8	0	.6

[$\bar{a}_{s1}$, $\bar{a}_{s2}$ correspond to slack variables]

7.

Final Tableau

$\bar{c}_B$	Basis vectors	c_j $\bar{x}_B$	-2 $\bar{a}_1$	-3 $\bar{a}_2$	-1 $\bar{a}_3$	-4 $\bar{a}_4$	0 $\bar{a}_{s1}$	0 $\bar{a}_{s2}$	0 $\bar{a}_{s3}$	$-M$ $\bar{a}_{A1}$	$-M$ $\bar{a}_{A2}$	$-M$ $\bar{a}_{A3}$
-2	$\bar{a}_1$	1.326	1	0	0	-0.492	-0.146	-0.077	-0.054	0.146	0.077	0.054
-3	$\bar{a}_2$	2.562	0	1	0	0.815	0.008	-0.154	-0.208	-0.008	0.154	0.208
-1	$\bar{a}_3$	0.138	0	0	1	0.585	-0.108	0.154	-0.092	0.108	-0.154	0.092
		-10.475	0	0	0	1.954	0.377	0.462	0.823	$M-.38$	$M-.45$	$M-.88$

[$\bar{a}_{s1}$, $\bar{a}_{s2}$, $\bar{a}_{s3}$ correspond to slack variables; $\bar{a}_{A1}$, $\bar{a}_{A2}$, $\bar{a}_{A3}$ correspond to artificial variables]

CHAPTER 7

1. Min $v = 50w_1 - 12w_2 + 30w_3$

$$\begin{aligned} 3w_1 - 2w_2 + 5w_3 &\geq 2 \\ 4w_1 - w_2 + 3w_3 &\geq 3 \\ 5w_1 - 2w_2 + w_3 &\geq 5 \end{aligned}$$

2. Min $v = 21w_1 + 38w_2$

$$\begin{aligned} 2w_1 + 3w_2 &\geq -3 \\ 3w_1 + w_2 &\geq -5 \\ w_1 + 4w_2 &\geq -2 \end{aligned}$$

3. Min $v = 19w_1 - 22w_2 + 38w_3$

$$\begin{aligned} 3w_1 - 2w_2 + w_3 &\geq -2 \\ 2w_1 - 3w_2 - w_3 &\geq -3 \\ w_1 + w_2 + 2w_3 &\geq 4 \\ -2w_1 - 3w_2 - 3w_3 &\geq -5 \end{aligned}$$

$w_1, w_2 \geq 0$

w_3 unrestricted in sign

4. $z_{min} = v_{max} = 40$

9. If c_1 is increased to 6, the solution is still optimal.
If c_1 is increased to 7, the optimal solution becomes

$$x_1 = \frac{20}{3}, \quad x_4 = \frac{2}{3}, \quad x_2 = x_3 = 0, \quad z = \frac{140}{3}$$

If c_2 is decreased from 8 to 5, the original optimal solution is still optimal.
If c_2 is decreased to 3, the optimal solution becomes

$$x_1 = \frac{20}{3}, \quad x_4 = \frac{2}{3}, \quad x_2 = x_3 = 0, \quad z = 20$$

10. The solution is still feasible and, hence, still optimal.

12. Amounts Shipped to Retail Stores

Warehouse								
1	3000	2000	1000			4000		2000
2				5000				
3							3000	
4				5000	8000			

The optimal cost is $300,000.
Note: There are other optimal solutions.

15. Maximal flow equals 14.

CHAPTER 8

1. $x_1^* = 6,\ x_2^* = 0,\ z^* = 29$

2. $x_1^* = 1,\ x_2^* = 2,\ z^* = 0$

4. (a) Local maxima: $(-0.215, 2.112)$, $(4,30)$.
(b) Global maximum is $(4,30)$, with $z^* = 34$.
(c) Local minima: $(1.55,-0.63)$, $(-2,-12)$.

12. $x_1^* = \frac{18}{13},\ x_2^* = \frac{12}{13},\ z^* = -\frac{468}{169}$

CHAPTER 9

11. $z^* = 27$

13. $x_1^* = x_2^* = 1,\ z^* = 5$

18. $z^* = 28$; Optimal assignment is 2 men to task A, none to task B, 3 to task C, and 1 to task D.

INDEX

DATE DUE